Eco-Social Contracts for Sustainable and Just Futures

Patrick Huntjens • Najma Mohamed
Katja Hujo • Manisha Desai
Editors

Eco-Social Contracts for Sustainable and Just Futures

Mobilising Collective Power to Deal with the 21st Century Polycrisis

 Springer

Editors
Patrick Huntjens
Inholland University of Applied Sciences
Delft, The Netherlands

Katja Hujo
UNRISD
Bonn, Germany

Najma Mohamed
UN Environment Programme World
Conservation Monitoring Centre
UNEP-WCMC
Cambridge, UK

Manisha Desai
Stony Brook University
Stonybrook, NY, USA

ISBN 978-3-031-99108-0   ISBN 978-3-031-99109-7   (eBook)
https://doi.org/10.1007/978-3-031-99109-7

This Springer imprint is published by the registered company Springer Nature Switzerland AG.
The registered company address is: Gewerbestrasse 11, 6330 Cham, Switzerland

If disposing of this product, please recycle the paper.

Synopsis

Eco-Social Contracts for Sustainable and Just Futures is a groundbreaking volume that reimagines how societies can respond to the interwoven crises of our time: climate collapse, biodiversity loss, widening inequality, and the erosion of public trust and democratic legitimacy. At its heart lies a fundamental realisation that can no longer be ignored: the social contract has been broken for billions of people, and with it, the bonds between people, planet, and power must be rewoven.

Grounded in diverse knowledge systems, from Indigenous cosmologies and feminist and care-based economics to regenerative development and post-growth thought, this book presents eco-social contracts as a bold and transformative vision for systemic renewal. Far more than policy tools, these contracts serve as a compass for profound cultural and institutional change, what Joanna Macy has described as hospicing the old systems that no longer serve life, while midwiving new ones rooted in care, reciprocity, and collective flourishing.

With contributions from leading scholars, policymakers, practitioners, artists, and activists across the globe, the volume bridges theory and practice to illuminate how communities are already advancing inclusive, regenerative, and dignity-affirming alternatives. Whether addressing food or climate justice, rethinking democracy, or embedding the Rights of Nature into law, each chapter offers a window into the transformations already underway, and the deeper shifts they call forth.

Increasingly recognised by the UN, global assessments such as the IPBES Transformative Change Assessment, and by a growing international community of civil society leaders, youth movements, and NGOs, eco-social contracts call for renewed solidarity, systemic equity across generations and communities, inclusive governance, and a fundamental transformation of economic systems. They challenge dominant economic logics and call for a reimagining of economies not as engines of extraction, destruction, and inequality, but as systems in the service of life, supporting people, planet, and the more-than-human world.

This vision moves beyond GDP and consumption as measures of progress. It centres relational, ecological, and cultural understandings of broad and shared prosperity. It honours the intrinsic value of natural systems, cultural lifeways, and all living beings, refusing to reduce life to metrics of utility or profit. It is a call for regeneration, rebalancing, and reweaving, where equity, dignity, and ecological integrity guide how we live and what we value.

This timely and courageous book captures the growing global momentum towards societal transformation. It offers both hopeful inspiration and hard-won lessons on unlocking collective agency in an age of ecological breakdown and social fragmentation.

For changemakers, students, and all those seeking hope, direction, and clarity in a time of global uncertainty, this book is both a call to action and a guide for transformation—inviting readers to imagine and co-create sustainable and just futures our hearts and minds know are possible.

Acknowledgements

The editors are grateful for financial support received from the following organisations:

Inholland University of Applied Sciences (Netherlands).

Stony Brook University's Office of Diversity, Inclusion, and Intercultural Initiative (United States).

The Atlantic Fellowship for Social and Economic Equity, based at the London School of Economics' (LSE) International Inequality Institute.

The United Nations Research Institute for Social Development (UNRISD) and the Robert-Bosch Stiftung for supporting this publication.

The members of the editorial team, Patrick Huntjens, Najma Mohamed, Katja Hujo, and Manisha Desai, express their gratitude to everyone who has contributed to this edited volume. We want to thank all participants of the Global Policy Seminar, co-convened by UNRISD and the Green Economy Coalition (GEC) in Bonn, Germany, in August 2023, which inspired this book project. We also want to thank the members of the Global Research and Action Network for a New Eco-Social Contract for actively participating and sharing their ideas, knowledge, and experiences during numerous events and the fruitful discussions held in the thematic working groups of the network.

We are deeply thankful to all authors who have enthusiastically contributed to this book project and responded to several rounds of revisions. We are also grateful to Paramita Dutta for her invaluable support in the finalisation of the manuscript, and to Jearelle Wolhuiter and Nicole Harris for the copy editing provided for some chapters.

Our gratitude also extends to the editorial team at Springer, who responded patiently to all our queries and supported the production of this book in various ways. To our colleagues who provided kind encouragement as we journeyed through the writing and editing process over the last year, and our families for the love, understanding, and support to ease our commitment to this work, thank you. More broadly, we acknowledge the courage of the communities and citizens that inspire and motivate us to imagine and to work for a world where we can live with reciprocity, with respect and care, for one another and for the planet we call home.

This book was written with the belief in the words of Dr. Martin Luther King Jr's that "*The arc of the moral universe is long, but it bends towards justice*", and a commitment to ideas and actions that could bend this arc.

Praise for *Eco-Social Contracts for Sustainable and Just Futures*

"Our success in addressing the triple planetary crises and securing a sustainable future for all requires that we transform our societies and our economies to be in harmony with nature. This volume provides a constructive and progressive contribution to this transformative journey through exploring new ways of relating to our environment. I hope it informs and inspires further urgent action at all levels for the benefit of nature and people."
—Inger Andersen, *Under-Secretary-General of the United Nations and Executive Director of the United Nations Environment Programme*

"The very essence of our future lies in recalibrating our focus. As this book convincingly argues, we must transition from systems that devalue life to economies that cherish sustainable well-being and comprehensive prosperity."
—Kumi Naidoo, *President of the Fossil Fuel Non-Proliferation Treaty Initiative, Former Secretary-General of Amnesty International and Executive Director of Greenpeace International*

"In a time where the world is going through multiple crises and people are losing trust, this book shows that collectively we have the power to fix our broken social contracts, to restore planetary health and to work together as a global community for inclusive and sustainable futures. The vision of eco-social contracts has guided UNRISD's work over the last years, culminating in this volume—an enlightening combination of theoretical and conceptual insights with empirical studies on how eco-social contracting and transitions happen on the ground. Important and inspiring!"
—Magdalena Sepúlveda Carmona, *Director, United Nations Research Institute for Social Development (UNRISD)*

"Transformation is not only possible—it is already happening. This book shows how we can recognize ourselves as agents of change, capable of engaging with the quantum potential of the world around us. When we do, new realities become thinkable and actionable."
—Karen O'Brien, *Co-Chair of Transformative Change Assessment (TCA) of the Intergovernmental Science-Policy Platform on Biodiversity and Ecosystem Services (IPBES) and Professor at University of Oslo*

"In a time of geopolitical and socio-economic fractures, ensuring social contracts that provide some sense of solidarity and cohesion is essential. This important book provides insights into possibilities for eco-social contracts that encompass social, environmental, economic, cultural, and institutional dimensions, recognizing the rights and duties of care for others and for nature that are essential for our common survival. These are not just 'feel-good' phrases: the varied contributions provide clear and viable pathways for this. Some light in the seeming darkness!"

—Jayati Ghosh, *Club of Rome, UN High-Level Advisory Board on Economic and Social Affairs, WHO council on the Economics of Health for All, and Professor of Economics, University of Massachusetts Amherst, USA*

"The rapidly worsening climate, nature and inequality crises confront us with an existential challenge. This book makes clear that a new eco-social contract must be built on structural justice, intergenerational equity, and inclusive governance that respects cultural and ecological diversity."

—Mary Robinson, *President of Ireland (1990–97), UN High Commissioner for Human Rights (1997–2002), and Chair of The Elders*

"The time has come to not only talk about a different social contract, but to act collectively in reshaping it—to ensure that it is inclusive, regenerative, and rooted in both justice and sustainability. This book shows how such a transformation is not only necessary, but possible—through bold ideas, shared values, and collective action."

—Carlos Alvarado Quesada, *President of the Republic of Costa Rica (2018–22), Professor of Practice in Diplomacy, The Fletcher School at Tufs University, and Richard von Weizsäcker Fellow at the Robert Bosch Academy*

"Young people feel the urgency of the climate crisis, but they lack the perspective of a structural solution. For a shared vision of the future, we need new forms of involvement. The eco-social contract offers an inspiring example of how we can build a sustainable future together."

—Daan Zieren, *Chair Youth Climate Movement (The Netherlands), part of the We Are Tomorrow Global Partnership (WAT-GP)*

"The climate crisis, and economic systems that facilitate inequality are disrupting lives, livelihoods, economic systems and even societal norms, in every part of the world. This new body of work, explores how renewed social contracts can serve as frameworks for just and system-wide transformations in the face of climate breakdown, biodiversity loss and inequality. It deserves to be read and discussed in all fora, and read by all government officials who signed the Paris Agreement and need tools to help adapt to a world where people, climate, and nature are in harmony."

—Manuel Pulgar-Vidal, *WWF Global Climate and Energy Lead, and Former COP20 President*

"As we seek a whole-of-society approach to implement the Global Biodiversity Framework, this publication offers us important insights into the role of 'eco-social contracts' i.e. collective agreements across different levels of governance and society that can help ensure we leave a better planet for future generations. Central to this idea is the reality that without nature we are nothing. Reimagining the relationship between people and the natural world is critical to tackle the interconnected challenges before us, be it climate change, nature loss and land degradation, and pollution and waste."

—Elizabeth Maruma Mrema, *Deputy Executive Director, UNEP and Executive Secretary, Secretariat of the Convention on Biological Diversity (2020–23)*

"This is a groundbreaking and timely contribution to the urgent global dialogue on sustainability and equity. This book brings together diverse perspectives—from Indigenous wisdom to progressive policy frameworks—to present a compelling vision for a new eco-social contract. It challenges us to rethink our relationship with nature, not as a resource to be exploited, but as a partner in a shared future. The chapters offer actionable insights for policymakers, activists, and academics, demonstrating that transformative change is not only possible but already underway in communities worldwide. This book is a must-read for anyone committed to building a just and sustainable world."

—Carina Bachofen, *Convenor, Green Economy Coalition*

"At a time when our social and ecological fabrics are both fraying, this book reminds us that renewal begins with re-connection. Re-connection with each other, with nature, and with what truly matters. *Eco-Social Contracts* are not abstractions, but living agreements about how we inhabit the Earth together. This is an inspiring guide for anyone who believes finance, policy and community can once again serve life."

—Hans Stegeman, *Chief Economist, Triodos Bank*

"With humanity teetering on the brink of interconnected ecological, social, and economic catastrophes, we desperately need new visions and fresh ideas. This thought-provoking new book answers the call with imagination, insights, and compelling evidence that a more just, beautiful, and happy world remains within reach. Eco-social contracts offer a powerful tool for catalyzing the urgently sought transformations."

—David Boyd, *UN Special Rapporteur on human rights and environment (2017–24) and Associate Professor at University of British Columbia*

"This book represents a crucial and timely voice to the urgent need for transformative change and global sustainability. The book reframes the human–nature relationship as one of mutual responsibility and collective well-being. It bridges ethical, political, and practical dimensions to guide transformative change by using real-world examples. The eco-social contracts provide new imaginative and promising ways which can inform visioning processes with the transformative potential to change mindsets and behaviours needed for transformative change across actors and sectors for a just and sustainable world."

—Sebastian Villasante, *Director EqualSea Lab, and Researcher Professor at University of Santiago de Compostela*

"Young people want to take action. The eco-social contract finally offers them a shared starting point: not endless talk about change or getting stuck in compromise, but moving together toward a just future in which nature, people, and food are meaningfully connected."

—Maria Geuze, *Former Director, Slow Food Youth Network*

"When both humanity and the planet are threatened by dramatic climate and environmental devastation, by war and possible nuclear annihilation and extreme and growing inequalities, and political leaders do not seem to be up to the task, 'we, the peoples' have serious work to do to meet the challenges. This publication with its comprehensive and solution-oriented approach is of great inspiration in the quest for a broad transformative movement towards a world based on the principles of a culture of peace and sustainability."

—Ingeborg Breines, *Norwegian peace educator, Former Director of the Women and a Culture of Peace Programme of UNESCO, and President of the International Peace Bureau*

"This book offers an inspiring vision of how new eco-social contracts can help regenerative futures. This resonates strongly with IUCN's vision of a just world that values and conserves nature and the conviction that governance arrangements have a very important role to play in this."

—Liliana Jauregui, *Director, IUCN NL*

"This groundbreaking volume offers a timely and visionary framework for reimagining our shared future in the face of intersecting ecological, social and governance crises. Eco-Social Contracts for Sustainable and Just Futures powerfully weaves diverse global perspectives into a compelling call for inclusive, regenerative and justice-oriented transformation. It is both a scholarly achievement and a practical guide for collective action that speaks to changemakers across disciplines, sectors and communities."

—Daryl Swanepoel, *Research Fellow at the School for Public Leadership at Stellenbosch University and the Chief Executive Officer of the Inclusive Society Institute, and Former South African Member of Parliament*

"The book offers a great many valuable insights, experiences and proposals for a path to a new, eco-social contract. Armed with knowledge the reader can consider how best to contribute towards creating the political will to mainstream these insights, experiences and proposals."

—Kerstin Leitner, *UN Resident Coordinator, China (1998–2003), Assistant Director-General, WHO, Geneva, in charge of Health and Environment (2003–2005), at present, Board Member UNA Germany*

Contents

About the Editors and Contributors

About the Editors

Patrick Huntjens is Professor of Social Innovation and Sustainability Transitions at Inholland University of Applied Sciences (2018–present), and Maastricht University (2021–2024), and member of the Global Research and Action Network for a New Eco-Social Contract (GRAN-ESC). An award-winning scholar and practitioner, he received the 2022 Nautilus Book Award (Gold Medal) for *Towards a Natural Social Contract* (2021) and was named Professor of the Year in the Netherlands (2021). With over 25 years of global experience across academia, policy, and practice, Patrick focuses on eco-social contracts, regenerative economies, sustainability governance, and transformative change, with a strong commitment to social and environmental justice. He is a lead author for IPBES, whose globally recognised work received the 2022 Gulbenkian Prize for Humanity and the 2024 Blue Planet Prize. Patrick has also served as a lead mediator in the Israeli-Palestinian water conflict and advised institutions including the World Bank, United Nations, and European Union.

Najma Mohamed leads UNEP World Conservation Monitoring Centre's work on nature-based solutions and is a Senior Atlantic Fellow in Social and Economic Equity and member of the Africa Europe Foundation's Women Leaders Network. Najma is member of the Global Research and Action Network for a New Eco-Social Contract (GRAN-ESC). An interdisciplinary scholar and practitioner, her work focuses on ideas and solutions that address climate change, fight inequality, and restore nature to achieve systemic change. From her early work on environmental justice in post-apartheid South Africa, she has maintained her long-standing interest in transformative and inclusive approaches to governance, policy, and practice. She edited the volume *Sustainability Transitions in South Africa* (Routledge 2019) exploring South Africa's sustainability transition through reflections on critical policy, economic, technological, social and environmental drivers of change. https://www.routledge.com/Sustainability-Transitions-in-South-Africa/Mohamed/p/book/9780367500382.

Katja Hujo is Head of the UNRISD Bonn office and leads the Transformative Social Policy Programme. Katja's academic work focuses on social policy, poverty, and inequality, as well as socio-economic development and the sustainability transition. She studied economics and political science at Eberhard-Karls-University Tübingen, Freie Universität Berlin (FUB) and National University of Córdoba, Argentina, and holds a doctoral degree in economics from FUB. At UNRISD, she is lead author and coordinator of the 2022 flagship report Crises of Inequality: Shifting Power for a New Eco-Social Contract and co-editor of Between Fault Lines and Front Lines: Shifting Power in an Unequal World (Bloomsbury June 2022). In partnership with the Green Economy Coalition, she represents UNRISD as co-convenor of the Global Research and Action Network for a New Eco-Social Contract (GRAN-ESC).

Manisha Desai is the Executive Director of Center for Changing Systems of Power and the Empowerment Trust Endowed Professor of Global Citizenship at Stony Brook University and part of the Global Network for Research and Action for a New Eco-social Contract. Her areas of research and teaching include gender and globalisation/development, transnational feminisms, global justice, particularly climate justice movements and human rights. She is the recipient of multiple awards, including the Sociologist for Women in Society's 2015 Distinguished

Feminist Award and the 2016 Faculty Mentor Award from the Compact for Faculty Diversity in the USA. She has served in many leadership capacities including as President of Sociologist for Women in Society.

About the Contributors

Alina Saba is an Indigenous activist of the Limbu Indigenous Peoples from the Eastern hills of Nepal. She works on Indigenous Peoples' rights and climate justice with International Funders for Indigenous Peoples (IFIP). She is a board member of the National Indigenous Women Forum (NIWF), Nepal. She has worked with social movements such as climate justice, the Indigenous Peoples movement, and human and labour rights in Nepal, Thailand, and globally. She holds an MA in Politics and Public Policy, University of Sheffield, UK. Her MA thesis was on "Identifying the challenges of addressing diversity in Nepal through federalism".

Ashish Kothari is a founder-member of Kalpavriksh and active in many people's movements in India and transnationally. He has taught at the Indian Institute of Public Administration and coordinated India's National Biodiversity Strategy & Action Plan. He serves on the boards of Greenpeace International & India, ICCA Consortium and as a judge on the International Tribunal on Rights of Nature. He helps coordinate Vikalp Sangam (www.vikalpsangam.org), Global Tapestry of Alternatives (www.globaltapestryofalternatives.org), & Radical Ecological Democracy (www.radicalecologicaldemocracy.org). He is the co-author of *Churning the Earth*, *Alternative Futures*, and co-editor of *Pluriverse: A Post-Development Dictionary*.

Caitlyn E. Sutherlin serves as the facilitator for the Global Alliance for the Rights of Nature (GARN) North America Hub, and co-facilitator of the GARN Academic Hub. She holds an M.S. in Environmental Science and Policy from Northern Arizona University and currently resides in the Upper Peninsula of Michigan in the USA where she is working towards her Ph.D. in Environmental and Energy Policy at Michigan Technological University. Her research looks at the intersection between local climate change adaptation and eco jurisprudence principles in California, Usulután, El Salvador.

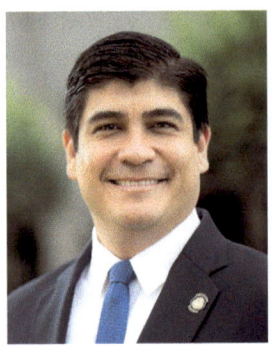

Carlos Alvarado Quesada served as Costa Rica's 48th President (2018–2022). Known for his commitment to democracy, human rights, and climate action, he launched Costa Rica's National Decarbonization Plan in 2019, the first of its kind since the Paris Agreement. Under his leadership, Costa Rica's response to the COVID-19 pandemic was highly successful and widely recognised. Alvarado received several recognitions, including TIME's 100 Next (2019) and the Planetary Leadership Award from the National Geographic Society (2022). He holds degrees in journalism, political science, and development studies and currently teaches diplomacy at the Fletcher School of Tufts University.

Carlos Emiliano Villaseñor Moreno is a political scientist at the Autonomous Technological Institute of Mexico (ITAM) and a current candidate for an M. Phil. in Conflict Resolution and Reconciliation with Trinity College Dublin exploring the climate change-conflict nexus. For the last four years Carlos has worked in the fields of just energy transitions and the right to energy with Ombudsman Energía México as well as climate mitigation policy through carbon pricing with the Mexican carbon platform, MÉXICO.

Catherine Haas is the Director of the Eco Jurisprudence Monitor (EJM), where she oversees the monitoring and evaluation of Rights of Nature and ecological law initiatives worldwide. She holds a Master's in Global Studies from the University of North Carolina at Chapel Hill, where her research focused on pluriversality and Indigenous ontology within the global Rights of Nature movement. Additionally, she serves as a Steering Committee member of the Global Alliance for the Rights of Nature (GARN) Academic Hub.

Chris Walker is a professor in the Dance Department at the University of Wisconsin–Madison and Special Advisor to the Provost on the Arts. He is the founding artistic director of the First Wave Hip Hop and Urban Arts Program. A contemporary dance artist and scholar from Jamaica, Walker's work investigates African Caribbean dance as a lens for environmental activism, social justice, and cultural preservation, bridging creative practice with academic inquiry.

Danielle Hersch-Castros is Programme Officer: Strategic Partnerships, African Climate Foundation (ACF), a philanthropic grant-making organisation that seeks to advance both climate and development objectives across the African continent. She obtained her PGDip in African Philanthropy and Resource Mobilisation from Wits Business School/The Centre on African Philanthropy and Social Investment and holds a BA in Brand Building and Management. Prior to working at the ACF, she worked at an intellectual property law firm where she provided counsel to Chinese enterprises navigating the African market. Danielle believes in the power of collaboration to drive meaningful change and is passionate about philanthropy that goes beyond mere financial contributions, shifting philanthropic practices towards long-term, sustainable grant making that values local perspectives and sees communities as capable, knowledgeable actors in development.

Davide Sofia is a Programme Executive at Friends of Europe, working in the Climate, Energy and Natural Resources expertise area and leading on the engagement of the European Young Leaders (EYL40) programme. Before joining Friends of Europe, he has been extensively involved in civil society organisations such as "Libera contro le mafie", working on active citizenship, fight against organised crime and corruption. He holds an MA in European and International Studies from the University of Trento, where he developed a dissertation on the economics of European integration and analysed the economic impact of legal and administrative obstacles in European cross-border regions.

Erin McCandless is Acting Director of the Qatar-South Africa Centre for Peace and Intercultural Understanding at the University of Johannesburg. A seasoned academic and policy advisor with three decades of experience in 25 countries, her current research explores rising complexity in conflict-crisis contexts, the role of emerging and non-Western actors in mediation and sustaining peace, principled movements for a just international order, and climate justice within fragile states. Her expertise includes leading multi-stakeholder applied research, conducting trainings, and facilitating strategy, visioning and policy processes with UN agencies, international organisations, civil society and activists.

Gabriele Koehler is a development economist and an UNRISD Senior Research Associate. From 1983 to 2010, she held positions as UN Resident Coordinator in Latvia, UNICEF Regional Social Policy Advisor South Asia, and Special Assistant to the UNCTAD Secretary-General. She researches the UN Agenda 2030, eco-social policies, and business and human rights. She is on the board of Women Engage for a Common Future (WECF), a member of the UNICEF National Committee Germany, and an independent advisor to the Peoples'20 network.

Georgios Kostakos is Co-founder and Executive Director of the Brussels-based Foundation for Global Governance and Sustainability (FOGGS), which focuses among other things on the need for a new globalisation narrative, the establishment of a Global Resilience Council, rethinking education for the digital era, and supporting SDG implementation. About half of Georgios' thirty-years+ work experience has been with the UN, including with the Executive Office of the UN Secretary-General, the High-level Panel on Global Sustainability, the UN Framework Convention on Climate Change (UNFCCC), and field missions for political affairs and human rights. The other half of his career has been with think tanks, academic institutions and as a consultant on global governance for sustainability, peace and resilience.

Hisayama Yuho is an Associate Professor for European Literature as well as the Director of Kobe Institute for Atmospheric Studies (KOIAS) at Kobe University, Japan. He received his B.A. (2006) and M.A. (2008) from Kyoto University (Japan), his first Ph.D. (2013) from TU Darmstadt (Germany) and his second Ph.D. (2021) from Kyoto University. Currently he works von Goethe, History of Ideas (*pneuma*, Geist, *qi /ki / ke*) and cross-cultural and cross-disciplinary atmospheric studies. He is the author of the book *Erfahrungen des* ki. *Leibessphäre, Atmosphäre, Pansphäre* (Freiburg und München 2014).

Karen O'Brien is a professor of human geography at the University of Oslo and co-founder of cCHANGE, an organisation that engages with society to promote and accelerate transformative change for a just and sustainable world. Her research focuses on the human and social dimensions of environmental change, with an emphasis on scaling transformations to sustainability. Karen's recent books include *You Matter More Than You Think: Quantum Social Change for a Thriving World* and *Climate and Society: Transforming the Future* (with Robin Leichenko). She was co-recipient of the BBVA Foundation's Frontiers of Knowledge Award for Climate Change in 2021 and was co-chair of the IPBES Transformative Change Assessment (2024).

Kazuhiko Ota is an Associate Professor in the Department of Policy Studies at Nanzan University and a Visiting Associate Professor at the Research Institute for Humanity and Nature (RIHN). He explores the recent partial slowdown in the Great Acceleration and its significance for sustainable transitions. His work spans food and agricultural ethics, environmental ethics, and serious games. Ota's major academic contributions include translating Paul B. Thompson's *The Spirit of the Soil* (1995), *From Field to Fork* (2015), and Giorgos Kallis's *LIMITS* (2019), as well as co-editing volumes on urban green space and hosting the Asia-Pacific Conference on Agricultural and Food Ethics (APSAFE). He is an editorial board member for the journals *Food Ethics* and 環境倫理 (*Environmental Ethics*), and leads Serious Board Game Jam initiatives, in which participants collaboratively design games to enhance dialogue and mutual learning about complex social and environmental issues.

Kiah Smith is a Sociologist and Senior Research Fellow at the Centre for Policy Futures, The University of Queensland. She currently leads the Fair Food Futures project that explores how civil society is re-imagining what a better food system might look like in response to multiple crises. Kiah has published widely on food security, the right to food, gender empowerment, sustainable livelihoods, resilience, financialisation, ethical trade, green economy, and food governance, and regularly provides expert sustainability advice to the United Nations and various civic food organisations. More information about Fair Food Futures—including the podcast—can be found at https://fairfoodfutures.com

Kumi Naidoo is a South African human rights and environmental justice activist, who currently is the President of the Fossil Fuel Non-Proliferation Treaty and the Payne Distinguished Lecturer at Stanford University. He is the former Secretary-General of Amnesty International (2018–2020) and also the first person from the Global South to lead Greenpeace International (2009–2015). His family started the Riky Rick Foundation for the Promotion of Artivism. Kumi is also the author of award-winning *Letters to My Mother: The Makings of a Troublemaker* and the host of the podcast Power, People and Planet.

Lauren Tarr lives on the shores of Lake Michigan in the midwest United States. Since 2020 she has been helping water the seeds of Ecological Jurisprudence through her work with the Global Alliance for the Rights of Nature. She is currently an Environmental Science PhD Candidate at the State University of New York College of Environmental Science and Forestry (SUNY ESF). She has a Masters of Public Administration from Syracuse University, and a Masters of Environmental Studies from SUNY ESF. Her work and studies focus on restoring positive human-environmental relationships.

Lysa John currently serves as the Executive Director of the Atlantic Institute, based in Oxford, which supports a global network of social change leaders. Lysa has a background in human rights, international development and social justice, with over 25 years of experience in large-scale change initiatives. Her professional journey includes supporting urban poor movements in India, leading transnational campaigns on governance accountability and serving as the Head of Outreach for the UN High Level Panel on the Post-2015 agenda. She has also held leadership roles as Global Campaigns Director at Save the Children and as Secretary General of the global civil society alliance, CIVICUS. Lysa has contributed to several networks in an advisory capacity, including the Global Partnership for Sustainable Data Development, the International Budget Partnership, the Open Government Partnership, and the World Benchmarking Alliance. She stays grounded through yoga, meditation, and gratitude for the worldwide network of friends, family, and fellow activists who nurture and inspire her.

Maja Groff, an international lawyer based in The Hague, is Convenor of the Climate Governance Commission, which seeks to propose high-impact global governance innovations adequate to meet the climate challenge. She serves as Co-Chair of the Coordinating Committee and Senior Treaty Advisor for the International Anti-Corruption Court (IACC). She has previously worked on the development and implementation/administration of multiple multilateral treaties. She was a Visiting Professor at Leiden University and is now Chair of the Planetary Governance Program, The New Institute. She was co-winner of a major international prize on global governance innovation (New Shape Prize), and is co-author of the book, *Global Governance and the Emergence of Global Institutions for the Twenty-first Century* (Cambridge). She serves on advisory boards for B for Good Leaders and ebbf, organisations devoted to ethical business.

Mary Robinson is Member of The Elders and former Chair; First woman President of Ireland and former UN High Commissioner for Human Rights; a passionate advocate for gender equality, women's participation in peace-building, human dignity and climate justice. Mary Robinson is a globally recognised voice on climate change and frequently highlights the need for drastic action from world leaders, as well as the intersectionality of the climate emergency: from intergenerational injustice to gender inequality and biodiversity loss. She is also honorary co-president of the Africa Europe Foundation (AEF), launched in 2020 to reset and bolster Africa-Europe relations, and co-Chair of AEF's Women Leaders Network. Mary is the author of two books, *Everybody Matters* and her most recent work *Climate Justice-Hope, Resilience, and the Fight for a Sustainable Future*.

Neema Pathak Broome is a member of Kalpavriksh and coordinates the Conservation and Livelihoods programme, as part of the programme she is engaged in supporting community-led efforts for conservation and well-being; self-strengthening of community governance, knowledge and management systems; and community-led or collaborative knowledge production. Neema is also part of the South Asia regional coordination team for ICCA Consortium and also the International Policy Coordinator for ICCA Consortium.

René Kemp conducts research into transitions to a circular economy, a low-carbon energy system and the governance of sustainability transitions. For the Dutch government he developed a model of transition management (with Jan Rotmans) and for the province of Limburg and Chemelot he currently investigates policies for achieving a transition to plastics recycling. In the last 10 years, he developed a keen interest in the humanisation of the economy through transformative social innovation and an alternative social contract. He has published in innovation journals, environmental and ecological economics journals, policy journals, transport and energy journals, sustainable development journals and transition studies journals. He even published an article in a biology journal (*Philosophical Transactions—B*) about how actors are myopically called in processes of co-evolution. In his research he pays equal attention to possibilities for steering and the limits of steering approaches. As a professor of innovation and sustainable development at Maastricht University he tries to do research that contributes to a better world. His research became more multidisciplinary and transdisciplinary and includes studies of citizenship and justice.

Saliem Fakir is Executive Director, African Climate Foundation. Prior to establishing the African Climate Foundation, Saliem served for 11 years as the Head of the Policy and Futures Unit of WWF South Africa. He was a Senior Lecturer at the Department of Public Administration and Planning and an Associate Director for the Centre for Renewable and Sustainable Energy at Stellenbosch University. For eight years, Saliem was the Director of the International Union for Conservation of Nature (IUCN) in South Africa. He has served on several Boards, currently serving as the chairman on the board of Atlantis Special Economic Zone Company SOC Ltd., and is a prolific writer, contributing to leading publications.

Shrishtee Bajpai is a researcher, writer, and activist working on themes of interspecies justice, earthy governance, and systemic transformations. She is a member of Kalpavriksh, an environmental action group in India, and coordinates the Alternatives Programme within the group. She helps coordinate the Vikalp Sangam (Alternatives Confluence) network that researches, documents, networks around systemic alternatives in India, is the core team member of Global Tapestry of Alternatives, and part of the Emerging Futures: Visionaries Programme of Joseph Rowntree Foundation. She also serves on the executive committee of Global Alliance for the Rights of Nature.

Teresia Teaiwa (1968–2017) was an I-Kiribati and African-American South Pacific Studies scholar and poet. A prominent educator, academic innovator, and poet, she directed the Va'aomanū Pasifika unit (Pacific Studies and Samoan Studies) at Victoria University in Wellington, Aotearoa-New Zealand, the first and only programme to offer a Ph.D. in Pacific Studies. Teresia served as the co-editor of the *International Journal of Feminist Politics* from 2008 to 2011. She was also a committed activist against militarisation and for sovereignty in the Pacific region.

Yogi Hale Hendlin is an environmental philosopher and public health scientist, working at the intersection of complex systems theory, the commercial determinants of health, environmental justice, and biosemiotics. Currently based at Erasmus University Rotterdam, an Assistant Professor in the Erasmus School of Philosophy and Dynamics of Inclusive Prosperity Initiative, Yogi has co-edited *Food and Medicine: A Biosemiotic Perspective* as well as *Being Algae: Transformations in Water, Plants*, and since 2021 has served as Editor-in-Chief of the journal *Biosemiotics*. Yogi's work on this chapter has been funded by the Research Institute on Humanity and Nature, in Kyoto, Japan, through the project "Diagnosis and Treatment to Multisolve for Ecological Civilization".

Abbreviations

ABS	Australian Bureau of Statistics
ACF	African Climate Foundation
AFNQL	Assembly of First Nations Quebec-Labrador
AGDA	Anwar Gargash Diplomatic Academy
AI	Artificial Intelligence
AIPP	Asia Indigenous Peoples Pact
ANC	African National Congress
APIB (in Portuguese)	Articulation of Indigenous Peoples of Brazil
ARIP	Adaptation and Resilience Investment Platform
ATF	Alternative Transformation Framework
AU	African Union
BJP	Bharatiya Janata Party
BMCs	Biodiversity Management Committees
BPL	Bijzonder Provinciaal Landschap
CAA	Citizenship Amendment Act
CAN Europe	Climate Action Network Europe
CAP	Common Agricultural Policy
CAR	Central African Republic
CBD	Convention on Biological Diversity
CBDR	Common But Differentiated Responsibilities
CCAs	Community Conserved Areas
CCUS	Carbon Capture, Utilisation and Storage
CESC	Climate Emergency Social Contract
CFNs	Civic Food Networks
COP	Conference of Parties
CPN-UML	Communist Party of Nepal (Unified Marxist-Leninist)
CSA	Community-supported Agriculture
CSDDD	Corporate Sustainability Due Diligence Directive

CSOs	Civil Society Organisations
CSRD	Corporate Sustainability Reporting Directive
DAOs	Decentralised autonomous organisations
DIY	Do-it-yourself
EEB	European Environmental Bureau
EGD	European Green Deal
EU	European Union
EZLN (in Spanish)	National Zapatista Liberation Army
FAO	Food and Agriculture Organization
FCRA	Foreign Contribution Regulation Act
FFF	Fridays for Future
FOGGS	Foundation for Global Governance and Sustainability
FRA	Forest Rights Act
GARN	Global Alliance for the Rights of Nature
GBF	Global Biodiversity Framework
GCAP	Global Call to Action Against Poverty
GCF	Green Climate Fund
GDP	Gross Domestic Product
GEF	Global Environment Facility
GHG	Greenhouse Gas Emissions
GND	Green New Deal
GoI	Government of India
GPDP	Gram Panchayat Development Plan
ICESCO	Islamic World Educational, Scientific and Cultural Organization
ICJ	International Court of Justice
IEADB	International Environmental Agreements Database
IFEES	Islamic Foundation for Ecology and Environmental Science
IGD	Institute for Global Dialogue
IMF	International Monetary Fund
INC	Indian National Congress
IoT	Internet of Things
IPBES	Intergovernmental Science-Policy Platform on Biodiversity and Ecosystem Services
IPCC	Intergovernmental Panel for Climate Change
IPES-Food	International Panel of Experts on Sustainable Food Systems
ISI	Institute for Security and International Studies
ITLOS	International Tribunal for the Law of the Sea
JET-P	Just Energy Transition Partnership
JOCA (in Spanish)	Youth for Climate Argentina
KNOCA	Knowledge Network on Climate Assemblies
LAHDC	Ladakh Autonomous Hill Development Council
LCA	Lifecycle Assessments

LGBTQI	Lesbian, Gay, Bisexual, Transgender, Queer and Intersex
LTK	Local Traditional Knowledge
LVVN (in Dutch)	Ministry of Agriculture, Fisheries, Food Security and Nature (Netherlands)
MENA	Middle East and North Africa
MGS	Maha Gram Sabha
MOCAF (in Spanish)	Mexican Network of Forest Peasant Organizations
NAPs	National Adaptation Plans
NBSAPs	National Biodiversity Strategies and Action Plans
NCDHR	National Campaign on Dalit Human Rights
NDCs	Nationally Determined Contributions
NDTV	New Delhi Television Limited
NEFIN	Nepal Foundation of Indigenous Nationalities
NGEU	Next Generation EU
NAP	National Adaptation Plan
NGN	NAP Global Network
NGOs	Non-Governmental Organisations
NHK	Japanese Broadcasting Corporation
NIWF	National Indigenous Women's Federation
ODI	Overseas Development Institute
OIC	Organisation of Islamic Cooperation
PCA	Permanent Court of Arbitration
PCC	Presidential Climate Commission (South Africa)
PES	Payments for Ecosystem Services
PESA	Panchayat Extension to Scheduled Areas Act
PRS	Panchayati Raj System
RoN	Rights of Nature
RRF	Recovery and Resilience Facility
SDGs	Sustainable Development Goals
SFSCs	Short Food Supply Chains
SMEs	Small and Medium Enterprises
TEK	Traditional Ecological Knowledge
TFA	Transformation Flower Approach
TG	Transformative Governance
TM	Transition Management
TSEI	Transformative Social-Ecological Innovation
TT	Transition Towns
UK	United Kingdom
UN	United Nations
UNDP	United Nations Development Programme
UNDRIP	United Nations Declaration on the Rights of Indigenous Peoples
UNDRR	United Nations Office for Disaster Risk Reduction

UNEP	United Nations Environment Programme
UNFCCC	United Nations Framework Convention on Climate Change
UNGA	UN General Assembly
UNRISD	United Nations Research Institute for Social Development
UNSC	UN Security Council
UNSG	United Nations Secretary-General
UPA	United Progressive Alliance
US	United States
VC	Village Council
VDB	Village Development Board
VS	Vikalp Sangam
WCED	World Commission on Environment and Development
WEGo	Wellbeing Economy Governments
WMO	World Meteorological Organization
WSSD	World Summit on Sustainable Development
WWII	World War Two

List of Figures

List of Tables

1

Foreword: The Transformative Potential of Eco-Social Contracts

Karen O'Brien

We live in a world of change. Many of us have difficulty keeping up with the accelerating pace of change driven by new technologies and contentious geopolitics. Yet the current direction of change is deeply troubling. Human activities are contributing to the degradation and loss of nature, the acceleration of climate change, threats to oceans and ice cover, and growing inequality. When we add to these the increasing risks of reaching irreversible biophysical tipping points, the current pace and direction of change is extremely dangerous.

We can do better. In fact, *it is possible to deliberately transform* ourselves and societies to realise a just and sustainable world (IPBES, 2024). We have the knowledge, tools, and resources to act. The question is: How do we unlock our transformative potential?

This is where social contracts come in—specifically, eco-social contracts. Social contracts are collective agreements that define the rights and responsibilities between people and community leaders or governments. They can be formally recognised through constitutions and legislation, or they can be informal, expressed through unseen and unwritten social and cultural norms that influence how communities are organised and what is prioritised. Building on social contract theory and practice, this book defines eco-social contracts as *"implicit or explicit collective agreements across multiple levels of governance, among members of society, aimed at addressing the interconnected*

K. O'Brien (✉)
Oslo University, Oslo, Norway
e-mail: karen.obrien@sosgeo.uio.no

© The Author(s) 2025
P. Huntjens et al. (eds.), *Eco-Social Contracts for Sustainable and Just Futures*,
https://doi.org/10.1007/978-3-031-99109-7_1

1

polycrisis of the twenty-first century, including inequalities, injustices, climate and ecological breakdown, and faltering trust in institutions. These agreements are rooted in cooperation and the recognition of shared norms and values oriented toward sustainability, equity, and justice. Importantly, eco-social contracts encompass social, environmental, economic, cultural, and institutional dimensions, and articulate the corresponding rights and duties of care for the environment and the well-being of others, including future generations and all forms of life on Earth" (Huntjens & Kemp, 2025).

The ideas, insights, elements, and examples of eco-social contracts discussed in the pages ahead present a powerful point of departure for creating pathways towards a just and sustainable world. The authors describe the many ways that such contracts are emerging in society, linking to theoretical and normative approaches grounded in moral and political philosophy. We learn that eco-social contracts are, in fact, being revived, reimagined, and renegotiated around the world.

This is promising at a time when social contracts are coming undone. We need new approaches to transforming views, structures, and practices. These must be deep, systemic changes—not superficial adjustments. The IPBES Transformative Change Assessment (IPBES, 2024) highlights the importance of *addressing the underlying causes* of biodiversity loss and nature's decline, which include the disconnection from and domination over nature and people; concentration of power and wealth; and prioritisation of short-term, individual, and material gains (IPBES, 2024). These underlying causes represent deep and persistent patterns in society that influence the drivers of interconnected global problems.

Unfortunately, at a time when we should be mobilising our collective wisdom, capacities, and resources to address interconnected problems, the underlying causes of nature's decline are becoming even more entrenched. Although millions of people are working hard to create a better world for people and nature, shifting systems and structures is not easy. It involves relating to people and nature differently and activating change through individual agency, collective agency, and political agency. This includes showing up fully and embodying a quality of agency that transforms systems and cultures across all scales. Above all, it is about generating new patterns based on a recognition that our deepest values are sources of individual change, collective change, and systems change (O'Brien, 2021).

The IPBES assessment identifies four principles of transformative change that contribute to shifting views, structures, and practices: equity and justice; pluralism and inclusion; respectful and reciprocal human–nature relationships; and adaptive learning and action (IPBES, 2024). Amplifying and scaling transformative change for a just and sustainable world involves consistently applying principles to specific structures and practices. This may include, for example, transforming the dominant economic paradigms that prioritise private interests over nature and social equity, or promoting more inclusive, accountable, and adaptive governance systems (IPBES, 2024).

This volume is timely because contemporary social contracts have not kept up with changing views of human–nature relationships. Recall that the United Nations Universal Declaration of Human Rights recognises the inherent dignity and of the equal and inalienable rights of all members of the human family, emphasising the foundation of freedom, justice, and peace in the world. This recognition was a milestone in 1948. However, 75 years later we need to broaden what we mean by the human family and expand our ethics of care. After all, scientific views of the world have changed rapidly over the past century, and many people have a deeper appreciation of the multiple values that people hold, including relational values of Indigenous and local knowledge systems and wisdom traditions that can help achieve a world living in harmony with nature.

For example, over one hundred years ago, quantum physics revealed that reality is not what it seems, and we increasingly recognise that social reality is not what it seems; we are, in fact, entangled with each other and with nature (Barad, 2007; O'Brien, 2021). Research on mirror neurons shows we are wired to connect with each other, and countless studies document how we are reconnecting with nature (Lieberman, 2013; Richardson, 2023). Research on social consciousness has revealed that meaning making can change over time, enabling us to expand our circles of care (Schlitz et al., 2010). Publication of the *Earthrise* photo taken on the Apollo 8 mission in 1968 showed us that we are all part of a much larger whole. Since then, more and more people recognise the importance of taking care of this whole.

Let's be clear: we have a long way to go. Many societies still do not recognise the rights of all people, species, and non-human entities. Some groups have been explicitly excluded from social contracts. In contrast, corporations have increasingly been included in formal and informal state-society agreements; they have been given rights, yet without responsibilities to citizens (O'Brien et al., 2009). Nature is still viewed by many as separate from humans and thus treated as a resource to be exploited or polluted for instrumental, opportunistic, and material gains. The sacred, spiritual dimensions of nature and human–nature relationships have also been discounted or ignored. Yet, Indigenous knowledge and wisdom traditions have long emphasised the significance of our relationship to nature. For example, Haudenosaunee Faithkeeper Oren Lyons points out that "What you people call your natural resources our people call our relatives". He reminds us that "we share the same river of life. What befalls me befalls you. And downstream, downstream in this river of life, our children will pay for our selfishness, for our greed, and for our lack of vision".

Our lack of vision has made us nearsighted. The focus on short-term gains has left us blind to the long-term consequences of our actions, both for people and nature. Many countries are backsliding on universal rights, while democratic institutions and global governance are increasingly threatened. Existing social contracts are crumbling. They are not capable of responding to the destruction of nature, the displacement of people, and the challenges of global sustainability. In short, many of today's shared agreements about governance reflect an outdated, human-focused worldview. However, like life, these agreements represent norms and relationships that are alive and changing. It is time for them to evolve into eco-social contracts that reflect a deeper, more integrated understanding of human–nature relationships.

It is time for a new social contract, and as the authors of this volume emphasise, it must be an eco-social contract. This volume provides hope and optimism that we can do better. The diversity of voices and perspectives in this volume underscores the potential for transformative change—and eco-social contracts provide a powerful, tangible framework for making this transformation a reality.

References

Barad, K. (2007). *Meeting the universe halfway: Quantum physics and the entanglement of matter and meaning*. Duke University Press Books.

Huntjens, P., & Kemp, R. (2025). The transformation flower approach for eco-social contracting: Comparative insights from eight case studies in the Global South and North. In P. Huntjens, N. Mohamed, K. Hujo, & M. Desai (Eds.), *Eco-social contracts for sustainable and just futures* (Chapter 16, pp. 283–312). Springer Nature.

IPBES. (2024). Thematic assessment report on the underlying causes of biodiversity loss and the determinants of transformative change and options for achieving the 2050 vision for biodiversity of the intergovernmental science-policy platform on biodiversity and ecosystem services (Transformative Change Assessment). In O'Brien, K., Garibaldi, L., & Agrawal, A. (Eds.), IPBES Secretariat. https://doi.org/10.5281/zenodo.11382215

Lieberman, M. D. (2013). *Social: Why our brains are wired to connect*. Broadway Books.

O'Brien, K., Hayward, B., & Berkes, F. (2009). Rethinking social contracts: Building resilience in a changing climate. *Ecology and Society, 14*(2), 12.

O'Brien, K. (2021). *You matter more than you think: Quantum social change for a thriving world*. Change Press.

Richardson, M. (2023). *Reconnection: Fixing our broken relationship with nature*. Pelagic Publishing.

Schlitz, M., Vieten, C., & Miller, E. M. (2010). Worldview transformation and the development of social consciousness. *Journal of Consciousness Studies, 17*(7–1), 18–36.

2

Eco-social Contracts for Sustainable, Regenerative, and Just Futures: Introduction and Overview

Manisha Desai, Katja Hujo, Patrick Huntjens, and Najma Mohamed

2.1 Introduction

In the wake of a global polycrisis—where climate change, inequality, and environmental degradation converge—this chapter invites you to reimagine the foundations of our shared future. Based on an analysis of the drivers of these crises, many of which are rooted in our global economic system, we propose that new alternative eco-social contracts must not merely aim to sustain what is but regenerate and repair what has been harmed. They must reimagine just, caring, and enriching relationships with each other and with

M. Desai (✉)
Stony Brook University, Stonybrook, NY, USA
e-mail: manisha.desai@stonybrook.edu

K. Hujo
UNRISD, Bonn, Germany
e-mail: katja.hujo@un.org

P. Huntjens
Inholland University of Applied Sciences, Delft, The Netherlands
e-mail: patrick.huntjens@inholland.nl

N. Mohamed
UN Environment Programme World Conservation Monitoring Centre, UNEP-WCMC, Cambridge, UK
e-mail: najma.mohamed@unep-wcmc.org

© The Author(s) 2025
P. Huntjens et al. (eds.), *Eco-Social Contracts for Sustainable and Just Futures*,
https://doi.org/10.1007/978-3-031-99109-7_2

institutions, as integral parts of nature. This chapter lays the groundwork for a transformative dialogue on alternative eco-social contracts, offering a fresh vision made up of philosophical or normative frameworks and imaginaries implemented through concrete policies and institutions for eco-social change. While acknowledging the Enlightenment origins of modern social contract theory, this book presents the ideals needed for eco-social alternatives from within a pluriverse of visions and values, including the wisdom of Indigenous traditions such as *Ubuntu, Sumak Kawsay*, and *Eco-Swaraj*. These traditions are not simply alternative perspectives but foundational and co-equal in shaping the pluralistic and regenerative futures this volume envisions. Recognising their epistemic legitimacy challenges the colonial hierarchies embedded in dominant political thought and affirms the need to centre diverse worldviews in the design of new social arrangements.

A pluriverse here is not meant to merely reflect "a collection of different knowledge systems existing side by side" but an active inquiry into what alliances and communities are needed to deal with the complex and seemingly intractable polycrisis while holding a common vision of the future: a vision for life-centred thriving on our planet (Culture Hack Labs, 2023). Fundamentally, the locus of pluriversal approaches is in addressing political-economic and cultural modes of dominance, including the anthropocentric gaze, towards a creative process that is emergent, dynamic, and self-organising. In this approach, citizens and communities are not merely placing their concerns and issues in the hands of a central authority (such as the state), but novel forms of organisation, collective action, commoning, economic systems, and so forth are employed to formulate and reflect eco-social ideas and actions (Culture Hack Labs, 2023; GTA, 2023).

By challenging outdated and elite social agreements that have long marginalised diverse voices and exploited vulnerable communities, the chapter proposes dynamic, inclusive social agreements where peoples' visions and values are visible and where nature is seen not merely as a resource but as a vital partner in our collective journey. This introduction sets the stage for a bold exploration of how restructured social, economic, and political systems, and values can nurture sustainable and just futures for all. Prepare to be inspired by new ideas that redefine the relationship between people and the planet and spark a movement towards genuine ecological, economic and social renewal.

2.2   Background and Motivation

In an era marked by escalating climate emergencies, deepening social inequities, and growing political instability, existing social contracts have proven woefully inadequate in safeguarding the rights and well-being of communities today and of future generations. A common characteristic of most twentieth-century-social contracts was its basis in an extractive political economy with an absence of rules to respect planetary boundaries, preserve biodiversity, and promote the sustainable use of natural resources (UNRISD, 2022: 231). Indeed, the polycrisis has put a spotlight on the absence of a contract for and with nature. Realising a sustainable, healthy, and just future for societies worldwide will therefore require a fundamental renegotiation and redesign of current social contracts and economies, which implies institutional change as well as multiple parties, multiple sectors, and multiple levels of governance acting and collaborating effectively for just transitions and inclusive care-based economies and societies (Huntjens, 2019, 2021).

Reimagining eco-social contracts requires a reconfiguration of not only the overarching goal of a social contract but also a fundamental restructuring of the current extractive political economy and how humanity views itself and its relationship with nature. Eco-social contracts are both a transformative goal and a negotiated process through which sustainable, regenerative, and just political economies and societies can emerge. They do not merely support regenerative outcomes but embody regenerative principles themselves, such as reciprocity, care, and interdependence, by redefining how we produce, consume, and relate to one another and the Earth through shared norms, rights, and responsibilities. A shift from the old social contracts to new eco-social contracts is urgently needed, and new discourses are emerging that propose new types of social contracts that address the anthropocentric, extractive, and exploitative foundations of our current economies and societies driving the ecological divide, while also being much more inclusive and democratic than past ones.

Transformative eco-social alternatives require a good understanding of the root causes of current crises. Despite considerable progress in human development over the past half-century, these gains have been uneven and increasingly fragile. The COVID-19 pandemic, followed by inflationary pressures and rising geopolitical tensions, has partially reversed development progress in many parts of the world, exacerbating poverty, inequality, and insecurity (World Bank, 2020; UN, 2021; Brignone et al., 2024; Yilmazkuday, 2024). These overlapping crises have deepened public mistrust in political institutions and intensified social fragmentation. In many Western liberal democracies, this disillusionment has contributed to a surge in exclusionary nationalism and a growing anti-immigrant backlash, posing what Kapelner (2024) calls the "democratic dilemma" of immigration policy: how to reconcile democratic values with increasingly restrictive political pressures. As a result, social contracts, the principles, values, and public institutions our societies are founded upon, are unravelling in various parts of the world, prompting UN Secretary-General António Guterres to state in his "Our Common Agenda" report that our social contract is broken (UN, 2021).

For the 1.8 billion young people who will live with the social and ecological consequences of economic decisions that discount the well-being of current and future generations, the social contract is broken. It is broken for the more than 800 million people who continue to go hungry daily while the world produces enough food to feed humanity. It is broken for Indigenous peoples who steward much of the world's biodiversity and yet remain on the margins of decision-making processes on the protection and conservation of nature. And for many others, including the millions pushed into poverty while wealth is concentrated in the hands of the super-rich, the social contract has always been broken. Governments have largely failed to uphold their end of the social contract: to guarantee safety, offer protection, uphold rights, fight inequality, and act in the best interest of *all* people (Mohamed, 2023).

The science is clear on the scale and urgency of the ecological and climate crises. Climate and environment risks make up the core of the global risks the world faces in the next decade (World Economic Forum, 2024)—the decade in which we have a once-in-a-generation opportunity to build a fairer, sustainable, and safer world. The impacts of climate change are already being felt as heat waves, wildfires, storms, and droughts are both more frequent *and* more intense, leading to billions worth of damage and loss of lives, livelihoods, and income (IPCC, 2022; Christian Aid, 2022). While our economies are fundamentally underpinned by the stability, health, and resilience of ecosystems (World Economic Forum and PWC, 2020), nature is disappearing at an alarming rate. Every national priority—jobs, welfare, poverty, production, infrastructure—will be hit as nature recedes with the most vulnerable and least responsible bearing the bulk of costs of biodiversity loss, pollution, and climate change (OECD, 2021).

Inequality is growing in every country and in every economy. Over the last decade the richest 1% have amassed around half of all the new wealth created (Oxfam, 2023), while extreme poverty and unemployment continue to rise. Rising costs of living is affecting households across the world, with people unable to afford essential needs, such as food, housing, water, and energy. At the same time, democracies are in decay, with the democratic institutions that form the "scaffolding of civic space" being eroded by both the decline in democratic values and the ascent of authoritarian regimes (Civicus, 2021). Trust is being lost as people are left increasingly vulnerable and unprotected, with mounting fears about economic security and their economic future (Edelman Trust Barometer, 2023).

COVID-19 has further eroded state capacity and legitimacy while inequality and poverty have increased. Scholars are talking about post-pandemic social contracts to emphasise how severe the disruption of social contracts was during the 2 years the pandemic ravaged lives and livelihoods and what is at stake for the global economy (Rodrik & Stantcheva, 2020). In the Global South, the pandemic has triggered debates about building more inclusive and rights-based social systems as foundations for a new social contract. However, recent reports on the *2030 Agenda for Sustainable Development*, the shared global blueprint for peace and prosperity for people and the planet, show that the achievement of the sustainable development goals agreed in 2015 are not on track or backsliding (UN, 2024).

The question then is how do we acknowledge the root causes of poverty, inequality, and unsustainable practices—primarily the extractive and exploitative neoliberal global political economy—and how can we renegotiate and rebuild new eco-social contracts that are founded on a care-based political economy that is more inclusive, resilient, sustainable, and just. The old social contracts were neither fully inclusive nor environmentally sustainable. In some countries it traded political voice and participation for welfare benefits and protection. In others it barely delivered on a minimum of social rights such as decent jobs, access to health and education, and social security. Since the 1990s, neoliberal hyperglobalisation has undermined social contracts worldwide, increasing inequalities and instability while dismantling welfare institutions and public policy. For a comparative overview of key characteristics distinguishing traditional social contracts from emerging eco-social contracts, see Table 16.1 in Chap. 16 (in this volume, Huntjens & Kemp, 2025).

While some contemporary strategies attempt to "green" the economy through technological innovation and more efficient resource use—within a framework of continued economic growth—such "green growth" approaches retain the fundamental logic of extractive capitalism. Eco-social contracts propose a deeper transformation: moving beyond the growth imperative itself. They call for regenerative, care-centred, and relational economies that prioritise the well-being of people and planet over endless material expansion, challenging the assumptions that money and consumption are suitable proxies for well-being and that human-made capital can fully substitute for nature.

Recent history shows that social contracts can be renegotiated, reformed, or overthrown when contexts change or when contracts lose legitimacy and support (see Chap. 14). According to Sen and Durano (2014: 5), a social contract "may be imposed from above, fought over from below, and always holding the potential for change". If they are to bring about transformative change, new eco-social contracts cannot be imposed from above, they need to be built bottom-up and reflect a broad consensus, derived through inclusive and democratic means and through empowerment of marginalised groups (UNRISD, 2022).

As argued in *Towards a Natural Social Contract* (Huntjens, 2021), and as visualised in Fig. 2.1, sustainability is no longer sufficient in the face of ecological collapse, structural inequality, and institutional failure. Fig. 2.1, titled "Beyond sustainability: designing regenerative cultures", offers a conceptual framing of how eco-social contracts can serve as transformative pathways that go beyond sustainability towards regenerative futures (Wahl, 2016). Many contributions in this volume (e.g. Chaps. 5, 6, and 16) echo a call to move beyond sustainability and towards restorative and regenerative approaches, where humans act as part of nature rather than merely upon nature. This implies not just mitigating harm but cultivating systems, institutions, and cultures that heal ecosystems, repair the social fabric, and restore trust in governance. Eco-social contracts, then, are not just about redistribution or recognition but about reweaving life-sustaining relationships through transformative political economy, governance, and practices at local and global levels. This aligns with the call for holistic eco-social imaginaries that centre life, ethics, and systems coherence—what Waddock (2024) describes as essential for

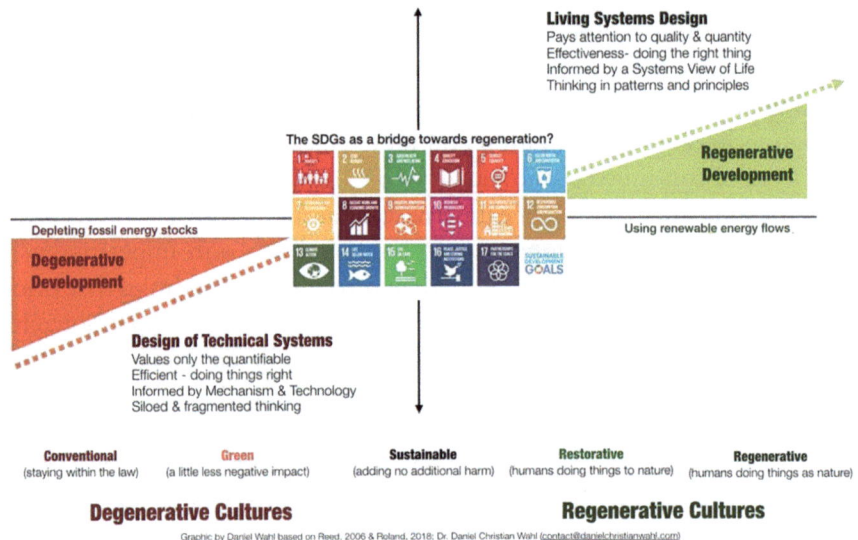

Fig. 2.1 Beyond sustainability: designing regenerative cultures. Source: Based on Reed (2006)

fostering life-affirming futures. In this spirit, eco-social contracts echo the dual imperative of hospicing the old and midwiving the new, a metaphor offered by Joanna Macy to describe the deep transitions required for regenerative transformation.

This regenerative vision represents a shift from technical efficiency—doing things right—to ethical effectiveness or doing the right things. It calls for new institutions and indicators that measure the health of ecosystems, the strength of social cohesion, and the resilience of communities. In this context, eco-social contracts are not only about redistribution or protection. They are designed to regenerate, helping to create the conditions in which all life, both human and more than human, can thrive.

To support this transformation, societies must prioritise broader measures of prosperity beyond GDP growth such as health and well-being, happiness, and ecological stability. This includes decarbonisation, renewable energy transitions, and investment in carbon-neutral technologies. It involves reducing pollution, waste production, and destruction of biodiversity and essential ecosystems. Redistributive mechanisms such as progressive taxation and universal social protection and basic income are also essential to address deep-rooted inequalities and make transitions just. In addition, adaptive systems are needed to withstand environmental and social shocks. These systems must ensure that infrastructure, cities, and communities are resilient, inclusive, and able to balance the needs of today with the rights of future generations.

2.3 What Is a Social Contract?

The social contract idea goes back to fundamental questions of political philosophy, reflected among others in Islamic, African, and Indigenous communitarian thinking (see Sect. 2.1). It is, however, most often associated with European Enlightenment philosophy as represented by Thomas Hobbes, John Locke, Immanuel Kant, and Jean-Jacques Rousseau, deliberating about political authority, social order, and equality. Social contract theorists define it as the moral and political obligations that free individuals accept voluntarily among themselves and vis-à-vis their government (sovereign) in order to escape the state of nature. It has also been defined as the explicit and implicit agreements between state and citizens defining rights and obligations to ensure legitimacy, security, rule of law, and social justice (UNRISD, 2022; see Chap. 14).

An eco-social contract has been defined in an ideal-typical sense as: "…implicit or explicit collective agreements across multiple levels of governance, among members of society, aimed at addressing the interconnected polycrisis of the 21st century, including inequalities, injustices, climate and ecological breakdown, and faltering trust in institutions. These agreements are rooted in cooperation and the recognition of shared norms and values oriented towards sustainability, equity, and justice. Importantly, eco-social contracts encompass social, environmental, economic, cultural, and institutional dimensions, and articulate the corresponding rights and duties of care for the environment and the well-being of others, including future generations and all forms of life on Earth" (Huntjens & Kemp, 2025, see Chap. 16). It is grounded in a fundamentally different worldview—one that is eco-social, eco-centric, or Earth-centric, rather than anthropocentric (see Chap. 5; Huntjens, 2019, 2021). It also reflects a shift in our understanding of humanity: from *homo economicus*, the self-interested rational actor, to *homo ecologicus*, a relational being embedded in and responsible for the web of life. This reorientation extends to our economic and cultural systems, favouring regenerative and well-being-oriented economies over linear, extractive models (Huntjens, 2021; Mohamed, 2023). In practice, eco-social contracting unfolds through dialogue and collaboration among citizens, communities, businesses, governments, and other actors who negotiate how to live well within planetary boundaries. Even if these actors never explicitly use the term "Eco-Social Contract", the process itself of cooperation, shared responsibility, and care across people and nature *is already* an act of eco-social contracting (Huntjens & Kemp, 2025, see Chap. 16).

While Jennings (2016) suggests that the social contract is primarily a way of looking at the world, we argue that it must also be a tool for systemic redesign—a foundation for forging new agreements that reflect ecological interdependence, historical justice, and the plurality of knowledges and worldviews. It is therefore a powerful approach to envisioning climate and ecological futures that are just and sustainable, connecting ideas (frameworks and imaginaries) with action (policies and institutions) that seek to achieve eco-social change. It is connecting movements that are working for social, ecological, and climate justice, leading to a new and evolving narrative space that exists at the intersection of diverse rights and justice-based approaches. These movements, as outlined in Fig. 2.2, encompass various claims that often overlap and work hand in hand.

Figure 2.2 synthesises the justice and rights-based foundations that underpin the vision of eco-social contracts. While social contracts have often been built on exclusions and unequal rights, new eco-social contracts must be grounded in intersectional alliances for social, climate, and ecological justice and should reflect the key dimensions of justice: procedural, distributive, and

Fig. 2.2 Rights and justice dimensions underpinning eco-social contracts. Source: Authors

restorative justice (Mohamed & Montmasson-Clair, 2022). Justice today requires not only redistribution and recognition but also representation: full parity of participation in social life, including the capacity to shape the rules that govern it (Fraser, 2008). These dimensions of justice are championed by diverse movements across the globe, whose voices and actions challenge existing systems and inspire regenerative, inclusive futures. Acknowledging these multiple dimensions helps ensure that eco-social contracts are not only theoretically robust but also grounded in lived realities.

The elements of new social contracts, as well as a commitment to developing them, have gained international recognition as a critical framework and transformative vision for addressing the interconnected challenges of social justice, environmental sustainability, and regenerative economies for a thriving future. This includes recognition in global and regional environmental and economic frameworks, such as the World Economic Forum's call for a "great reset" post-COVID-19 (Schwab, 2020), proposing an integrated approach to global economic recovery that aligns with sustainability and social justice principles. Regional policies such as the European Green Deal (see Chap. 9) and the Inflation Reduction Act in the United States both reflect eco-social contract ideals by linking climate action with social equity and economic transformation. Social movements, such as the Rights of Nature movements, have supported the advancement of legal rights for nature (see Chap. 11), advancing the eco-social contract by embedding nature's rights into governance. Others, like the youth-led Fridays for Future and Extinction Rebellion movements, champion eco-social contract principles by demanding systemic changes to ensure climate justice and equitable futures address historical rights and intergenerational justice. Indigenous advocacy groups have also emphasised integrating traditional ecological knowledge into governance, a cornerstone of the eco-social contract.

The concept is also gaining traction in academic discourse, with scholars proposing new eco-social agreements as key to developing a societal vision for alternatives to traditional growth-focused paradigms, linking environmental, social, and economic systems cohesively. Ideas and movements have laid the intellectual foundations for eco-social contracts by promoting systemic change, equitable governance, and the integration of ecological and social considerations into policy and theory. This is equally visible in key intergovernmental processes (see Table 2.1).

While there is growing and widespread support for new eco-social contracts, as the chapters in this book show, these social pacts are also sites of social and political struggle and contestation, of sometimes unequal power relations, of political choices that are not democratically debated (Bet & Saujot, 2024, see Chaps. 11 and 15), and of voices and knowledges that are neither valued nor visible. Eco-social contracts will have to be reimagined with this in mind.

2.3.1 Diversity of Social Contracts

Beyond the social contract associated with post-war Western welfare capitalism, different types of social contracts can be identified across the non-Western world, for example, the communitarian approaches dedicated to the common good such as *Ubuntu*—"I am because we are"—in Southern Africa, *Sumak Kawsay* (more commonly known by the Spanish *buen vivir*), and *Eco-Swaraj*

Table 2.1 An overview of international recognition of key elements of eco-social contracts

The Earth Charter (Earth Charter Commission, 2006)	An early declaration emphasising global interdependence, ecological integrity, and social justice, serving as a moral foundation for the eco-social contract
Sustainable Development Goals (SDGs) (United Nations General Assembly, 2015)	The *2030 Agenda for Sustainable Development* embodies elements of an eco-social contract, emphasising the integration of environmental protection (Goals 13, 14, 15), social equity (Goals 1, 5, 10), and economic transformation (Goal 8)
The Paris Agreement (United Nations, 2015)	The Agreement highlights the importance of social and ecological resilience in combating climate change, advocating for a just transition to sustainable societies
Food and Agriculture Organization's (FAO) *Social Protection Framework* (FAO, 2017)	FAO's strategic framework emphasises "Four betters", which reflect eco-social contract principles: better production, better nutrition, better environment, and better life. It also has a long-standing commitment to delivering its mandate to "free humanity from hunger" and ensure access to food for all by mainstreaming social protection, environmental sustainability, and climate resilience in its programmes
Reimagining Our Futures Together: A new social contract for education (UNESCO, 2021)	This publication advances constructive conversations and inspiring innovative approaches to renew a social contract for education—one that embodies justice, cooperation, and the transformative power of knowledge
Resolution on the human right to a healthy environment (United Nations General Assembly, 2022)	The United Nations General Assembly recognised access to a clean, healthy, and sustainable environment as a universal human right, aligning with eco-social contract principles

(*continued*)

Table 2.1 (continued)

United Nations Research Institute for Social Development (2022) *Crisis of Inequality: Shifting Power for a New Eco-Social Contract*	The report outlines the need for structural reforms to overcome entrenched inequalities and ecological degradation. It shows how inequalities and crises reinforce and compound each other, leading to extreme disparity, vulnerability, and unsustainability, and argues the social contract has broken down to the great detriment of people and planet
Kunming-Montreal Global Biodiversity Framework (GBF) (Convention on Biological Diversity, 2022)	The GBF supports the achievement of the SDGs and sets out an ambitious pathway to reach the global vision of a world living in harmony with nature by 2050. It aims to "catalyze, enable and galvanize urgent and transformative action by Governments, and subnational and local authorities, with the involvement of all of society"
World Health Organization (Lemmens et al., 2022)	The WHO advocates for policies that address social determinants of health and environmental risks, aligning with the broader objectives of an eco-social contract. Through the framing of a social contract, the report explores roles and responsibilities to promote access to health
International Labour Organization (ILO) established *Global Coalition for Social Justice* in 2023	The ILO promotes the concept of a just transition, ensuring social equity and fair labour practices during the global shift towards a green economy, a core tenet of the eco-social contract. It also launched the *Global Coalition for Social Justice* (ILO, n.d.), a ground-breaking initiative aimed at intensifying collective efforts to urgently address social justice deficits
Transformative Change Assessment, Intergovernmental Science-Policy Platform on Biodiversity and Ecosystem Services (IPBES, 2024)	The Assessment emphasises the systemic changes needed to halt biodiversity loss and achieve the 2050 Vision for Biodiversity. The eco-social contract is described in the assessment as a transformative and actionable vision that redefines the relationship between people, governments, and nature, emphasising shared responsibility and collaboration among all societal actors to create a just and sustainable future
United Nations Environment Programme and International Science Council (2024) *Foresight Report: Navigating New Horizons—A Global Foresight Report on Planetary Health and Human Wellbeing*	The report frames new social contracts as central to addressing global challenges and one of the eight key shifts and signals of change needed for a world where people and the planet can thrive

Source: Authors

(Desai, 2022) or faith traditions like liberation theology (see Chap. 4) which offer ways to imagine and develop societal and economic models based on justice and reciprocity between people and between people and nature, to post-colonial social contracts concerned with nation-building, state legitimacy, and social cohesion (see Chap. 14).

The concept of *Sumak Kawsay* (living well or living in harmony) focuses on collective well-being, viewing humans, communities, and nature as deeply interconnected and interdependent. It is a holistic vision which sees humans as stewards of nature and assumes a dynamic balance between material and nonmaterial needs such as recognition and affection (Barié, 2014). Originating in Indigenous communities' struggle for autonomy and power in Latin America (Altmann, 2014), the Living Well paradigm is now the normative foundation for national development strategies in the constitutions of Bolivia and Ecuador via rights for *Pacha Mama* (Mother Earth) and also increasingly referred to in alternative development discourses in Peru and Colombia. *Sumak Kawsay* can be considered an explicit eco-social contract through its formalisation in constitutions, laws, and policies as well as governance reforms that grant more rights to Indigenous communities and nature, as reflected in constitutional amendments on plurinationality, interculturality, and rights of mother earth. The *Sumak Kawsay* concept in the constitutions of Bolivia and Ecuador is a potentially transformative vision, even if overturned in practice. It resonates with understandings and value propositions in Indigenous and communitarian cosmovisions in other cultures such as *Ubuntu* or *Eco-Swaraj* (see Chap. 10) and is considered an alternative vision to mainstream development approaches.

Similarly, feminist scholars have provided several conceptualisations to realise new eco-social alternatives. For example, following the lead of Mies, Bennholdt-Thompson, and Von Werlhof (1988) and Mies and Shiva (2014), recent feminists of social reproduction (Di Chiro, 2019; Solon, 2019) call for centring care in economic and social policies and institutions to ensure sustainable societies (see Chap. 6). Feminist political ecologists such as Clement et al. (2019) and Sato and Alarcón (2019) reframe the concept of commons to active processes of commoning not only of natural resources but also social and cultural resources to produce societies that live in reciprocal and dynamic relations with all other living and non-living entities. Abolitionist feminists such as Gilmore (2017) highlight the importance of the unfinished process of liberation for all, reparations, and place making as a way to enact new eco-social contracts.

There is a growing movement engaging afresh with the eco-social contract as a frame for new thinking about the agreement between multiple stakeholders in society about the mutual rights and responsibilities that uphold a just and sustainable social order (McCandless, 2021). Some focus on key social challenges, as seen in movements championing for new social contracts that are anti-racist (Christian Aid and Partners, n.d.) and gender-just (ITUC, 2022) and that promote decent work (ITUC, 2021) and inter-generational justice (Weiss, 1992; Gardiner, 2011; Krznaric, 2020). Others harness the increased awareness and understanding of the climate and eco-logical emergency to define a new eco-social contract that addresses the cli-mate crisis (Willis, 2020), the ecological divide (Huntjens, 2021; Mohamed & Huntjens, 2023), and transitions to a low carbon and sustainable world (Messner, 2015)—a missing dimension in much of social contract thinking and practice to date. Together these approaches connect the sustainability, inequality, and democracy movements in a call for new eco-social contracts.

2.3.2 Elements of New Eco-Social Contracts

It is crucial to distinguish between ideal understandings of a social contract (the norms and values underpinning its vision and objectives, which vary according to different worldviews and ideologies) and real-world experiences (the actual institutions and policies that are implemented and their effects) (UNRISD, 2022; see Chap. 14). This volume will discuss both the normative and empirical side of eco-social contracts. Given the shortcomings of existing social contracts and their failure to live up to their promises, it is necessary to redesign our social contracts if we aim to achieve sustainability for all; they must become eco-social contracts, incorporating the ecological dimension and creating new contracts for the planet and future generations. They also need to deliver for persons that have been excluded from previous social con-tracts or included on unfavourable terms. The negotiations for new eco-social contracts should include the following elements (See also Chap. 14 and UNRISD (2022):

Historical injustices: New eco-social contracts must recognise ongoing coloni-ality and how it marginalises communities in the Global North and South and be committed to decolonising political, economic, and sociocultural relations, institutions, and structures. This includes but is not limited to reparations for past and ongoing injustices, a commitment to restructure institutions and policies informed by Indigenous and other excluded knowledges, values, and practices. Such decolonial considerations must inform policies to combat the climate crisis, ecological destruction, and social inequalities through just transitions.

Ethics and values: New eco-social contracts need to be based on ethics and values that (1) reflect a cultural shift from an anthropocentric paradigm to a balanced eco-social paradigm, redefining the relationship between people, governments, and nature, emphasising shared responsibility and collaboration among all societal actors to create a just and sustainable future; (2) cultivate a sense of stewardship and collective responsibility for nature/ecosystems and society; (3) encourage shifts in value orientations or constellations, norms, and behaviours towards greater sense of being part of nature; (4) promote environmental education and awareness to empower citizens to make informed, sustainable choices in line with values; and (5) recognise diverse visions of the human-nature relationship that cultivate an eco-social paradigm.

Transformed economies and societies: Transforming economies means moving away from extractive and exploitative neoliberal capitalist modes of production and consumption, progressing through transitional models like green and circular economy approaches, and ultimately moving towards regenerative economies—such as well-being, solidarity, and care-based models—proposed by scholars and social movements (Huntjens, 2021; Mohamed, 2023). This means ensuring broad economic well-being and equitable access to opportunities, ecological sustainability, and the well-being of current and future generations. These are economies that not only reduce harm but actively restore ecosystems, repair social relations, and rebuild trust in governance.

Social justice and equity: New eco-social contracts must recognise that previous social contracts have often functioned as elite bargains and have been built upon an unequal inclusion and recognition of groups marginalised and oppressed because of their identities, place in social, economic, and political hierarchies, and trapped in contexts of war and conflict. These groups are disproportionately impacted by ecological degradation and inequities and must have fair access to natural resources and services. New eco-social contracts must therefore ensure equal access and response-ability of all for productive and reproductive labour. Everyone needs to have access to quality social services and social protection, in particular, persons that have been excluded in previous social contracts, for example, informal workers or migrants or persons with disabilities.

Legal and institutional recognition of earth system governance frameworks: Eco-social contracts must move beyond anthropocentric legal and governance frameworks to develop laws and policies that explicitly protect ecosystems and regulate their use. New frameworks such as Rights of Nature, that is, granting legal personhood or similar rights to natural entities such as rivers, forests, and mountains, to safeguard ecosystem health, or ecological jurisprudence should be considered to transform current governance systems so they are aligned with the vision and values of eco-social contracts.

Peace and solidarity: New eco-social contracts require bottom-up approaches to transformative change bringing together social movements and progressive alliances between science, policymakers, and activists. It must overcome the mindset of "us against them", fostering instead a spirit of "all united against" global challenges such as climate change, inequalities, and social fractures. It needs to foster peace and non-violent, respectful behaviours vis-a-vis human beings and nature, including animals, plants, and ecosystems.

This edited volume attends to the critical pathways needed to build a world where the above elements are central to ensuring that all life flourishes and identifies possible avenues to break from the dominant social paradigm built on the ecological divide, one of the greatest fault lines of our time (Mohamed & Huntjens, 2023). The shift from anthropocentric to eco-centric social contracts requires a fundamental reassessment of the purpose, goal, and vision of our societies and economies and what this means for the relationship between people, between people and power, and between people, nature, ecosystems, and planet.

This new vision and discourse is nurtured and disseminated in the Global Research and Action Network for a New Eco-Social Contract established in 2021 (www.ecosocialcontract.net), coordinated by the United Nations Research Institute for Social Development (UNRISD, 2021) and the Green Economy Coalition (GEC), and with more than 370 participating organisations joining in the first 4 years. This edited volume contains contributions from network members and beyond—researchers, practitioners, and activists working for social, economic, and environmental justice, as well as other scholars and practitioners with relevant work on the subject.

The book aims to broaden understanding around the rationale, ideas, and actions for reimagining "new" eco-social contracts as transformative pathways to achieve a just and sustainable world. It aims to explore framings of a new eco-social contract as a pathway to build and strengthen the social mandate for social and ecological transformations and explore the governance innovations, the policies, processes, and movements, that can advance a just transition to a climate-resilient, sustainable, and equitable future.

2.3.3 How to Get there: Collective Action, Political Participation, and Governance Reform

Transformative change towards eco-social contracts is not just about what needs to change but critically about how such change can be implemented and governed and emerge from the bottom up. The governance frameworks needed for eco-social contracts pose both a normative and practical challenge to governance (Jennings, 2016; Huntjens, 2021). Transformative approaches to governance that integrate ecological, social, and economic priorities, fostering a sustainable and equitable future for all, are required. These should be catalysed by participatory mechanisms and processes that recognise and respect the values, perspectives, and rights of all actors while also addressing power differences among them. It should build a societal mandate by enabling people's participation and engagement in the design of policies, projects, and investments, ensuring that the outcomes, trade-offs, costs, and benefits of interventions are distributed equitably and with transparency. And it requires an enabling environment—civic and democratic space and an engaged and informed citizenry with access to diverse participatory mechanisms to engage with and hold power to account.

Inclusive, transparent, and accountable multi-level governance systems, processes, and institutions are therefore central to eco-social contracts. Recent governance literature and practical innovations such as the Transformation Flower Approach, discussed in detail in Chap. 16 by Huntjens and Kemp, highlight the importance of participatory, value-driven, and adaptive governance to enable deep systemic change (Huntjens & Kemp, 2022; IPBES, 2024; Huntjens & Kemp, 2025).

Reinvigorated eco-social contracts that establish greater linkage and ambition for economic, environment, and social change require greater citizen voice, deliberation, and agency and address unequal power dynamics. Cross-community deliberative processes, such as citizens assemblies and juries, participatory constitutional development, and citizen-led policy dialogues, are being employed to explore the various elements of twenty-first-century eco-social contracts, making global and national-specific proposals for political reform, new governance structures, and policy change. These eco-social contracts in the making are grounded in collective diagnosis, debate, and decision-making, culminating in a broad consensus.

The contributions in this book reimagine eco-social contracts as societal agreements that ensure a just coexistence between humans and between humans and the natural world at local, national, global, and transnational levels. It explores the processes and mechanisms through which these agreements are negotiated and governed, including the purpose and principles of new eco-social contracts, the people and groups who participate (and are excluded) in setting the priorities and vision for social contracts, and the processes whereby these agreements are negotiated. It links ideas to action in that it applies a critical analysis to understand the "institutional structures, mechanisms and tools that can create spaces for multi-stakeholder diagnosis, views and values" (Mohamed, 2020: 59) to craft eco-social contracts. It adopts a position that robust and well-resourced institutional mechanisms and processes are needed to highlight a whole-of-society contract encompassing a pluriverse of visions, values, and priorities. This process of inclusive governance creates a "demand pull" on social and political agreements that can make otherwise "top-down" decision-making more inclusive, accountable, transparent, and sustainable. Meaningful spaces that include the voices of citizens, workers, and social movements in international agreements, national policy, and legal frameworks, and practices and projects on the ground, are therefore needed.

Negotiating eco-social contracts is thus a two-step process—first it is a goal in and of itself–an imagination and articulation of the world we want that results in an "agreement out of which free and equal human beings create common rules and constraints that allow them [and the Earth] to flourish together in ways they could not flourish alone" (Jennings, 2016: 37). Second, it is a process by which all people are brought into decision-making processes ensuring that multiple values, visions, and priorities are considered. A new eco-social contract could help define the goals and shape of the world we want and also revive legitimacy and accountability to governance systems needed for the twenty-first-century transition we are now a part of. It is a tonic for declining public trust (Hopkins et al., 2020) and a way for people to articulate what they want from their government. Transparent and accountable systems and institutions of governance, rooted in trust, must enable citizen deliberation, action, and voice to be integrated in decision-making. Participatory and inclusive governance is a central part of the overarching ambition of building a movement for a just and sustainable world.

Finally, new eco-social contracts need to be implemented through transformative policies and institutions, with strong accountability mechanisms ensuring that agreed objectives are realised.

Several aspects of the old social contract thus point the way towards its reconceptualisation, that is, reminding those in power of their conditional positionality—they are, after all, servants of the general will; expanding democratic practices to hold power to account; and an exploration of democratic approaches which favour direct participation instead of representation (Culture Hack Labs, 2023). However, rising reactionary and nationalist politics, declining public trust, fraying multilateralism, loss of political legitimacy, financial constraints and shrinking democratic and civic space all point to democratic decline and challenges to enacting regenerative and just eco-social contracts (Edelman Trust Barometer, 2023; UNDP, 2022; CIVICUS, 2021, 2024).

The diversity of ideas and actions in this book together propose eco-social contracts that build institutions and processes and foster alliances and solidarity for a transition to green and fair societies and economies. This also connects the inequality, climate, and environmental agendas with efforts to revive and revitalise democracies and political decision-making to create the collective solidarity needed for a critical mass sufficient, not just to accelerate this transition but to ensure it is just, democratic, and sustainable.

2.4   Structure of the Book

The chapters in the book highlight important ways in which actors around the world are addressing the polycrisis through negotiating new eco-social contracts along different dimensions and with varied results.

In Part I authors discuss the urgency of a fundamental transformation in our normative and theoretical understanding of our relations with each other and our environment. Chris Walker demonstrates the importance of aesthetic and artistic creations in informing how we care for all living entities on earth as our kin. Najma Mohamed draws our attention to Islam's liberation eco-theology and how it has inspired a movement to respond to climate and ecological crises in countries across the world and how it can align with liberatory movements working for eco-social justice. In a similar vein Najma Mohamed and Patrick Huntjens call for a new pact with nature, one that is based on reciprocity, partnership, and connectedness rather than our current paradigm of nature as a resource for wealth and profit accumulation and unsustainable consumption. Drawing upon feminist and Indigenous insights, Manisha Desai calls for moving away from sustainable development to sustainable societies in which care-centred economies are part of the shift from an anthropocentric to eco-centric understanding of ecosystems that is committed to liberatory radical politics of coalition building.

Part II moves to the realm of governance at different levels and the changes needed to align with the normative transformations called for above. Maja Groff, Georgios Kostakos, and Patrick Huntjens identify the changes needed at the global governance level in tandem with national stewardship to ensure eco-social justice. Davide Sofia and Patrick Huntjens focus on the European Green Deal and the ways in which it represents a big step forward in addressing ecological and social issues at a regional level while it still needs more transformatory policies to serve as a model of an eco-social contract. Shristee Bajpai, Ashish Kothari, and Neema Pathak Broome show how local movements draw upon local and post-colonial practices of democracy to craft a radical local democracy that enables them to challenge neoliberal development and enact *Eco-Swaraj*, a multidimensional eco-social conception. Drawing upon the Eco Jurisprudence Monitor, Lauren Tarr, Catherine Haas, and Caitlyn Sutherlin examine the possibilities of rights to nature to further eco-social governance. Drawing upon the findings of the Fair Foods Project in Australia, Kia Smith describes how a new eco-social contract can be extended via food justice networks and policies at local and state levels.

Authors in Part III examine the processes and negotiations involved in designing eco-social contracts. Katja Hujo synthesises research and conceptual thinking on how to fix our broken social contract, maps and discusses various examples where social contracts have been made more inclusive and sustainable, and lays out key principles to steer new eco-social contracts in a transformative direction. Salim Fakir and Danielle Hersch-Castros focus on the role of philanthropy in promoting eco-social contracts in tandem with other similar ideas of eco-social justice in the context of Africa. Patrick Huntjens and Rene Kemp apply the Transformation Flower Approach to eight comparative case studies in the Global North and South, showing how eco-social contracting can help reweave values, institutions and practices. The chapter concludes that while power asymmetries and institutional inertia remain major barriers, the approach enables diverse actors to co-create visions, identify leverage points, and build coalitions that open new governance pathways towards more regenerative and just futures. Erin McCandless draws upon intergovernmental climate policy processes to define how eco-social contracts can be crafted that address environmental sustainability and peace building.

Finally, in Part IV, authors highlight how eco-social contracts are actually being implemented on the ground at various levels. Carlos Villaseñor looks at governance within social movements in Latin America to understand how democratic decision-making or lack thereof can hamper movements' eco-justice work. Saba and Koehler focus on the formation of eco-social contracts in India and Nepal at the national level, the crucial role of civil society actors in advocating for new eco-social contracts, and how vulnerable they are to changes in political power. Yogi Hale Hendlin, Yuho Hisayama, and Kazuhiko Ota show the diversity of the ecovillage movement in Japan which ranges from shifts in corporate headquarters to rural settings to *satoyama* farming communities. Their preliminary findings suggest that ecological and social justice concerns reinforce each other in these communities. Finally, Najma Mohamed explores participatory mechanisms that are being used to address economic transformation and that hold the potential to centre people in advocating for economies that are not only greener and fairer but also democratic and inclusive.

In addition to the comprehensive chapters, the book features powerful narratives from thinkers and leaders at the beginning of each part, setting the tone for the chapters that follow. These narratives include contributions from renowned artists, activists, and policymakers—Teresia Teaiwa, Mary Robinson, Kumi Naidoo, and Lysa John—in addition to Karen O'Brien's foreword and an epilogue by Carlos Alvarado Quesada.

2.5 Conclusion

As this introductory chapter draws to a close, it leaves you with a call to explore the transformative potential of eco-social contracts. The ideas presented here are both a critique of broken commitments and a roadmap for building a future where social justice, environmental care, regenerative economies, and inclusive governance converge.

New eco-social contracts cannot be a repeat of the old social contract. They must rethink the nature and goals of a political economy that will prioritise fighting inequality. They must restore trust. They must uphold and advance human rights. They must be responsive to the climate and ecological emergency. They must be negotiated with people and with the well-being of future generations in mind. And they must reimagine the relationship between humans and nature, ensuring that people and the planet thrive together. In the words of choreographer and dancer Chris Walker, the crisis "demands that we move, that we innovate, that we step forward—not despite the weight, but with it, finding a rhythm that does not break us, but remakes us" (Chap. 3).

In the chapters that follow, you will discover a rich tapestry of scholarly insights, grassroots experiences, and innovative policy and governance proposals that challenge conventional paradigms and offer real pathways to change. Let this conversation ignite your curiosity and commitment to a world where every individual and ecosystem is valued and empowered. Your journey towards a just, sustainable future begins here—an invitation to reimagine the bonds between society and nature and to actively participate in shaping thriving societies on a healthy planet.

References

Altmann, P. (2014). Good life as a social movement proposal for natural resource use: The indigenous movement in Ecuador. *Consilience: The Journal of Sustainable Development, 12*, 82–94.

Barié, C. G. (2014). Nuevas narrativas constitucionales en Bolivia y Ecuador: el buen vivir y los derechosde la naturaleza. *Latinoamérica: Revista de Estudios Latinoamericanos, 59*, 9–40.

Bet, M., & Saujot, M. (2024). *The social contract: A framework for rethinking our common ground*. Institute for Sustainable Development and International Relations. Accessed April 21, 2024, from https://www.iddri.org/en/publications-and-events/blog-post/social-contract-framework-rethinking-our-common-ground

Brignone, D., Gambetti, L., & Ricci, M. (2024). *Geopolitical risk shocks: When the size matters*. ECB Working Paper No. 2024/2972. European Central Bank. Available at SSRN: https://ssrn.com/abstract=4919668

Christian Aid. (2022). *Counting the cost 2022. A year of climate breakdown*. Christian Aid. Accessed February 15, 2025, from https://www.christianaid.org.uk/sites/default/files/2022-12/counting-the-cost-2022.pdf

Christian Aid and Partners. (n.d.). *New feminist and anti-racist social contracts for people and the planet*. Christian Aid. Accessed February 15, 2025, from New feminist and anti-racist social contracts | Christian Aid

CIVICUS. (2021). *What is civic space and why is it shrinking?* Accessed February 15, 2025, from https://www.youtube.com/watch?v=Fm-KeAbhVCk

CIVICUS. (2024). *Rights reversed: A downward shift in civic space*. Accessed April 15, 2025, from RightsReversed.2019to2023.pdf

Clement. F., Harcourt, W., Joshi, D. & Sato, C. (2019). Feminist political economy of the commons and communing. *International Journal of the Commons 13*(1), 1–15.

Conference of the Parties to the Convention on Biological Diversity (CBD). (2022). *Kunming-Montreal global biodiversity framework*. CBD.

Culture Hack Labs. (2023). *Alliance for ecological and social transitions: Co-creating the Pluriverse*. Report prepared for the Green Economy Coalition. Green Economy Coalition.

Desai, M. (2022). *Going beyond the social. Communitarian imaginaries as inspirations for rethinking the eco-social contract*. United Nations Research Institute for Social Development (UNRISD) Issue Brief 12. UNRISD.

Di Chiro, G. (2019). Care not growth: Imagining a subsistence economy for all. *British Journal of Politics and International Relations, 21*(2), 303–311.

Earth Charter Commission. (2006). *The Earth charter*. Accessed April 30, 2025, from https://earthcharter.org/wp-content/uploads/2020/03/echarter_english.pdf

Edelman Trust Barometer. (2023). *Global report 2023*. Accessed February 15, 2025, from https://www.edelman.com/sites/g/files/aatuss191/files/2023-01/2023%20Edelman%20Trust%20Barometer%20Global%20Report_Jan19.pdf

Food and Agriculture Organization (FAO). (2017). *FAO social protection framework: Promoting rural development for all.* FAO. Accessed April 30, 2025, from https://socialprotection.org/sites/default/files/publications_files/i7016en.pdf

Fraser, N. (2008). Social justice in the age of identity politics: Redistribution, recognition, and participation. In *Geographic thought.* Routledge

Gardiner, S. M. (2011). *A perfect moral storm: The ethical tragedy of climate change.* Oxford University Press.

Gilmore, R. (2017). Abolition geography and the problem of innocence. In G. T. Johnson & A. Lubin (Eds.), *Futures of Black Radicalism* (pp. 225–240). Verso.

GTA. (2023, June). *Global tapestries of alternatives, Weaving solidarity and hope -stories of regeneration* (Vol. 3). Global Tapestry of Alternatives Core Team. https://global-tapestryofalternatives.org/reports:pandemic:03

Hopkins, C., Greenfield, O., & Mohamed, N. (2020). *Is the moment for a new social contract here? Green economy coalition.* Green Economy Coalition. Accessed February 15, 2025, from https://www.greeneconomycoalition.org/news-and-resources/is-the-moment-for-a-new-social-contract-here

Huntjens, P. (2019). *Sociale innovatie voor een duurzame samenleving: Op weg naar een natuurlijk sociaal contract.* Lectorale boek, IMPACT Lectoraat Sociale Innovatie in het Groene Domein, Hogeschool Inholland. https://doi.org/10.48544/90f2b3be-ab4d-4acf-a9c3-13effea07bcf

Huntjens, P. (2021). *Towards a natural social contract: Transformative social-ecological innovations for a sustainable, healthy and just society.* Springer. https://doi.org/10.1007/978-3-030-67130-3

Huntjens, P., & Kemp, R. (2022). The importance of a natural social contract and co-evolutionary governance for sustainability transitions. *Sustainability*, MDPI, *14*(5), 1–26. https://doi.org/10.3390/su14052976

Huntjens, P., & Kemp, R. (2025). The transformation flower approach for eco-social contracting: Comparative insights from eight case studies in the Global South and North. In P. Huntjens, N. Mohamed, K. Hujo, & M. Desai (Eds.), *Eco-social contracts for sustainable and just futures* (Chapter 16, pp. 283–312). Springer Nature.

International Labour Organization. (n.d.) *Global coalition for social justice.* Accessed April 30, 2025, from https://social-justice-coalition.ilo.org/

International Trade Union Confederation (ITUC). (2021). *New social contract: Five workers' demands for recovery and resilience.* Accessed September 21, 2024, from https://www.ituc-csi.org/new-social-contract-five-demands

IPBES. (2024). *Thematic assessment report on the underlying causes of biodiversity loss and the determinants of transformative change and options for achieving the 2050 vision for biodiversity of the intergovernmental science-policy platform on biodiversity and ecosystem services (Transformative change assessment)* (O'Brien, K., Garibaldi, L., & Agrawal, A. Eds.). IPBES Secretariat. https://doi.org/10.5281/zenodo.11382215.

IPCC. (2022). *Climate change 2022: Impacts, adaptation, and vulnerability. Contribution of Working Group II to the sixth assessment.* Report of the

Intergovernmental Panel on Climate Change. Cambridge University Press. https://doi.org/10.1017/9781009325844

ITUC. (2022). *A new social contract for a gender-transformative agenda. Workers and trade unions major group sectoral position paper to the HLPF (high level political forum)*. ITUC.

Jennings, B. (2016). *Ecological governance. Toward a new social contract with the earth.* West Virginia University Press.

Kapelner, Z. (2024). Anti-immigrant backlash: The democratic dilemma for immigration policy. *Comparative Migration Studies, 12*(1), 12.

Krznaric, R. (2020). *The good ancestor: How to think long term in a short-term world.* Random House.

Lemmens, T., Ghimire, K. M., Perehudoff, K., & Persaud, N. (2022). *The social contract and human rights bases for promoting access to effective, novel, high-priced medicines*. Oslo Medicines Initiative technical report. WHO Regional Office for Europe.

McCandless, E. (2021). *Social contracts: A pathway for more inclusive societies*. Research Paper. Pathfinders for Peaceful, Just and Inclusive Societies, New York University Centre for International Cooperation.

Messner, D. (2015). A social contract for low carbon and sustainable development. *Technological Forecasting and Social Change, 98*, 260–270.

Mies, M., & Shiva, V. (2014 [1993]). *Ecofeminism.* Zed Books.

Mies, M., Bennholdt-Thompson, V., & Von Werhof, C. (1988). *Women: The last colony.* Zed Books.

Mohamed, N. (2020). *Inclusion matters. Policy insights and lessons from the green economy coalition's national dialogues.* Green Economy Coalition.

Mohamed, N. (2023). *Building new social contracts: An overview of participatory mechanisms for economic governance.* Green Economy Coalition.

Mohamed, N., & Huntjens, P. (2023). *Dismantling the ecological divide: Toward a new eco-social contract.* United Nations Research Institute for Social Development (UNRISD) Issue Brief 15. UNRISD. https://cdn.unrisd.org/assets/library/briefs/pdf-files/2023/ib15-a-new-contract-with-nature.pdf

Mohamed, N., & Montmasson-Clair, G. (2022). Normative framework to assess the just transition to a net zero carbon society. In N. Xaba & S. Fakir (Eds.), *A just transition to a low carbon future in South Africa low carbon futures.* Mapungubwe Institute for Strategic Reflection.

OECD. (2021). *The inequality-environment nexus: Towards a people-centred green transition.* OECD Green Growth Papers, 2021-01. OECD Publishing.

Oxfam. (2023). *Survival of the richest how we must tax the super-rich now to fight inequality.* Oxfam Briefing Paper. Oxfam GB.

Reed, B. (2006). Shifting our mental model: "Sustainability" to regeneration. *Rethinking Sustainable Construction*, 1–18.

Rodrik, D., & Stantcheva, S. (2020, June 11). *The post-pandemic social contract.* Project Syndicate. The Post-Pandemic Social Contract by Dani Rodrik & Stefanie Stantcheva - Project Syndicate.

Sato, C., & Alarcón, J. M. S. (2019). Toward a postcapitalist feminist political ecology's approach to the commons and commoning. *International Journal of the*

Commons, 13(1), 36–61. JSTOR, Accessed September 4, 2025, from https://www.jstor.org/stable/26632712.

Schwab, K. (2020, June 3). *Now is the time for a great reset*. World Economic Forum. Accessed April 30, 2025, from https://www.weforum.org/stories/2020/06/now-is-the-time-for-a-great-reset/

Sen, G., & Durano, M. (2014). *The remaking of social contracts: Feminists in a fierce new world*. Zed Books.

Solon, P. (2019). Is Vivir Bien possible? Candid thoughts about alternatives. In K. Bhavnani et al. (Eds.), *Climate futures* (pp. 254–262). Zed Books.

UN. (2015). *The Paris agreement*. United Nations. https://unfccc.int/sites/default/files/english_paris_agreement.pdf

UN. (2021). *Our common agenda – Report of the secretary-general*. United Nations.

UN. (2024). *The sustainable development goals report*. United Nations.

UNDP. (2022). *Human development report 2021–22: Uncertain times, unsettled lives: Shaping our future in a transforming world*. United Nations Development Programme.

UNESCO. (2021). Reimagining our futures together. A new social contract for education. UNESCO.

UNGA. (2022). *The human right to a clean, healthy and sustainable environment*. Resolution passed at the Seventy-sixth session, Agenda item 74 (b). United Nations.

United Nations Environment Programme (UNEP) and International Science Council. (2024). *Navigating new horizons: A global foresight report on planetary health and human wellbeing*. Kenya.

United Nations General Assembly (UNGA). (2015). *Transforming our world: The 2030 agenda for sustainable development*. Resolution adopted by the General Assembly on 25 September 2015. United Nations.

UNRISD. (2021). *A new eco-social contract. Vital to deliver the 2030 agenda for sustainable development*. UNRISD Issue Brief 11. United Nations Research Institute for Social Development.

UNRISD. (2022). *Crises of inequality: Shifting power for a new eco-social contract*. United Nations Research Institute for Social Development.

Waddock, S. (2024). Holistic eco-social imaginaries for a life-centered future. *Sustainability Science, 19*, 2119–2134.

Wahl, D. C. (2016). *Designing regenerative cultures*. Triarchy Press.

Weiss, E. B. (1992). In fairness to future generations and sustainable development. *American University International Law Review, 8*(1992), 19.

Willis, R. (2020). A social contract for the climate crisis. *IPPR Progressive Review, 27*(2), 156–164.

World Bank. (2020). *Poverty and shared prosperity: Reversals of fortune*. World Bank.

World Economic Forum (Ed.). (2024). *The global risks report* (20th ed.). World Economic Forum.

World Economic Forum and PwC. (2020). *Nature risk rising. Why the crisis engulfing nature matters for business and the economy*. World Economic Forum.

Yilmazkuday, H. (2024). Geopolitical risks and energy uncertainty: Implications for global and domestic energy prices. *Energy Economics, 140*, 107985.

Part I

Normative and Theoretical Approaches on Eco-Social Contracts

To Island

Teresia Teaiwa

Shall we make "island" a verb?
As a noun, it's so vulnerable to impinging forces.
Let us turn the energy of the island inside out.
Let us "island" the world!
Let us teach the inhabitants of planet Earth how to behave as if we were all living on islands!
For what is Earth but an island in our solar system?
An island of precious ecosystems and finite resources.
Finite resources.
Limited space.
The islanded must understand that to live long and well, they need to take care.
Care for other humans, care for plants, animals; care for soil, care for water.
Once islanded, humans are awakened from the stupor of continental fantasies.
The islanded can choose to understand that there is nothing but more islands to look forward to.

From: Teresia Kieuea Teaiwa, compiled and edited by Katerina Teaiwa, April K. Henderson, and Terence Wesley-Smith. 2022. Sweat and Salt Water: Selected Works. Honolulu: University of Hawaii Press.

Continents do not exist, metaphysically speaking.

It is islands all the way up, islands all the way down.

Islands to the right of us, islands to the left.

Yes, there is a sea of islands.

And "sea" can be a verb, just as "ocean" becomes a verb of awesome possibility.

But let us also make "island" a verb.

It is a way of living that could save our lives.

3

Seaweed King: Weaving Narratives of Loss and Renewal in the Anthropocene

Chris Walker

3.1 Introduction

Seaweed King is an interdisciplinary solo performance that explores resilience, resistance, and ecological care in the face of environmental and cultural exploitation. Drawing from Black and Indigenous knowledge systems, the work weaves personal and collective narratives into a call for reflection and action.

As Mia Mottley warns, "Our world knows not what it is gambling with…" (Mottley, 2021). Island nations like those in the Caribbean contribute little to climate change yet bear its most severe consequences—rising sea levels, intensified storms, and environmental degradation.

Through choreography, costume, and audience engagement, *Seaweed King* transforms ecological crisis into an embodied conversation. Inspired by Sylvia Wynter's critique of Western humanism (Wynter, 2003), the work reimagines humanity's relationship with the Earth, enacting a praxis of care and renewal. This chapter traces the creative and collaborative process, examining how the work engages with the principles of an eco-social contract—where art becomes both reckoning and response.

C. Walker (✉)
University of Wisconsin–Madison, Madison, WI, USA
e-mail: cawalker2@education.wisc.edu

© The Author(s) 2025
P. Huntjens et al. (eds.), *Eco-Social Contracts for Sustainable and Just Futures*,
https://doi.org/10.1007/978-3-031-99109-7_3

3.2   Gathering the Strands: A Collaborative Process

This chapter follows the evolving process of *Seaweed King*, tracing the movement of ideas, materials, and embodied research across disciplines and creative partnerships. The work is analysed both as a choreographic practice and as an intervention into climate discourse. At its core, the *Seaweed King* emerges from collaboration—between choreographer and dancer, costume and movement, and artist and audience. It interrogates performance as a site for ecological reflection and collective responsibility, weaving together movement traditions, material culture, and Caribbean intellectual thought.

The following sections guide the reader through this journey: beginning with an evocation of the *Seaweed King's* thematic core through poetic imagery and reflective storytelling, moving into the creative and collaborative process—where traditional dance vocabularies are reimagined through exchange and experimentation—and exploring the symbolic resonance of costume and audience participation. Finally, the chapter positions *Seaweed King* as an artistic gesture of the new eco-social contract—one that demands solidarity between people, planet, and future generations (United Nations, 2021), considering how performance itself becomes a site for reckoning and a call to renewal.

3.3   *Seaweed King*: A Dance of Ecology and Resistance

The *Seaweed King* waits, hidden under the discarded layers—fabrications of a world taken for granted, voices silenced beneath the weight of forgotten things. To emerge from this pile is to feel both heavy and light, as if carrying an unending load, yet finding grace in each precarious step. The dancer is still and listens closely under the pile, sensing each footfall, each murmur from the audience who mistakes him for mere refuse. "No one expects a human to be there," he says. And isn't that the point?

This performance—this presence of the *Seaweed King*—asks not for applause but for a reckoning. Not a "How can this be fixed?" but a "Where do you stand?" In the slow, careful unravelling of each movement, there is a pulling back, a soft unwinding of what we have wound so tightly around us. Here, the body speaks; here, there is only the dance of bearing what cannot be discarded, of tending to what others have let slip from their grasp. Kevin Ormsby, the first person to perform the work, describes the urgency: "How do I get to that next piece, that next thing I have to protect?"

And in this gathering—of fabric, of species, of memory—the *Seaweed King* becomes something larger, a caretaker whose movements are as rooted in the sea as they are in time itself. The audience is invited to witness, not as observers but as participants in a shared world. The rhythm shifts, the weight grows, and yet there is no retreat, only a careful, pressing forward.

What does it mean to be a burden-bearer in a world unravelling? Or perhaps, as the *Seaweed King* shows us, we might ask instead: What does it mean to become entwined with what we carry, to find a new rhythm within the weight? Like kelp growing slowly in the dark, we rise through layers, through tides, embodying the memory of what once was and what still could be. And in this motion, a question gathers strength with each turn:

If the weight is already upon us, how will we choose to carry it?

3.4 The Framing and Evolution of *Seaweed King*

Seaweed King is deeply personal, shaped by histories, knowledge systems, and resistance strategies that have enabled Black and Indigenous communities to endure and thrive in the face of environmental and social violence. Caribbean intellectuals like Sylvia Wynter (1970) influenced my artistic identity, particularly as I began exploring ways to create contemporary art that was rooted in the traditions, myths, and dance vocabularies of Jamaica and the Caribbean.

Wynter's analysis of Jonkonnu as a process of cultural resistance resonated with me early on (Wynter, 1970). She describes how African-descended Jamaicans transformed imposed colonial forms into assertions of survival and identity. This idea sparked a deeper inquiry into creative expression as a tool for reclaiming and reframing relationships with history, community, and the environment. In the same way Jonkonnu reconfigured ritual, mask, and satire as strategies of survival, *Seaweed King* asks: how might movement, costumes, and performance embody strategies of ecological care and resistance? How can the work engage with the interconnected environmental and social crises of today?

The creative and collaborative process of developing *Seaweed King* became an ongoing exploration of these questions. The movement research draws on dances rooted in the African Caribbean wake complex, traditions that mediate loss, honour ancestors, and affirm the continuity of life. Shuffling steps from "*Ettu*"—a death observance ritual from southeastern Jamaica—were reimagined as progression movements, their varying speeds and postures reflecting resilience and labour. Circular rocking motions, often accompanying hymn singing at wakes, were slowed and exaggerated to depict *Seaweed King* emerging from a whirlpool. Masquerade-inspired jumps, layered with costuming, added depth to the interplay between cultural memory and ecological symbolism. These movements—drawn from ancestral traditions but reinterpreted through contemporary performance—allowed me to honour their origins while imagining new futures rooted in resilience and interconnectedness.

Wynter's framing of Indigenous knowledge systems as adaptive frameworks for survival and environmental care is also reflected in the costumes and props (Wynter, 2003). Inspired by the Pitchy Patchy figure in Jamaican Jonkonnu, which is traditionally constructed from discarded materials, *Seaweed King* employs a similar ethos of reclamation and transformation with the use of upcycled fabric strips. The bilum bag, a tote bag traditionally woven from natural fibres by Papua New Guinea (PNG) women, is a central symbol in *Seaweed King*. Bilum, in the Tok Pisin language of PNG, means "womb," symbolising care, sustenance, and the responsibility of carrying life. When worn by a man, it embodies ancestral ties and collective stewardship, signifying that to carry is to care, to remember, and to sustain. In *Seaweed King*, the bilum becomes a vessel—a ritual archive where fabric strips, each marking environmental destruction, accumulate until the bag appears to expand without limit. Its use reflects Wynter's notion of resilience through Indigenous practices, bearing the weight of ecological harm while sustaining life.

By contrast, the fishing net entangles the performer, representing both the destruction wrought by colonial systems and the potential for renewal through deliberate care. Plastic bags, filled with packaged air, serve as a cautionary commentary on the commodification of essential resources, a critique that aligns with Wynter's emphasis on challenging the coloniality of power and the reduction of nature to a mere resource for profit (Wynter, 2003). These material choices bridge the past with an urgent call for collective care and regeneration.

My grandparents' home in Tredegar Park was where I first learned how human actions upstream create downstream consequences. The banks of the Rio Cobre, marked by industrial intervention and colonial extraction, revealed the ways in which ecological harm is inherited. Their backyard—once a sanctuary of abundance, filled with fruit trees, healing plants, and memory—was later washed away by a storm. This personal loss reinforced a truth central to *Seaweed King*: the effects of human intervention in natural systems do not unfold in isolation—they accumulate, compound, and alter the future.

The themes of loss and renewal in *Seaweed King* mirror this downstream effect of human intervention, but they also suggest that through collective care, repair is possible. The legacy of Black and Indigenous ecological knowledge is not only one of survival but of transformation. This legacy informs the movement vocabulary, material choices, and symbolic gestures in *Seaweed King*, offering a framework for imagining regenerative futures. The performance enacts the resilience that Wynter highlights as essential to reimagining our relationship with the Earth—one where care is not just a survival strategy but a creative, generative act.

Note: Veiled in bilum and net, *Seaweed King* surfaces, embodying the call to remember, repair, and protect. Chris Walker in performance, Eat Little/Live Long, Brooklyn NY. David McDuffie Photographer

3.5 Materials in Motion: The Birth of *Seaweed King*

The character of *Seaweed King* and its costumes were conceived as part of a deeply integrated creative process, where movement and material continuously informed each other. The character emerged organically during the development of *What Lives Beneath*, a project led by visual artist and designer Laura Anderson Barbata, which brought together shared visions of ecological storytelling and environmental activism, centred on movement explorations with designed objects, masks, and costume pieces. Drawing from her expedition to Papua New Guinea, Laura introduced bilum bags—rich in cultural and ecological significance—as central elements of the project's visual and conceptual design, directly shaping the costumes and themes of *Seaweed King*.

The bilum bag represents resilience and cultural survival, carrying both the burden of environmental harm and the knowledge systems that sustain life. Beyond its role as a carrier, the bilum also masks the performer, erasing individuality and reinforcing the anonymity of the burden-bearer. One bilum is worn as a loincloth, grounding the performer in earthly and ancestral ties, while another shifts from mask to a living archive, holding both the weight of loss and the possibility of renewal. This transformation shifts the bilum from a symbol of life-giving care to a stark emblem of burden and resilience.

As a choreographer, I worked alongside dancers to experiment with these materials, allowing their textures and movements to shape the energy and identity of the characters. In one rehearsal, I began manipulating fabric strips that Barbata had brought—upcycled test fabrics that would have otherwise been discarded. Their flowing, undulating kelp-like qualities immediately suggested connections to natural elements like soft corals and sea grasses. As I draped more fabric strips over my body, their sargassum-like mass and weight transformed my movement and gestures, slowing them down, giving them new direction. A bilum bag, slung across my shoulders, amplified this feeling. In that moment, *Seaweed King* was born, unplanned and emerging as a new character defined by its deep relationship with these materials.

To complement the symbolic load of the bilum bag, and adding layers to visual and conceptual storytelling, Barbata designed a fishing net jumpsuit. The net enveloped the performer's body, tangling the fabric strips and making their manipulation intentionally difficult. This disruption required slower, more deliberate gestures, influencing the character's movement vocabulary to embody struggle and resilience. The physical impact of the net—its ability to mark the skin with welts and bruises—became a deliberate metaphor for the harm inflicted by overfishing and ecological exploitation, mirroring the fragility of ecosystems when mishandled or neglected.

From the outset, the development of *Seaweed King* integrated costume and character into a unified entity, where every gesture and material element contributed to the narrative of resilience and care. This iterative process extended to the involvement of other dancers. Tiffany Merritt-Brown was the first to explore the character outside of my own embodiment, using improvisational prompts to interact with the fabric strips. Her interpretations revealed new layers of symbolism, expanding the character's potential and refining movement discoveries. Kevin Ormsby later brought *Seaweed King* to life for a live audience, further developing the character through meticulous precision and emotional intensity.

In performance, *Seaweed King* became a living, breathing narrative of ecological harm and renewal. Barbata's costume elements, the destructive net, upcycled fabric strips, and bilum bags come together to form a symbolic narrative—a live sculpture—contributing to a layered work of artivism that embodies the complexities of environmental stewardship and resilience.

Reflecting Barbata's broader practice, the costume blends cultural reverence with ecological critique, positioning *Seaweed King* for its evolution from a character in *What Lives Beneath* into a stand-alone work that delivers a unique call for ecological care and cultural reconnection.

3.6   A Performance in Three Parts: Rituals, Archetypes, and Acts of Care

Performed as a solo by the choreographer in Madison, Wisconsin, in 2018, *Seaweed King* has evolved over time with reconstructions for Kevin Ormsby in 2019 and 2023. These revisions reflect the work's ongoing exploration of ecological responsibility and its adaptability to different performers and contexts. The piece is divided into three sections: rituals, archetypes, and acts of care; each requires different performance skills and technical approaches. There is an awakening and learning for *Seaweed King* from the journey through the archetypes, returning transformed. This process models for the audience their own capacity for change, inspiring them to reflect on their impact and envision a renewed relationship with the Earth.

3.7   Rituals

The work begins with an audience engagement ritual developed collaboratively with Laura Anderson Barbata. Audience members are invited to write the names of extinct species, lost habitats/settlements, or beloved ancestors on strips of fabric. Reflective of Barbata's art intervention practice—inviting participation without asserting authority—audience members choose to actively take part by interacting with fabric strips or remain observers, engaging through visual storytelling. Video clips of environmental degradation alongside a recital of Caribbean poet Dennis Scott's *Uncle Time* play on repeat in the background (Scott, 1973). The poem's vivid imagery portrays time as a relentless, unyielding force that exposes the consequences of human actions, resonating with the core themes of loss, accountability, and resilience. In *Seaweed King*, this inevitability, embodied in the audience ritual where they engage with the weight of loss and responsibility, mirrors the poem's meditation on the inescapable consequences of time. Their personal markings, each representing a unique story of loss or memory, are placed onto a seemingly innocuous pile on the stage.

By transforming abstract environmental issues into intimate, personal reflections, this intervention lasting 15–20 min connects audiences both physically and emotionally to its themes. This connection is crucial for moving viewers from awareness to action, creating visceral experiences that resonate deeply. As the fabric strips accumulate, the audience collectively builds the figure of *Seaweed King*, contributing to a form that is rooted in communal memory and a collective acknowledgement of what has been lost.

Unbeknownst to the audience, the pile of strips holds the still figure of a dancer, embodying *Seaweed King* in a suspended performance of stillness. For the dancer, this moment is about tuning—opening of the senses, feeling the weight of the fabric and contact with it, smelling the ink from the writings, feeling the small channels of breeze, listening to the sounds—the audience's whispers, and the subtle movements in the space. There is an intimacy that grows in this stillness, an amplified awareness that connects the body of the performer to the environment of the performance space.

The dancer's presence, initially hidden, is the quiet resilience of the natural world—a force that, though subdued by human impact, remains part of our shared environment.

The pile begins to move, the audience audibly reacts, and the body appears—a deliberate slowness, a refusal to rush through the act of revelation. This slow pace is meant to challenge the audience's expectation of movement, asking them to sit with the weight of the moment. The action-based compositional structure of the opening sequence reflects a series of actions of a living, breathing entity that grows, morphs, and evolves as *Seaweed King* becomes aware of his surroundings. The upcycled fabric strips, salvaged from what would have been discarded, take on new roles, both elevating and obstructing *Seaweed King's* emergence from the depths, reminding us that even in destruction lies the potential for transformation.

Seaweed King's slow emergence is rooted in an ancient relationship to time. It echoes Dennis Scott's *Uncle Time*—the slow, deliberate movement of a wise elder (Scott, 1973). This reverence for time makes the opening of the work both sacred and unsettling. The expectation of the proscenium stage invites a specific kind of performance—one of immediacy and resolution. However,

Seaweed King interrupts these expectations, deliberately cultivating discomfort to engage the audience more profoundly. This intervention models a pathway for transformation, urging viewers to confront their own impatience, their need for action, and their discomfort with stillness. Time, especially in the context of environmental degradation, cannot be hurried or taken for granted.

The audience engagement invites and confronts, drawing them into the complex terrain of ecological loss and responsibility. The opening act implicates the audience in the subject matter, revealing multiple perspectives on their involvement. For some, this ritual may evoke a sense of shared responsibility, uniting each participant in a communal act of remembrance and care. The strips, fragments of what the world has already lost, encourage a feeling of collective burden and, ideally, foster a deeper commitment to protect what remains.

Yet there is another layer to this act—a possibility that implicates the audience in a less comfortable way. By casting their tokens of loss onto the pile, participants risk symbolically offloading their burdens onto the environment itself, expecting the *Seaweed King* to bear the weight alone. This act mirrors the often-unconscious way society distances itself from its ecological impact, treating nature as a receptacle for its waste and sorrow. The audience sees *Seaweed King* embodying the weight of these burdens, his effortful shedding and gradual revelation. This dual implication forces each viewer to consider their role: are they contributing to a collective effort or merely transferring their losses onto a world already overburdened?

Ideally, the opening experience introduces a sense of urgency, making clear that addressing the climate crisis demands active, shared responsibility—not passive or symbolic participation. It invites audiences to see themselves not apart from but as integral to the Earth's systems of balance and renewal. Within this system, every part—be it geological, biological, or atmospheric—contributes to a larger balance. Even materials buried deep within the planet's layers eventually resurface, transformed by natural processes, ready to support new cycles of life. It is from this deep that *Seaweed King* emerges.

3.8 Archetypes

Humans, too, have a role in this interdependence. Just as plants, rivers, and ecosystems contribute to a healthy Earth, so do human actions affect the planet's cycles and stability. The emergence of *Seaweed King* as a humanoid being in the second section of the work is in direct response to this crisis. The character rises from the aftermath of a catastrophic event, stepping into a world marred by debris and ruins.

When *Seaweed King* first stands tall, the spatial design evokes the aftermath of a catastrophe. Strips of seaweed spread out around him like a chaotic, sprawling organism, consuming space. *Seaweed King* embodies a series of archetypes—coloniser, environmentalist, star/influencer, fishmonger, and Indigenous person—each revealing a different side of humanity's relationship with the environment. His first action is to survey the land. The coloniser moves with the angular rigidity, embodying a system that takes without giving, dominating without understanding. The movement vocabulary of the coloniser, the surveyor, the capitalist—is that of figures that have historically represented exploitation and disruption. The movement is militant, with grid-like patterns that reflect precision and command. The body is upright, angular, almost mechanical, with directional prescribed gestures—what I think of as an "upright death." There is no reverence to the earth in the vocabulary; the torso stays upright showing a cold sense of authority and extraction. It is the destructive force of colonialism—how it fractures ecosystems, commodifies nature, and disturbs the delicate balance of interconnected systems.

From the rigid figure of the coloniser, *Seaweed King* transitions into the environmentalist, a youth archetype embodying urgency and advocacy for the planet. This shift begins with a subtle costume change: the netting worn as a necktie by the coloniser is repurposed as a scarf for the environmentalist, symbolising a move from dominance to care. The environmentalist's gestures become open and beseeching—arms outstretched, chest clutched, and direct pointing to audience members—encouraging shared ecological responsibility and collective care. This transformation, marked by the repurposing of costume elements, reframes "king" as an advocate for the planet's future.

By contrast, the star archetype, representing privilege and excess, uses plastic bags filled with air—a cautionary critique of commodification and environmental imbalance, echoing Wynter's warning against exploitative systems. This character uses entire bags of air to speak, discarding them carelessly, privileging performance and spectacle over substance. The plastic waste left behind underscores the imbalance caused by unchecked consumption highlighting the devastating environmental cost of privilege. The influencer/star archetype highlights the allure and potential of visibility, a modern form of power often associated with celebrity culture. The influencer/star archetype critiques the superficiality of influence for its own sake, exposing the environmental and social costs of excess.

The fishmonger archetype embodies an intimate understanding of ecological systems and cycles. Rooted in the rhythms of daily life, this figure understands conservation through practice. This character redefines "king" as a caretaker deeply connected to the rhythms of the Earth, someone who recognises that true power lies in sustaining rather than exploiting. Through the fishmonger, the *Seaweed King* invites audiences to honour the wisdom of those who work within the cycles of nature, rather than against them, and challenges us to see labour and care as sovereign acts.

Finally, *Seaweed King* embodies the Indigenous person archetype, moving with a grounded, patient grace that reflects ancestral wisdom and humanity's place within the delicate web of life. In this context, the archetype redefines "king" as a steward—a figure of humility and reverence for Earth's ecosystems. It underscores the necessity of listening to and learning from Indigenous perspectives, which offer critical models for ecological harmony and resilience. This archetype also carries the profound weight of grief, perceiving the Earth's suffering—the poisoned rivers, scarred landscapes, and irreparable harm caused by humanity. At the height of their performance, they release a wail, a cry of anguish and reckoning, calling audiences to confront the urgency of collective responsibility and repair.

Note: *Seaweed King*, moving through ecological debris, caught between anguish and awakening. Chris Walker in performance, Eat Little/Live Long, Brooklyn, NY. David McDuffie, Photographer

3.9   Acts of Care

As the *Seaweed King* embodies each human archetype, he reflects the contradictions most of us carry in our relationship with the Earth. He moves between destruction and advocacy, waste and conservation, and dominance and stewardship—his journey mirrors our own struggles to reconcile these conflicting roles. By the time he returns to himself, the *Seaweed King* sees the evidence of exploitation scattered everywhere, and it compels him to act. Underscoring the tension between immediate comforts and long-term consequences, his first act is cleaning the plastics left behind by the influencer, securing them in the bilum bag. His work becomes about repair—tending, gathering, and securing what is broken. Each movement is deliberate and purposeful, reflecting a shift towards care and responsibility. This is the call I hope resonates with the audience: to see themselves in his actions and to consider what role they play in this shared work of restoration.

Seaweed King concludes with an act of care—marked by bending, a reaching, a gathering of what remains, and becoming a thread that stitches together what has been unravelled. It is work, yes, but it is also ritual, renewal, and reckoning, like *Uncle Time*'s insistent cunning labour, his "web" of consequence spun across generations (Scott, 1973). It is the work of time itself. It is sprawling and messy, urgent, and meticulous, strategic in its focus, and tender in its execution. Each piece that *Seaweed King* collects is held with reverence, a gesture that honours loss while insisting on the possibility of repair. Kevin Ormsby captures it perfectly when he calls it "an act of protecting the species." The fabrics collected during this section are, in his words, "extensions of us as human beings," holding the names of extinct species, lost settlements, and cherished loved ones. This act of gathering is both tedious and profound, deeply rooted in care.

I designed the section to embody the labour of navigating the direct consequences of ecological degradation with vocabulary cycling through increasing technical difficulty amid the weight and imbalance of the continuously expanding bilum bag. The ground is always shifting; nothing can be left behind. This deliberate instability amplifies the tension between control and precariousness, mirroring the ecological challenges we face, where every action feels fragile but carries immense consequence. Kevin Ormsby conveys this tension, describing the physicality of twisting, stretching, and contorting to retrieve fabric fragments while perched precariously. *Seaweed King* turns and balances, and between shuffling and rolling, falling, and suspending, there are moments of stillness—pauses to notice even the smallest fragments. Kevin likens this to negotiating with an ever-changing environment: "You have to find the next thing with immediacy," he explains, "you have to be that agile—that alert" (Ormsby pers. comm., 2019) The act of filling the bag becomes a potent ritual. As the bag expands beyond what seems possible, it absorbs the overwhelming weight of human impact on the Earth—a visceral acknowledgement of the planet's burden.

Building on its earlier symbolism, navigating off-centre turns, leaps, and balances with the bilum bag loaded with fabric reflects resilience in the face of adversity, the uneven weight of ecological loss. The echoes of Dennis Scott's *Uncle Time* reverberate in these moments, as the poem describes how Uncle Time, with "sea-win' laughter," is "scraping away de lan'" (Scott, 1973). This relentless and quiet erosion mirrors the inexorable pull of time, framing *Seaweed King's* labour as both necessary and inevitable. The body contorts and struggles to carry the weight, much like our struggle to bear the weight of our environmental impact. Ormsby's reflection brings this poetic resonance to life: "You feel the weight, but there's a lightness too, because you've found a new stance. A new space of understanding" (Ormsby pers. comm., 2019). The performance does not offer resolution; it offers reckoning. The audience—writing their grief and their losses on fabric strips—ties themselves to the work and to the weight.

Despite the heaviness of the themes, there is a sense of joy. *Seaweed King* finds a new rhythm, and as he lifts the burden onto his shoulders, we are reminded of Scott's words: "Uncle Time is a ole, ole man" (Scott, 1973). Time moves forward; the work is infinite, collective, and urgent. A transformation happens—not one of release but one of adaptation. *Seaweed King* does not shed the weight but instead learns to carry it differently. The final step, drawn from *Ettu*, carries a playful release, symbolising the emergence of life amidst death.

It is a joy born from survival that acknowledges the precariousness of our existence but refuses to be consumed by fear. *Seaweed King*, in its final moments, reminds us that there is beauty in the struggle and that even as we face the destruction of our environment, as Scott's suggests, "when 'im touch yu, weep…" (Scott, 1973). Weep for the Earth, for what has been lost and for what might still be saved. Then, bend down, pick up the fragments, and, like *Seaweed King*, find a new rhythm and begin again. The weight is ours and so is the dance.

3.10 Conclusion: Caribbean Leadership in an Eco-Social Future

At its heart, *Seaweed King* is both a process and a praxis. It embodies a rupture from entrenched anthropocentric ideologies, proposing an alternative relationship with the planet, one rooted in care, interdependence, and collective responsibility. The grief carried by *Seaweed King* is inescapable—an acknowledgement of ecological loss and cultural erosion. Yet grief must not stagnate; it must transform into action, for within the depths of loss lies the potential for growth, reparation, and new forms of being.

Seaweed King enacts Wynter's call for decolonial praxis, urging a fundamental shift in how we inhabit the Earth. The performance bridges theory and action, transforming ecological crisis into an embodied conversation that disrupts passive spectatorship. This is art as radical engagement, an invitation not only to reflect but to act, adapt, and intervene. In this process, the audience is not merely an observer—they are placed at the centre, implicated as agents of ecological and social transformation.

Mia Mottley's warning remains clear: "Our world knows not what it is gambling with, and if we don't control this fire, it will burn us all down" (Mottley, 2021), sounding a call that echoes Aimé Césaire's charge more than 70 years ago in *Discourse in Colonialism*. Césaire cautioned that civilisations unable to solve the problems they create are decadent and destined for collapse. *Seaweed King* emerges within this lineage of warning and reckoning, evoking the unfinished work of confronting colonial devastation and imagining new ways of being in relation to the earth. *Seaweed King* offers a critique of these systems and their downstream impacts, as well as a pathway forward. By reframing grief as a call to action, the performance insists—like Césaire—that renewal is possible but only if we commit to building a world that values interdependence over exploitation. This too is the charge of the eco-social future—not only to reckon with crisis but to innovate through it, to step into the fire and move, carrying the weight not as an anchor, but as momentum—a fragile hope for renewal.

A striking example can be seen in the Pelagic Sargassum crisis in the Caribbean. Once a seasonal and ecologically beneficial presence, Sargassum blooms have become overwhelming. Massive accumulations disrupt marine ecosystems, devastate fisheries, and suffocate coastal communities. The burden of this crisis has fallen largely on those whose economies are directly tied to the sea, revealing a stark reality: nature alone cannot correct the damage inflicted upon it. Human responsibility and intervention are necessary to restore balance.

The Caribbean has long been a site of eco-social experimentation, where environmental justice is deeply entwined with histories of resistance and survival. The United Nations Environment Programme's 2021 Sargassum *White Paper* frames the ongoing influxes of Sargassum as both a crisis and a potential catalyst for new forms of regional resilience (UNEP-CEP, 2021). Across the region, nations are exploring innovative ways to repurpose Sargassum, transforming an ecological crisis into economic opportunity. Scientific research in Barbados and the Dominican Republic has successfully repurposed Sargassum into biofuels and other sustainable materials (Reingold, 2024), demonstrating how Caribbean nations are transforming an environmental crisis into opportunities for climate resilience and economic innovation.

Nations like Guyana exemplify this commitment to climate accountability and sustainable development. With 87% of its landmass covered in forests, Guyana plays an outsized role in global climate solutions, storing approximately 19.5 gigatons of carbon dioxide equivalents. President Irfaan Ali's *Low Carbon Development Strategy 2030* protects these forests (Government of Guyana, 2021) while also investing in renewable energy, climate-resilient agriculture, and coastal defences—showcasing a model where environmental responsibility is woven into national development. Guyana has also committed to reducing its carbon emissions by 70% by 2030, recognising that action must be both systemic and sustained.

This urgency is also evident in real-world eco-social transformation across the Caribbean. Organisations like Jamaica's youth-led Climate Change Youth Council push for climate education, advocacy, and policy reform, empowering local communities to act. Meanwhile, the *Barbados National Energy Policy 2019–2030* outlines the nation's commitment to achieving 100% renewable energy and full electrification of its transport sector by 2030 (Government of Barbados, 2019). This transition is already underway, exemplifying the principles of an eco-social contract—one that integrates ecological sustainability with social well-being, fosters equitable access to clean energy, and promotes environmental stewardship among citizens, businesses, and the state.

These efforts reveal that the eco-social contract is not just aspirational—it is already unfolding, forged through resilience, adaptation, and collective ingenuity.

Seaweed King emerges from this landscape as a demand for action. The work aligns with the ethos of these Caribbean movements, illustrating that change is a shared responsibility. Whether through performance, policy, or community engagement, *Seaweed King* stands within this moment of reckoning, insisting that the climate crisis is not an abstraction. It demands that we move, that we innovate, and that we step forward—not despite the weight but with it, finding a rhythm that does not break us but remakes us.

References

Government of Barbados. (2019). *The Barbados National Energy Policy 2019–2030*. Ministry of Energy and Water Resources.

Government of Guyana. (2021). *Low carbon development strategy 2030 (LCDS2030)*. Office of the President.

Mottley, M. (2021). *Our world knows not what it is gambling with*. Speech delivered at the United Nations Climate Change Conference (COP26), Glasgow.

Ormsby, K. (2019, 2022). *Personal communication with the author, 2019, 2022*.

Reingold, J. (2024, December 19). *Turning the Sargassum crisis into a seaweed industry*. Pulitzer Center.

Scott, D. (1973). Uncle time. In *Uncle time and other poems*. University of Pittsburgh Press.

United Nations Environment Programme – Caribbean Environment Programme (UNEP-CEP). (2021). *Sargassum white paper – Turning the crisis into an opportunity*. UNEP-CEP.

Wynter, S. (1970). Jonkonnu in Jamaica: Towards the interpretation of folk dance as a cultural process. *Jamaica Journal, 4*(2), 34–48.

Wynter, S. (2003). Unsettling the Coloniality of being/power/truth/freedom: Towards the human, after man, its overrepresentation—An argument. *The New Centennial Review, 3*(3), 257–337.

4

Restoring Planetary Balance: Exploring Muslim Eco-Social Covenants for the Earth

Najma Mohamed

4.1 Introduction

Humanity is increasingly challenged by the fallout from human aggressions and transgressions against the Earth and its inhabitants. Extreme heat, polluted air and biodiversity loss are some impacts of the interlinked climate and ecological crisis. Even as we celebrate the achievements of a half-century of "green" policymaking, humanity continues to bankroll the extinction of life on Earth. But the social and ecological impacts of overextraction, pollution and human-induced climate change are not borne equally. Our economies are driving the accumulation of wealth for the few in highly industrialised countries, while the global majority bear the burden of the triple planetary crisis: biodiversity loss, climate change and pollution (Wilkinson & Pickett, 2022; Chancel et al., 2023).

For a growing community of scholars that seek to understand the root causes of this crisis, the social and ecological malaise we face today emerges from a crisis of worldview and presents a spiritual and moral, rather than a technological, conundrum. As discussed throughout this book, dominant worldviews that have shaped political economic and social systems today often only think in technological fixes to address the crisis, shaping much of

N. Mohamed (✉)
UN Environment Programme World Conservation Monitoring Centre, UNEP-WCMC, Cambridge, UK
e-mail: najma.mohamed@unep-wcmc.org

© The Author(s) 2025
P. Huntjens et al. (eds.), *Eco-Social Contracts for Sustainable and Just Futures*,
https://doi.org/10.1007/978-3-031-99109-7_4

our aspirations for action today. Yet, as Özdemir (2008) maintains, this crisis has metaphysical and philosophical roots, since there is a relationship between human treatment and conceptualisation of nature. To right the relationship between humans and nature, we must therefore "re-examine our understanding of nature, and our place within, and relationship to, the natural world" (Baker & Morrison, 2008: 36). Tackling this crisis will require shifting from anthropocentric values, principles and beliefs to ecocentric pathways of connection and relatedness that situate humanity within the broader community of life (Mohamed & Huntjens, 2023). The planetary crisis thus requires not only scientific, policy and technological fixes but a shift in worldview which should address the unjust relationship between people and nature, as well as growing inequalities between the haves and the have-nots.

The beliefs, values and knowledges embedded in cultures and religions provide environmental imaginaries which can inform and inspire eco-social action (Grim & Tucker, 2014). Yet knowledges emerging from non-Western peoples and traditions are often given less credence and delegitimised as a foundation for ecological ideas and practices. This epistemic injustice results from, to paraphrase Shiva (1993), a "monoculture of minds" that silences diverse ecological knowledges in the same way that monocultures destroy the conditions for sustainability. Fricker (2007: 1) defines epistemic injustice as a form of injustice "done to someone specifically in their capacity as a knower", excluding and silencing knowledges and understanding and calling into question their validity and contribution. For instance, the systematic exclusion and under-representation of Indigenous knowledges in climate and biodiversity planning and policy (Byskov & Hyams, 2022), despite the crucial role of Indigenous people as stewards of the world's biodiversity, are telling. This holds equally true for other environmental imaginaries and knowledges that have the potential to counter the exploitation of people and nature, Islam's ecological imaginary included (Mohamed, 2012). These alternative imaginaries can offer ways of living in connection, in reciprocity and in justice with people and nature, and are vital to envision eco-social contracts. This chapter seeks to contribute to the reimagining of eco-social contracts premised on diverse ways of knowing and living with nature.

4.2 A Pluriverse of Ecological Knowledges

Ecological knowledge, defined as "the knowledge, however acquired, of relationships of living beings with one another and with their environment", and constituted by a knowledge-practice-belief complex, is embedded within a worldview (Berkes, 2008: 5). While worldviews can be based on religious, spiritual or secular beliefs and ethics, of interest in this chapter is making visible knowledges that nurture and respect life on Earth and how these could give voice to more diverse perspectives on the human-nature relationship.

In *Pluriverse: A Post-development Dictionary*, Kothari et al. (2019) curate over 100 essays that weave together a "tapestry of narratives, imaginaries and visions" that both theorise the human-nature relationship *and* seek to achieve eco-social transformations. These imaginaries present ways of understanding the world and offer pathways for just and sustainable living. By shedding a light on this pluriverse of environmental imaginaries, we not only address the epistemic injustice that has systematically excluded diverse eco-social visions (Trisos et al., 2021) but also seek to remedy and redress this exclusion. For many in the world, these environmental imaginaries are derived from religious beliefs.

Religious discourse offers one of the most cogent forms of "cultural conversation outside the modern story of economic growth and technological fixes" (Oelschlager, 1996: 47). Religious resources for the environment are ideational, practical *and* political, since they present not simply an ecological ethic but also offer an eco-justice ethic in which social and ecological justice are inextricably linked: a liberation eco-theology (Mohamed, 2012). Liberation eco-theology, rooted in liberation theology, privileges a commitment to eco-social justice, harnessing faith as a vehicle for eco-social transformation. It incorporates the liberative dynamic within religious traditions by placing social justice at the centre of eco-social action. It builds on the strong tradition of liberation theology within religious tradition, which includes standing at the forefront of actions to abolish slavery and apartheid, to fight poverty and to act against the sins of pollution and environmental destruction (Boff, 1995; Gottlieb, 2003). More than adding social justice to ecology, it is about "raising the emancipatory potential of environmental ideas and to engage directly with the larger landscape of debates over modernity, its institutions, and its knowledges… which match the nuanced beliefs and practices of the world" (Peet & Watts, 1996: 37). This places environmental imaginaries *within* people's meaning systems, including diverse ways of knowing and living with nature (Arora & Stirling, 2020).

Islam plays a pivotal role in shaping the worldview of close to two billion people today. Its theology of nature, or eco-theology, has inspired a growing movement across the Muslim world (Foltz, 2000; Mohamed, 2024a) voicing its concern about and acting in response to the climate and ecological crises. Muslims are drawing on the teachings of Islam in the pursuit of a sustainable and just world, showing that in being true to these teachings, they must live in justice with all of creation. This chapter seeks to address the epistemic injustices in ecological knowledge by presenting the eco-justice ethics of Islam as an environmental imaginary which is seeding a movement to restore planetary balance.

4.3   A Liberation Eco-theology of Islam

Eco-theology is concerned with the relationship between religious beliefs and practices and how this impacts the human relationship with nature. Concern for the environment is deeply rooted in all fields of Islamic teaching and culture, and many concepts and principles in the Qur'an, which Muslims believe to be the word of God revealed to humanity, carry substantive implications for eco-social action. Islam's eco-theology defines humanity's relationship with God and with creation, their fellow human beings and the natural world. Izzi Dien (2000: 81) suggests that the ingredients for an eco-theology of Islam are "dissected parts of Islamic theology, law and ethics", while Gade (2019) makes the case that Muslim environmentalism is simply the practice of the religion. Islam's eco-theology presents the planetary crisis as a failure of human trusteeship resulting in "corruption (*fasād*) in the land and seas" (Qur'an Al-Rūm 2008, 40:41).

4.3.1   An Outline of Islam's Eco-theology

Islam's eco-theology is neither ecocentric (nature centred) nor anthropocentric (human centred) but theocentric (God-centred). The oneness of God, *tawḥīd*, infuses the environmental worldview of Islam with the recognition that the design of the world is purposive and functions in accordance with the will of God. The universe is believed by Muslims to have been created by God, with humanity tasked with living in submission to the laws of God, including not transgressing the primordial balance in which the Earth was created (Qur'an Al-Rahman 2008, 55:7–9). *Tawḥīd* has profound implications for ethical conduct since humans are enjoined to act in obedience to divine will: this obedience extends to living in harmony with nature. Thus, the Muslim should "not only feed the poor but also avoid polluting running water. It is pleasing in the eyes of God not only to be kind to one's parents, but also to plant trees and treat animals gently and with kindness" (Nasr, 1997: 9).

Muslims believe that the universe has been created in balance (*mīzān*) (Qur'an Al-Qamar 2008, 54:49). Humanity is invited to contemplate natural phenomena in a quest for meaning, to look to nature, observe its order and deduce the oneness of God (Özdemir, 2003). This observation imparts what harmonious and balanced living *within* planetary boundaries should entail (Parvaiz, 2003; Haque et al., 2021). It is mentioned in numerous instances in the Qur'an that humankind should observe the balance of the Earth and not cause corruption (*fasād*) therein after it has been set in order. Muslim scholars interpret the triple planetary crisis as an impairment of this balance for which humanity will be called to account (Ahmad, 1997). In the words of Abdul-Matin (2010: 131): "By treating the natural world as though it were our dumping ground, we risk disturbing the delicate balance (*mīzān*) that exists in nature". Khalid (2003: 315) calls *tawḥīd* the "primordial testimony to the unity of all creation and to the interlocking grid of the natural order", the creation of one God, in which all of creation possesses "co-integrated rights" (Haque et al., 2021).

Muslims consider that all creation (*khalq*) on Earth form part of an interconnected system originating from one God in which *all life* serves a purpose, making "the world one telic system, vibrant and alive, full of meaning" (Al-Faruqi, 1987: 25): a world in which all creation is regarded as communities (Qur'an Al-An'ām 2008, 6:38). The position of nature in Islam can be condensed into three domains, with the foremost being as signs of God. Second, nature has an ecological value as an integral part of the whole ecosystem. Finally, nature has an instrumental value that is necessary to sustain all life. This is expressed in Qur'anic verses that state the natural world has been "subjected" (Qur'an Al-Jāthiyah 2008, 45:13) by God for life to flourish. This does not entail human domination or exploitation. Rather, humankind bears the trust of stewardship and must maintain the primordial equilibrium between the needs of humanity and "the rights of other creatures to live out their lives on this earth" (Setia, 2007: 132).

In Islam, human stewardship, *khilāfah*, must be carried out in accordance with the teachings of the religion, which seeks to guide humanity to live in justice with all life. The Islamic worldview is that God created humankind and that, through a covenant with God (Qur'an Al-Ahzāb 2008, 33:72), humanity accepted the trust of being a steward or representative of God on Earth. The term *khalīfah* has been translated as representative, steward, guardian and vicegerent and covers every aspect of life. Humanity's just exercise of this trust, including interactions with the natural world, must be in harmony with the wishes of its Grantor, God (Sachedina, 1999).

Women and men in Islam are thus vicegerents, provided with bounties that should be enjoyed within limits. Nasr (1997: 8) states that "to be human is to be aware of the responsibility which the state of vicegerency entails". *Khilāfah* incorporates responsibility and accountability for conduct towards fellow human beings, fellow creatures and the planet itself. Humanity, if it betrays the duties of trusteeship, not only endangers the security of creation but also forfeits the right to its own physical and spiritual security in this world and the next. *Khilāfah* is both a responsibility and a trial by which human beings will be evaluated in terms of who has done the most good, acted according to God's purpose and served all of creation (Llewellyn, 2003). This metaphysical exaltation is linked to a weighty moral burden—to adhere to a code of action reflecting the best social behaviour and highest ethical values (Haq, 2001).

What happens when humanity fails? In the words of the Qur'an: "Corruption (*fasād*) prevails in the land and the sea because of all the evil that the hands of humanity have earned – so that God may cause them to taste something of that which they have done so that they may return in penitence to God" (Qur'an Al-Rūm 2008, 30:41). *Fasād* features prominently in Islamic eco-theology as a term which has been used to describe the climate and ecological crises (Ghoneim, 2000). Translated as destruction, corruption or mischief, *fasād* is said to apply to the realm of the environment as it does to any other part of life. It is the result of transgressing the limits of human behaviour as ordained by God. Exegetists extend *fasād* to the planetary crisis, which is regarded primarily as a failure of human trusteeship. Nature thus becomes the index of how well a particular society has performed its responsibility towards God (Ouis, 2003).

Islam's eco-theology compels Muslims to act and speak for the Earth as a matter of religious belief and practice. At almost 2 billion strong, Muslims are joining the movements advocating for climate and ecological justice, displaying the transformative force of their faith, which propels them to live in justice with all of creation. For, "indeed, God loves those who are just" (Qur'an Al-Māʾidah 2008, 5:42). This movement, like civil society globally, faces diminishing civic space for action, including in many Muslim majority countries governed by authoritarian regimes that do not govern in accordance with and largely do not display the liberative dynamics of Islam.

4.3.2 An Eco-theology of Justice

One of the distinguishing features of liberation theology is its emphasis on lived spirituality. Liberation theology "arises out of believers' outrage and protest against injustice… and a commitment to transforming it toward justice, freedom, and dignity" (Solberg, 2008: 310). Liberation eco-theologies put forward both a political and spiritual vision centred on the recognition that it is not only the poor, oppressed and disenfranchised who largely bear the burden of climate and environmental breakdown but the Earth itself which is suffering at the hands of humankind.

Islam's eco-theology connects ecological and social justice, placing exacting ethical demands on Muslims to strive to live in justice with *all* life. The Qur'an stresses the inextricable connection between belief and the struggle for social justice, obligating believers to stand in solidarity with the oppressed: "God has sent His Messengers and revealed His Books so that people may establish justice *[qist]*, upon which the heavens and the earth stand" (Al-Jawziyyah, 1973: 373). Further, Islam maintains this relationship between belief and social justice (Kerr, 2000) with its message of ethical monotheism culminating in the establishment of a sociopolitical community in Medina. The covenant of Medina, formulated in 622 CE, has been described as the first written constitution in world history (Bangash, 2011), establishing a community based on the rule of law where the rights and obligations of *all* inhabitants were equally recognised and upheld.

Islam's liberation eco-theology also develops a critique of prevailing economic and development paradigms (Al-Hamid, 1997; Khalid, 2003; Al-Jayyousi, 2015; Haque et al., 2021). Al-Hamid (1997), for instance, compares the social and environmental destruction wrought by ill-conceived international development interventions to the Qur'anic verse mentioning those who purport to do good but who in fact cause corruption or *fasād* in the land (Qur'an Al-Baqarah 2008, 2:204–205). Unjust and inappropriate developmental and economic systems are seen as drivers of climate and ecological injustice. Instead of prosperity and well-being, these systems have brought widespread indebtedness and have exacerbated multiple and intersecting inequalities. Relentless economic growth, without consideration of the ecological or social footprint of our societies and economies, is described by Al-Jayyousi (2015) as "the economic costs of corruption".

Liberation thus covers multiple dimensions of human life besides the spiritual. This approach to "[u]niversal justice, in Islam, means justice to all entities for it takes the stance that it is not enough for us to be humans, but rather to be *just* humans" (Haque et al., 2021: 205). Islam's eco-theology recognises that social and economic injustices play a pivotal role in understanding the planetary crisis.

4.4   Charting Pathways to Eco-social Action

The burgeoning green movement among Muslims, rooted in Islam's eco-theology, spans multiple efforts. This includes the establishment of practical initiatives that underpin the growth of an indigenous environmental movement compatible with and rooted in Islamic norms (Foltz, 2005). This section examines one expression of Muslim environmentalism, i.e. the charters, declarations and statements that represent Islam's eco-theology, to assess how these reflect the liberation eco-theology of Islam. The *Islamic Declaration on Sustainable Development*, the *Islamic Declaration on Global Climate Change* and *Al-Mizan: A Covenant for the Earth* have been selected. While there have been earlier statements issued as part of multi-faith messages, these three declarations offer an expression of the collective vision of Islam's eco-theology in response to the climate and ecological crisis. What is the history and background to these declarations?

The Islamic Declaration on Sustainable Development (2002): The declaration was championed by the Organisation of Islamic Cooperation (OIC), in response to a call from the United Nations for all international and regional organisations to prepare action programmes for the World Summit on Sustainable Development (WSSD), held in Johannesburg in 2002. The OIC is a membership organisation of 57 Muslim-majority countries spread over 4 continents. Established in 1969, its fundamental purpose is to strengthen solidarity among member states. Climate and environment represent one of the themes of cooperation in the OIC. To this end, the OIC convenes regular intergovernmental conferences of environment ministers. The first conference in Jeddah in 2002 led to the production of the *Islamic Declaration on Sustainable Development*. The Islamic Educational, Scientific and Cultural Organization (ICESCO), an inter-governmental organisation operating under the aegis of the OIC, in cooperation with the United Nations Environment Programme (UNEP), engaged with member states and key organisations to develop the declaration which was shared at the WSSD.

The *Islamic Declaration on Global Climate Change* (2015): The declaration, launched at the Islamic Climate Change Symposium in Istanbul, Turkey, in

August 2015, emerged from a collaboration between organisations and individuals working on climate change and environment in the Muslim world. This includes organisations such as the Islamic Foundation for Ecology and Environmental Science (IFEES) and Islamic Relief. Research and consultation for the drafting of the declaration included religious scholars, academics and environmentalists. The draft was circulated worldwide for input and then discussed for adoption by about 80 Islamic scholars, academics, UN representatives, representatives of many faiths and non-governmental organisations (NGOs). The final text was launched at the International Climate Change Symposium on 18 August 2015. Key organisations that participated in the symposium included the OIC, ICESCO and the International Islamic Fiqh Academy, a subsidiary organ of the OIC focused on the application of Islamic scholarship to contemporary challenges. The declaration is regarded as one of the "most notable contributions to environmental justice", presenting both "a discussion on climate change and action for change" (Turnbull, 2021). In it, Muslim leaders called on governments meeting later that year to develop the Paris Agreement on climate change to "bring their discussions to an equitable and binding conclusion" (United Nations Framework Convention on Climate Change, 2015).

Al-Mizan: A Covenant for the Earth (2024): The impetus for the development of *Al-Mizan* came from the Eighth Islamic Conference of Environment Ministers held in 2019. It emerged amid a call for a strategy to enhance the role of cultural and religious factors in climate and environmental protection. This resulted in a 5-year process convened by UNEP's Faith for Earth Coalition that led to the development of a declaration representing the Muslim response to the climate and ecological crisis. *Al-Mizan* is an Arabic word that means equilibrium or balance and is associated and mentioned in the Qur'an with *al-qist*, i.e. equity, fairness and justice. It "describes a cosmic equilibrium in which all interconnected and interdependent beings are integrated in harmony" (Llewellyn et al., 2024: 70). The declaration was created by a drafting team of international religious and environmental scholars and activists. The first draft was shared with more than 300 institutions and partners for consultation. *Al-Mizan: A Covenant for the Earth* was launched at the UN Environment Assembly in Nairobi, Kenya, in 2024 and has been translated into multiple languages. It covers the eco-theology of Islam and offers practical guidance for Muslim environmental stewardship and seeks to act as a "platform to link environmental issues with Islamic teachings and embrace Islamic views on nature" (Llewellyn et al., 2024: iv).

The analysis below explores five dimensions of these three eco-social contracts to assess how they reflect the liberation eco-theology of Islam. Table 4.1 offers a structured comparison of these three contracts, showing how varying

Table 4.1 An analysis of three eco-social contracts representative of the eco-theology of Islam

Domain of analysis		*Islamic Declaration on Sustainable Development* (First Islamic Conference of Environment Ministers, 2002)	*Islamic Declaration on Global Climate Change* (International Islamic Climate Change Symposium, 2015)	*Al-Mizan: A Covenant for the Earth* (Llewellyn et al., 2024)
Rationale and development	Why was the document produced?	In response to the UN call for programmes to advance people-centred sustainable development at the WSSD in 2002	In response to collective efforts by governments and stakeholders to develop a global response to climate change in the run-up to the Paris Agreement on climate change in 2015	In response to the need for a platform to inspire Muslim climate and environmental action
	What is the focus of the document?	It presents an action programme rooted in the Islamic perception of environment and development	It outlines the climate challenge and its causes and develops a call to action, rooted in principles of Islam's eco-theology	It draws on science to assess the crisis, provides a political economy-rooted critique of the causal factors and presents Islam's eco-theology
	Who was consulted in its production?	ICESCO and UNEP engaged OIC member states and regional and international organisations at the First Islamic Conference of Environment Ministers (2002)	It was drafted by scholars and academics and included an extensive consultation process. Outputs were considered by 80 organisations at a symposium where the final declaration was formulated	Convened by UNEP'S Faith for Earth Coalition, a core team made up of religious and environmental scholars drafted the covenant, which received feedback from 300 organisations and partners. This was integrated in the final text launched at the sixth session of the United Nations Environment Assembly (UNEA-6) in 2024

Framing			
What is identified as the main causes of the climate and ecological crisis?	Poverty, debt, conflict, overpopulation, lack of technology and the lack of skills and capacity needed for sustainable development actions	Human-induced corruption and devastation and disruption of the balance of the climate and earth systems, due to "relentless pursuit of consumption" and unbridled economic growth	Economic growth fuelled by the production and consumption patterns of wealthy nations, coupled with high levels of inequality and wealth accumulation
Are connections made between the climate and ecological crises?	Overarching focus on sustainable development and no consideration or mention of climate. Environmental and natural resources framing indicates a material value given to nature	Consequences of "human disruption of this balance" [in nature] identified as global climate change and environmental collapse	Both global climate change, driven by an "energy-hungry consumer civilisation", and the loss of biological and cultural diversity are regarded as the result of *fasād-fil-ardh* (corruption of the Earth)
Are the links between social, ecological and environmental inequalities recognised?	Yes. While this is a largely technocratic analysis of constraints, several of the solutions presented are eco-social, and the need for the industrialised world to "step up support" for development is identified	Yes. It recognises that the risks of climate change are "unevenly distributed" and will be borne by the "poor and disadvantaged" in every country	Yes. Profit-oriented economic growth, conspicuous consumption and colonial histories of exploitation are named as causal factors of the crisis. Actions focus on addressing these "root causes" and not furthering "existing inequalities and gross injustices"

(continued)

Table 4.1 (continued)

Domain of analysis		Islamic Declaration on Sustainable Development (First Islamic Conference of Environment Ministers, 2002)	Islamic Declaration on Global Climate Change (International Islamic Climate Change Symposium, 2015)	Al-Mizan: A Covenant for the Earth (Llewellyn et al., 2024)
Eco-theology of Islam	How is the relationship between God, people and nature framed?	The "lieutenancy" of humans is core, with humanity "mandated to build civilisation" and seen as "responsible for harnessing and protecting the environment" as a religious duty	The primary framing is on God as the creator and on humanity as accountable for their actions to the creator. Humanity must also work for the "greater good of all species"	The oneness of God is core. There is an emphasis on nature as "signs of God", with the earth created in cosmic and "dynamic balance and equilibrium"—And the wilful disturbance of this regarded as a "travesty of our faith"
	How is human stewardship formulated?	Trusteeship is presented through an anthropocentric reading of the Qur'an in which the environment is a "gift donated to man by God" and where humanity is duty-bound to not disturb the balance in nature	Stewardship is central, with a focus on accountability and just actions to remedy the corruption caused by humanity. The prophet Muhammad, in his ethic and care for nature, is identified as a role model	Stewardship is framed as an honour but also a test of moral accountability to be exercised throughout "space and time, incorporating all species, individuals and generations of God's creatures"
	How is the natural world envisioned?	Nature is regarded as a trust, a blessing and gift and the natural resource base for life on earth. Muslims are duty-bound to protect nature for the public good	Everything in creation has value, and the earth has been created for all beings to thrive. Humans are but a part of "a multitude of living beings" and have no right "to abuse creation or impair it"	The "inherent value of the natural world as God's creation" and the extension of the "qur'anic cosmology of justice" to all of creation entrench the belief that the "good of all beings" matter. The value of nature "as communities" is seen to precede its "value as commodities"

Eco-social policies and actions			
What are the key themes of the solutions proposed?	Solutions are framed around securing funding sources, increasing investment and accelerating technology transfer. Social development as well as respect for "civilisational heritage" and education are proposed	Solutions are framed around target audiences and include phasing out and divestment from fossil fuels, reducing consumption, investing in economic reform, committing to 100% renewable energy and exploring new development models	Actions encompass the expression of Islam's eco-theology in spiritual practices such as prayer, charity and fasting but also in social, political and economic life, e.g. economic and development pathways
Do these actions connect with dimensions of social justice?	Yes. While technocratic in nature, the solutions speak to justice, such as economic and trade justice and the "effective participation of developing countries in decision-making"	Yes. It identifies the uneven distribution and impact of climate risks, and the first call to action is to "well-off and oil-producing states" to fulfil their moral obligation of phasing out greenhouse gases and reducing consumption	Yes. The crisis is framed as systemic environmental and economic challenges, and solutions proposed include just and equitable transitions that address those who "disproportionately suffer"
What are examples of eco-social actions being proposed to address inequalities?	The declaration sets out to achieve "justice advocated by Islam between people and between all social categories". It calls for development programmes for the "least developed countries", a just system for world trade; and the preservation of cultural rights	The call to action focuses on the broader objectives of climate action. It highlights investing in a green and circular economy, "elevating the conditions of the world's poor", 100% commitment to renewable energy and a focus on climate adaptation for the most vulnerable countries and groups	Systemic solutions include fossil fuel divestment, changing development paradigms and placing nature and people, not profit, at the heart of decision-making

(continued)

Table 4.1 (continued)

Domain of analysis		Islamic Declaration on Sustainable Development (First Islamic Conference of Environment Ministers, 2002)	Islamic Declaration on Global Climate Change (International Islamic Climate Change Symposium, 2015)	Al-Mizan: A Covenant for the Earth (Llewellyn et al., 2024)
Eco-social activism	How is this vision situated on the spectrum of environmentalism?	It puts forward a technocentric approach, both in its framing of the problem and the identification of solutions that are largely related to the application of finance, technology and science	The theocentric starting point of Islam's eco-theology is strongly articulated, and while there are elements of anthropocentric-oriented technological solutions, there is a much stronger focus on both the rights of creation and the rights of people. It tends towards an ecocentric approach	It has a strong ecocentric framing. There is an emphasis on justice, with human trusteeship seen as promoting "the welfare of all God's creatures". Crimes against nature are seen as "analogous to crimes against humanity, and are no less grave"
	Does the vision recognise the value of alliance building?	In keeping with the technocentric focus, the key alliances identified are with international and regional organisations and institutions	It calls on all groups to join in "collaboration, cooperation and friendly competition", including working with other religions	A series of pledges encompasses a society- and economy-wide plan of action. It notes that actions require activism and an alliance of religious traditions that "represent more than 80% of the earth's human population"
	Are any key stakeholder groups or alliances mentioned?	The international community, in the shape of the UN, is identified to mediate a "just system of world trade" and to "rein in practices, policies and conducts that affect badly the environment" and promote "unsound consumption patterns"	A call to action is aimed at nations and states, corporations, financial institutions and the business sector, as well as people and leaders. It also includes a call for all Muslims to "tackle habits, mindsets and the root causes of climate change, environmental degradation and the loss of biodiversity in their particular spheres of influence"	Pledges are targeted at a wide range of actors in the Muslim world and key stakeholders, including Muslim jurists, educators, economists, ecologists, governments, corporations, policymakers, planners, conservation agencies, non-governmental and faith-based organisations and groups and local communities

interpretations of Islamic principles inform different visions of ecological responsibility, social justice and governance. First, it will look to the objectives and process of developing the vision. Second, it will consider the framing of the climate and environmental crises. Third, it will identify how it presents the key features of Islam's eco-theology. Fourth, it will delve into the eco-social actions being proposed, and, finally, it will assess whether the statements identify key opportunities for activism and alliance building. While these declarations play a key role in setting the terms of the debate, activating eco-social transformations will require not only a change in mindset but also actions that address the political and economic structures that the world is embedded in. This is the challenge that the world faces. So it is therefore not surprising that oil-rich Muslim-majority countries might endorse these declarations but not necessarily change their practice and policies on fossil fuel extraction.

What do these three illustrations of an Islamic eco-social contract reflect? All three statements were launched at key intergovernmental events, but the declarations on sustainable development and climate change were developed in response to a call by the international community for civil society, including faith-based organisations, to present their visions for action. *Al-Mizan* was developed over a longer period, enabling wider consultation, which resulted in a broader vision and a more holistic diagnosis of the crisis and presentation of solutions. The rationale for all three presents a liberatory intent to decolonise ecological knowledge and to "preserve civilisational heritage and knowledge" (First Islamic Conference of Environment Ministers, 2002), to defend cultural and religious identity, worldviews and knowledges and to make visible an ethos that presents a vision of living in harmony and connectedness with the "diverse communities of life" (Llewellyn et al., 2024: 6).

The framing of climate and ecological breakdown as rooted in the paradigms of "economic development and human progress" (International Islamic Climate Change Symposium, 2015, 2) is especially strong in *Al-Mizan* and the *Islamic Declaration on Global Climate Change*. Both address political and economic inequalities and explicitly mention "well-off nations and oil-producing states" (International Islamic Climate Change Symposium, 2015: 6) as needing to shoulder a bigger responsibility for phasing out fossil fuels and reduce conspicuous consumption given the extent of their contribution to the climate and ecological crisis. *Al-Mizan* also posits that the way out of this *fasād* must be a just transition that focuses on "affected workers and poor and vulnerable communities" (Llewellyn et al., 2024: 4). The *Islamic Declaration on Sustainable Development* does not adopt this political economy framing. Instead, it identifies poverty and indebtedness as constraints on sustainable development, with little mention of the political and economic histories of colonialism that are key to understanding prevailing patterns of inequalities, including climate and environmental inequalities (Sultana, 2022).

Islam's theocentric eco-theology is represented in all three eco-social contracts, which centres God as the Owner and Fashioner of all creation; the inherent value and sanctity of all creation; and the dynamic balance and equilibrium on Earth: "a climate in which living beings—including humans—thrive" (International Islamic Climate Change Symposium, 2015: 4). Trusteeship is framed in *Al-Mizan* and the *Islamic Declaration on Global Climate Change* as a relationship of ultimate accountability to God for actions in this life. Living in justice with people and with all creation on Earth is included in the ambit of good actions. The *Islamic Declaration on Sustainable Development* regards the role of trusteeship as achieving public good and emphasises the duty of humanity to take care of the environment it received "as a gift" (First Islamic Conference of Environment Ministers, 2002). It centres the human right to a healthy environment, in contrast to the other two eco-social contracts that present an eco-theology rooted in "responsible human existence among other creatures" (Gade, 2019: 3). A rights-based approach can be discerned in all three, even though there is a divergence between a human rights-based vs co-integrated or multi-species rights approach.

The political economy analysis adopted in *Al-Mizan* and the *Islamic Declaration on Global Climate Change* results in the identification of eco-social policies and actions that centre equity, justice and inclusion. *Al-Mizan* mentions solutions like decentralised renewable energy and inclusive and participatory governance and shares how Islamic practices and traditional institutions like *zakāh* (an obligatory wealth tax) and *himā* (a traditional conservation practice) are both rooted in justice. The *Islamic Declaration on Global Climate Change* focuses on transparency and accountability in the phasing out of fossil fuels and transitioning to 100% renewable energy and emphasises new models of development that prioritise well-being and resilience of all, including "the poor and disadvantaged communities of every country, at all levels of development" (International Islamic Climate Change Symposium, 2015: 2) that bear the brunt of climate inequality. These statements offer the strongest representation of the liberatory intent of Islam's eco-theology. It not only places responsibility on those who have contributed the most to climate and environmental breakdown but also proposes solutions that connect climate, social and ecological justice.

Eco-social activism and alliance building are dealt with extensively in *Al-Mizan*. The authors mention that since human-induced global climate change has "drastically interfered with the equilibrium… of the Earth's interconnected systems", they are obliged to support and endorse the "proposed Fossil Fuel Non-Proliferation Treaty" (Llewellyn et al., 2024: 3–4). All three documents direct their calls to action not only at all Muslims but to the world at large, with a vision to collaborate and cooperate.

Declarations that express alternative ecological imaginaries are only the first step in giving visibility to this worldview. How can this imaginary inspire action? The eco-theology of Islam has been implemented in many areas of Muslim religious and social life. This includes the greening of educational and religious institutions and guidance for integrating sustainability in key festivals and celebrations, like the pilgrimage and month of fasting (Mohamed, 2024a). There are now several Muslim faith-based climate and environmental organisations operating globally and nationally that raise awareness of and action on this eco-theology. For example, Muslim charities are connecting the social and green agendas and taking action in how they design their interventions, recognising that the issues they focus on including poverty, food and water insecurity, health and social inequality will be impacted by this crisis (Mohamed, 2024b).

This eco-theology movement of Islam is still in its early phases, having emerged as much of the environmental movement in response to the climate and ecological crisis. It has much to do to revive and raise awareness of this ecological imaginary and what this means for action. As a "religion concerned with action on core ethical principles that are deeply congruent with a love of the planet", Abdul-Matin proposes that the key tasks of Islam's growing environmental movement are to tell its stories, get educated and connect with others with the aim of becoming better representatives of God on the planet (2010: 14).

4.5 Alliances for Eco-social Change

Muslims draw upon religious teachings to shape their values, beliefs and attitudes towards life, including their response to the climate and ecological crisis. As Foltz states, "For many of the world's billion plus Muslims, the solution to this crisis must be an Islamic one—an environmentalist Islam rediscovered from the sources of the faith" (2000: 72). Eco-social contracts or covenants, rooted in Islam's liberation eco-theology, should focus not only on enlivening Islam's ecological imaginary *within* the Muslim world but also among the pluriverse of environmental imaginaries and movements which privilege a commitment to work for justice and sustainability. This concluding section points to areas of alliance building, some of which are already underway.

Making visible the "tapestry of alternative" worldviews, both old and new, to repair the relationship between people and nature, is a first port of call (Kothari et al., 2019). Islam's eco-theology is already well represented in the religion and ecology movement but is neither well understood nor represented in the pluriverse of environment and development alternatives and in alliances that are working for shifts in eco-social paradigms. This is being addressed through the work of organisations such as Ummah for Earth, 2 Billion Strong and Faithfully Sustainable, faith-based organisations that centre a commitment to climate and ecological justice and connect ideas and actions that link the eco-theology of Islam with grassroots and community-based action.

A second area for alliance building might offer a remedy to the under-representation of Islam's eco-theology. Economic relations, according to Islam, should be deeply rooted in achieving both justice and well-being, with the goal being to achieve balanced and equitable development, as well as the well-being of all creation (Haque et al., 2021; Kader, 2021). This aligns well with the frameworks inspiring the global movement for economic reform, which proposes models that recognise the dependency of societies and economies on nature, reformulate the purpose of economies and develop new measures and institutional structures that advance well-being and prosperity of all life. Linked to this are a wide range of alliance-building examples in sectoral economic transitions, most notably in the food (Yasin, 2014; Aziz et al., 2023), fuel (Dawes, 2021) and finance (Securities Commission Malaysia and World Bank, 2019) arenas. These highlight how issues of sustainable consumption and production, fair trade and ethical finance are being addressed in the Muslim world, showing how Islam's exacting ethical requirements, which connect social and environmental rights, could ally with some of the key movements working for economic reform.

A third arena for alliance building, already underway, is aligning with the movement championing the Rights of Nature. The growing focus on co-integrated rights and the intrinsic value of the Earth and all creation (Haque et al., 2021) in Islam's eco-theology can join forces with the movement advocating for the legal recognition of nature's rights. Stop Ecocide International, representing a global movement that wants to position the destruction of the environment as a crime against humanity, lauded the position of Muslims on ecocide as expressed in the *Al-Mizan* charter. The charter regards ecocide as "analogous to crimes against humanity and are no less grave" and calls for "this kind of corruption in the Earth…to be recognised, litigated, and penalised in national and international legislation" (Llewellyn et al., 2024, 35). These are only three avenues for alliance building, but the burgeoning scholarship on the eco-theology of Islam outlines many more.

Islam's liberation eco-theology formulates environmental imaginaries that link climate, ecological and social justice and provides fertile ground for building alliances to challenge dominant imaginaries. It presents a lived spirituality which situates believers as active participants in achieving social and ecological transformation, with the potential to connect and collaborate with the growing movement for climate and ecological justice (Gottlieb, 2003). And while structural barriers to eco-social change exist, as identified in the declarations analysed above, these demonstrate that the liberatory vision of Islam's eco-theology connects the ecological and the social change agendas and that Muslims can ally and join the movements working for eco-social contracts.

References

Abdul-Matin, I. (2010). *Green Deen: What Islam teaches about protecting the planet.* Berrett-Koehler Publishers.

Ahmad, A. (1997). *Islam and the environmental crisis.* Ta-ha Publishers.

Al-Faruqi, I. R. (1987). *Islamization of knowledge: General principles and workplan.* International Institute of Islamic Thought.

Al-Hamid, A. (1997). Exploring the Islamic environmental ethics. In A. R. Agwan (Ed.), *Islam and the environment* (pp. 39–69). Synergy Books International.

Al-Jawziyyah, I. Q. (1973). *'Ilām al muwaqqi'īn 'an rabbil 'ālamīn [Announcement for those who sign on behalf of the Lord of the worlds]* (Vol. 4). Dārul Jīl.

Al-Jayyousi, O. (2015). Rethinking degrowth: Islamic perspectives. *Degrowth.* Accessed August 2, 2024, from https://degrowth.info/de/library/rethinking-degrowth-islamic-perspectives-2

Arora, S., & Stirling, A. (2020). Don't save 'the world' – embrace a pluriverse! *STEPS Centre.* Accessed August 2, 2024, from https://steps-centre.org/blog/dont-save-theworld-embrace-a-pluriverse/

Aziz, N., Bakry, N., Mz, M. H., & Armia, M. S. (2023). The paradigm of modern food products and its relevance with the concept of food in the Quran. *Heliyon, 9,* e21358. Accessed August 2, 2024, from https://www.ncbi.nlm.nih.gov/pmc/articles/PMC10663733/pdf/main.pdf

Baker, S. C., & Morrison, R. (2008). Environmental spirituality. Grounding our response to climate change. *European Journal of Science and Technology, 4*(2), 35–50.

Bangash, Z. (2011). *The covenant of Madinah and the Inclusivist Islamic state*. The Institute of Contemporary Islamic Thought.

Berkes, F. (2008). *Sacred ecology*. Routledge.

Boff, L. (1995). *Ecology and liberation. A new paradigm*. Orbis Books.

Byskov, M. F., & Hyams, K. (2022). Epistemic injustice in climate adaptation. *Ethical Theory and Moral Practice, 25*, 613–634. https://doi.org/10.1007/s10677-022-10301-z

Chancel, L., Bothe, P., & Voituriez, T. (2023). *Climate inequality report 2023*. Report number: Study 2023/1. World Inequality Lab. Accessed August 2, 2024, from https://wid.world/wp-content/uploads/2023/01/CBV2023-ClimateInequality Report-2.pdf

Dawes, Z. (2021). ISNA urges adoption of fossil fuel non-proliferation treaty. *Good Faith Media*. Accessed August 2, 2024, from https://goodfaithmedia.org/isna-urges-adoption-of-fossil-fuel-non-proliferation-treaty/#:~:text=The%20 Islamic%20Society%20of%20North,8%20press%20release

First Islamic Conference of Environment Ministers. (2002). *Islamic declaration on sustainable development*. World Summit on Sustainable Development. Accessed September 9, 2024, from https://muslimenvironment.wordpress.com/2010/10/25/islamic-declaration-on-sustainable-development-johannesburg-august-september-2002/

Foltz, R. (2000). Is there an Islamic environmentalism? *Environmental Ethics, 22*(1), 63–72.

Foltz, R. (2005). *Environmentalism in the Muslim world*. Nova Science Publishers.

Fricker, M. (2007). *Epistemic injustice. Power and the ethics of knowing*. Oxford University Press.

Gade, A. (2019). *Muslim environmentalisms. Religious and social foundations*. Columbia University Press.

Ghoneim, K. (2000). The Quran and the environment. *Islam Online*. Accessed August 2, 2024, from https://islamonline.net/en/the-quran-and-the-environment/

Gottlieb, R. (2003). *Liberating faith: Religious voices for justice, peace, and ecological wisdom*. Rowman and Littlefield Publishers.

Grim, J., & Tucker, M. E. (2014). *Ecology and religion*. Island Press.

Haq, S. N. (2001). Islam and ecology: Toward retrieval and reconstruction. *Daedalus, 130*(4), 141–177.

Haque, N., Masri, A.-H., & Banaei, M. (2021). *Ecolibrium: The sacred balance in Islam*. Beacon Books.

International Islamic Climate Change Symposium. (2015). *Islamic declaration on global climate change*. Istanbul, Turkey. Accessed August 4, 2024, from https://www.ifees.org.uk/about/islamicdeclaration/

Izzi Dien, M. (2000). *The environmental dimensions of Islam*. Lutterworth Press.

Kader, H. (2021). Human well-being, morality and the economy: An Islamic perspective. *Islamic Economic Studies, 28*(2), 102–123.

Kerr, D. (2000). Muhammad: Prophet of liberation – A Christian perspective from political theology. *Studies in World Christianity, 6*(2), 139–174.

Khalid, F. (2003). Islam, ecology and modernity: An Islamic critique of the root causes of environmental degradation. In R. Foltz, F. Denny, & A. Baharuddin (Eds.), *Islam and ecology: A bestowed trust* (pp. 299–321). Harvard University Press.

Kothari, A., Salleh, A., Escobar, A., Demaria, F., & Acosta, A. (2019). *Pluriverse: A post-development dictionary*. Tulika Books.

Llewellyn, O. (2003). The basis for a discipline of Islamic environmental law. In R. Foltz, F. Denny, & A. Baharuddin (Eds.), *Islam and ecology: A bestowed trust* (pp. 185–247). Harvard University Press.

Llewellyn, O., Khalid, F., et al. (2024). *Al-Mizan: Covenant for the Earth*. The Islamic Foundation for Ecology and Environmental Sciences.

Mohamed, N. (2012). *Revitalising an eco-justice ethic of Islam by way of education*. PhD, Stellenbosch University. Accessed August 4, 2024, from https://scholar.sun.ac.za/server/api/core/bitstreams/e8bc9d38-6d96-48ff-82a6-af98c844f857/content

Mohamed, N. (2024a). Muslims going green locally and globally. *The Platform*. Accessed March 9, 2025, from https://platform.ilke.org.tr/analyze/muslims-going-green-globally-and-locally#!

Mohamed, N. (2024b). On being Muslim and Green. *The Forum, 6*, 55–57.

Mohamed, N., & Huntjens, P. (2023). Dismantling the ecological divide. Towards a new eco-social contract. In I. Kempf, K. Hujo, & R. Ponte (Eds.), *Global study on new eco-social contracts* (pp. 22–30). United Nations Research Institute for Social Development.

Nasr, S. H. (1997). Islam and the environmental crisis. *al'Ilm, 17*, 1–23.

Oelschlager, M. (1996). *Caring for creation: An ecumenical approach to the environmental crisis*. Yale University Press.

Ouis, S. P. (2003). Global environmental relations: An Islamic perspective. *The Muslim Lawyer, 4*(1), 1–7.

Özdemir, I. (2003). Toward an understanding of environmental ethics from a qur'anic perspective. In R. Foltz, F. Denny, & A. Baharuddin (Eds.), *Islam and ecology: A bestowed trust* (pp. 1–37). Harvard University Press.

Özdemir, I. (2008). *The ethical dimension of human attitude towards nature. A Muslim perspective*. Insan Publications.

Parvaiz, M. A. (2003). Scientific innovation and al-Mīzān. In R. Foltz, F. Denny, & A. Baharuddin (Eds.), *Islam and ecology: A bestowed trust* (pp. 393–401). Harvard University Press.

Peet, R., & Watts, M. (1996). *Liberation ecologies: Environment, development, social movements*. Routledge.

Qur'an. (2008). *The gracious Quran. A modern-phrased interpretation in English*. Translated by Ahmad Zaki Hammad. Lucent Interpretations.

Sachedina, A. (1999). Ethics of environment in Islam. In M. Abu-Sway & A. Sachedina (Eds.), *Islam, the environment and health* (pp. 40–56) Islamic Medical Association of South Africa.

Securities Commission Malaysia and World Bank. (2019). *Islamic green finance: Development, ecosystem and prospects*. World Bank Group.

Setia, A. (2007). The inner dimensions of going green: Articulating and Islamic deep-ecology. *Islam & Science, 5*(2), 117–150.

Shiva, V. (1993). Monocultures of the mind. *The Trumpeter, 10*(4). Accessed August 2, 2024, from http://trumpeter.athabascau.ca/index.php/trumpet/article/view/358/563

Solberg, M. (2008). The good news of liberation theologies. *Dialog, 47*(4), 310–311.

Sultana, F. (2022). The unbearable heaviness of climate colonialism. *Political Geography, 99*.

Trisos, C., Auerbach, J., & Katti, M. (2021). Decoloniality and anti-oppressive practices for a more ethical ecology. *Nature Ecology and Evolution, 5*, 1205–1212.

Turnbull, E. (2021). What Islamic contributions have been made to climate change action and how useful are they in promoting environmental justice? *Journal of Financial Crime, 28*(4), 1032–1043.

United Nations Framework Convention on Climate Change. (2015). *Islamic declaration on climate change*. Accessed August 2, 2024, from https://unfccc.int/news/islamic-declaration-on-climate-change

Wilkinson, R., & Pickett, K. (2022). From inequality to sustainability: Deep Dive Paper 01. *Earth4All*. Accessed August 2, 2024, from https://www.clubofrome.org/wp-content/uploads/2022/05/Earth4All_Deep_Dive_Wilkinson_Pickett.pdf

Yasin, D. (2014). Tayyib: The foundation of ethical eating & conscious consumption. *Medium*. Accessed August 2, 2024, from https://medium.com/ummah-wide/tayyib-the-foundation-of-ethical-eating-conscious-consumption-327117907493

5

A New Pact with Nature: From Social to Eco-Social Contracts

Najma Mohamed and Patrick Huntjens

5.1 From Social to Eco-social Contracts

Environmental degradation is advancing around the world, and scientists have warned that we are heading towards a major planetary catastrophe (Brondizio et al., 2019). In 2009, scientists identified nine planetary boundaries—key ecological systems and processes that regulate the stability and resilience of the Earth and sustain all life. These boundaries define a safe operating space within which humanity can thrive, supporting development and well-being. However, this safe space is increasingly being breached. To date, humanity has crossed six of these nine planetary boundaries, pushing global systems towards greater instability (Richardson et al., 2023; Stockholm Resilience Centre, 2024). As nature recedes, every social and economic priority—jobs, food, water, education, health, and the future of humanity—is at risk. This has spurred the recognition that we must fundamentally change the relationship between humankind and nature and make peace with nature (UNEP, 2021). With nature in crisis, new social contracts must take an eco-social turn (UNRISD, 2016).

N. Mohamed (✉)
UN Environment Programme World Conservation Monitoring Centre, UNEP-WCMC, Cambridge, UK
e-mail: najma.mohamed@unep-wcmc.org

P. Huntjens
Inholland University of Applied Sciences, Delft, The Netherlands
e-mail: patrick.huntjens@inholland.nl

P. Huntjens et al. (eds.), *Eco-Social Contracts for Sustainable and Just Futures*,
https://doi.org/10.1007/978-3-031-99109-7_5

73

There is a growing movement engaging afresh with the social contract as a frame for imagining a world that is both just and sustainable. This is visible in groups championing twenty-first-century social contracts that are anti-racist (Christian Aid and partners, 2024) and gender-just (ITUC, 2022) that promote decent work (ITUC, 2021) and intergenerational justice (Extinction Rebellion, 2024). Others reimagine social contracts that address the climate and ecological emergency. Messner (2015) and, more recently, Willis (2020) highlight the need for a social contract that responds to the climate crisis. Jennings (2016), Huntjens (2019, 2021) and Mohamed and Huntjens (2023) call for a reformulation of the human-nature relationship so that people and the planet thrive together.

UNEP and the International Science Council (2024: 64) highlight that a new social contract is "[k]ey to a better future", requiring "shared values that unite us rather than divide us". Transformative change, in paradigms and collective goals with the "environment at the heart of a social contract", should provide the foundation for new social contracts (UNEP and the International Science Council, 2024: 64). UNRISD (2021: 246) calls for a new eco-social contract for people and planet that would be "instrumental in reconfiguring a range of relationships that have become sharply imbalanced—those between state and citizens, between capital and labour, between the global North and the global South, between humans and the natural environment". As several chapters in this book show, there is a pluriverse of knowledges, traditions and wisdoms that recognise the interconnectedness of all life. These worldviews, exemplified in Indigenous knowledge systems that emphasise reciprocity and relationality (Kimmerer, 2013), provide fertile ground for the emergence of eco-social contracts but have largely been sidelined and excluded in social contract formulation.

The idea of a social contract can be traced back to political philosophy and is reflected, among others, in Greek, Islamic, African and Indigenous communitarian thinking (UNRISD, 2021). Yet today it is most often associated with the Enlightenment era of the seventeenth and eighteenth centuries and the work of the philosophers Thomas Hobbes, John Locke and Jean-Jacques Rousseau. At the core of a social contract lies the idea that it is the responsibility of the state to be competent, accountable and transparent in protecting the security, welfare and dignity of all people within that state. A broken social contract is therefore, in essence, a failure of governance.

But nature has been the blind spot of social contract theory and has "had little or no intrinsic value for most (but not all)" modern theorists (O'Brien et al., 2009). As a result, existing social contracts are largely anthropocentric and reinforce the divide between humans and nature. While social contract theorists do not usually set out to exclude nature, it is an omission rooted in the dominant social paradigm of anthropocentrism. In *Le Contrat Naturel* (*The Natural Contract*), Michel Serres challenges us to "look outside the narrow frame of the social contract to what, to our peril, it excludes: nature" (Serres, 1995: 1).

The anthropocentrism that serves as the foundation of prevailing social and economic paradigms is among the primary causes of one of the greatest societal fault lines of our time: the ecological divide (Scharmer, 2013; Scull, 2017; Bogert et al., 2022). The ecological divide reflects a disconnect between the self and nature, where humanity fails to recognise its dependence on and interconnectedness with the natural world. Communitarian and other visions of a new social contract challenge the neglect and exploitation of nature, advocating for a holistic and integrated understanding of the human-nature relationship. "Unlike the human-centred approach of the global North…[they] offer a relational worldview that goes beyond the social to encompass economic, political and cultural relations based on reciprocity, respect and equity with all living and non-living entities" (Desai, 2022: 1).

Eco-social contracts reframe the relationship between people and nature by embracing ecological knowledges and values that recognise the interconnectedness of all life and translate these values into practical action. Therefore, eco-social contracts cannot simply replicate the old social contract. They must fight inequalities. They must restore trust. And they must reimagine the relationship between humans and nature.

Demand for a different kind of world, one that is more sustainable, inclusive and fair, is increasing. Communities, workers and social groups are seeking to bridge the ecological divide by using language, concepts and visions that consider personal and communal well-being, planetary health and the Rights of Nature. This requires a reconfiguration of the overarching goal of a social contract but also a fundamental restructuring of how humanity views itself and its relationship with nature. The shift from an ego-centric to an ecocentric social contract requires a fundamental reassessment of the purpose, goal and vision of our societies and economies and what this means for the relationship between humans and nature.

This chapter provides an overview of key paradigm shifts needed in the human-nature relationship for eco-social contracts to emerge. Defined by the *Concise Oxford Dictionary* (1999), a paradigm shift is "a fundamental change in approach or underlying assumptions". The key paradigm shifts proposed in this chapter include changes in *worldview*, in *visions of nature* and in *understanding of human behaviour* and *societal well-being* (Fig. 5.1).

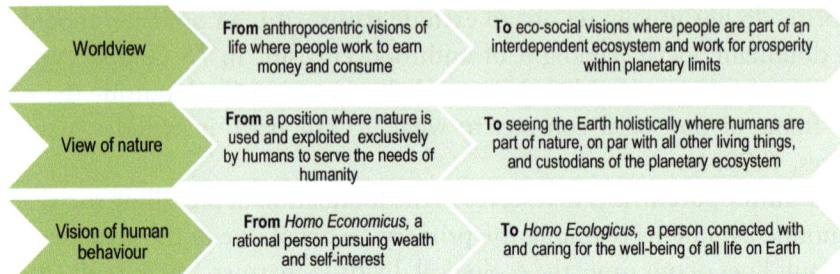

Fig. 5.1 Key shifts from an old to a new eco-social contract. Source: Authors' own elaboration

5.2 Key Paradigm Shifts to Bridge the Ecological Divide

In this section we describe the significant paradigm shifts and transformations from an anthropocentric (human-centred) to an ecocentric or biocentric (life-centred) worldview as crucial for eco-social contracts (see Fig. 5.1).

These are central to reshaping our approach to governance, policy and prevailing societal norms that define humanity's relationship with nature. It requires a comprehensive transformation in how we view our place in the world, make decisions and live our lives. By embracing ecocentrism, humanity can move towards more sustainable, equitable and resilient futures.

5.2.1 From Anthropocentric to Eco-social Worldviews

The schism between humans and nature, rooted in the dominant anthropocentric worldview, stems from three prevailing beliefs (Huntjens, 2021). Certain interpretations of religious traditions, including some readings of Judeo-Christian texts, have historically been used to justify human dominion over nature. In particular, the biblical notion of "dominion" has been interpreted in ways that reinforce human superiority. This perspective persisted and gained further traction during the Enlightenment (Plumwood, 2005), where, despite an emphasis on science and rationality, many thinkers continued to uphold the idea of humanity as separate from and above nature, with the earth and its creatures viewed primarily as resources for human use (Patel & Moore, 2020). However, contemporary eco-theologians and liberation theologians challenge these interpretations, emphasising responsible and accountable stewardship and the interconnectedness of creation instead.

Second, an unquestioning belief in the liberal market economy and infinite growth has shaped economic policy and planning (Huntjens, 2021). Third, within this dominant market-driven paradigm, people have been regarded primarily as consumers, leading to societies centred on individualism, materialism and short-term thinking (Huntjens, 2021). This has resulted in what Jennings (2016) describes as "a social contract of consumption".

Existing social contracts, rooted in anthropocentrism, reflect the ecological divide. Paradigms—deeply ingrained ways of thinking and doing—are heavily influenced by underlying worldviews. A worldview comprises the set of beliefs, values and assumptions that shape how individuals and societies perceive and interact with the world. The dominant social paradigm is steeped in an anthropocentric worldview of the human-nature relationship, such as human mastery, control and exploitation of nature. So far, social contracts are either human rights based or based on human interest with objectives such as maintaining social order, protecting rights and promoting social justice with limited or no recognition of the Rights of Nature (Huntjens, 2021). This human-centredness is deeply ingrained and reflected for instance in the dominance of market-based approaches. The underlying paradigm of Western liberal political thought also centres the individual.

An anthropocentric worldview considers nature primarily in terms of its utility to humans, emphasising human needs, desires and benefits. However, this perspective does not affect all humans equally; power asymmetries and inequalities—between social groups, countries and regions—mean that some humans exploit and oppress others, treating them as mere resources for personal or economic gain. In this sense, the exploitation of nature and of marginalised human communities often goes hand in hand, reinforcing cycles of dominance and extraction.

This worldview has been particularly dominant in the Global North since the Industrial Revolution (which began in the mid-eighteenth century in Britain) and has driven technological advances and economic development—often at significant social and environmental cost. Though initially rooted in Western thought, it has since been exported globally, shaping economic and political systems worldwide. Within an anthropocentric paradigm, nature is viewed as a resource to be used for human benefit, with environmental concerns often treated as secondary or external to economic and social priorities. Additionally, a critical distinction exists between those who exploit natural resources for survival and those who do so for wealth accumulation—further underscoring the uneven impacts and ethical dimensions of this worldview.

An eco-social perspective values all forms of life and the ecosystems that sustain them, recognising their intrinsic worth beyond mere human use. This approach emphasises the interconnectedness of all life and the importance of maintaining the health and integrity of ecosystems. Ecocentrism supports the idea that humans are part of the broader ecological community and should live in harmony with nature. To move from a utilitarian to a reciprocal relationship between people and nature, eco-social contracts must be based on eco-social visions of the human-nature relationship.

When a critical mass within a society begins to adopt a new worldview, this can drive a paradigm shift. The shift towards Rights of Nature can be seen as an example of this, from an exploitative view of nature to one that recognises the interconnectedness and intrinsic value of all life forms as well as intergenerational rights. Ecocentric paradigm shifts also entail a profound cultural transformation, where societal values and norms evolve to embrace sustainability, equity and respect for nature: an ethos reflected in Indigenous teachings of relational accountability and gratitude (Kimmerer, 2013). Education and public awareness are vital in fostering this cultural shift.

5.2.2 From *"Homo Economicus"* to *"Homo Ecologicus"*

This shift in worldview is closely linked to a shift in our view of humanity, which requires a different model of human behaviour, including a transition from ego-awareness to eco-awareness. In addition, a transformational change in our model of human behaviour is required, from "*homo economicus*" to "*homo ecologicus*" (Dryzek, 1996; Bosselmann, 2004; Becker, 2006; Cecchi, 2013; Huntjens, 2021).

Homo economicus is defined as a rational person who pursues wealth for his or her own self-interest and was first mentioned by John Stuart Mill, the English philosopher, in the nineteenth century. In economic theory, it is a model of human behaviour that assumes people will make choices in their own self-interest. This assumption of rationality—also called the theory of rational behaviour—is a simplification economists make to create a useful model of human decision-making. Modern behavioural economists challenge the traditional notion of purely rational decision-making, arguing that while humans are capable of rational thought, their choices are often influenced by cognitive biases, emotions and limited information. Decision-making can be distorted by misinformation, complexity and an inability to fully account for long-term or indirect consequences—such as environmental feedback loops and distant impacts on nature. Rather than being entirely irrational, human reasoning is shaped by both logical processes and psychological factors that can lead to suboptimal or short-sighted choices. Likewise, the concept of *homo economicus* has received substantial criticism from other disciplines. Critics argue that it misrepresents how social agents operate (Bourdieu, 2005), assumes individual rationality as the unquestioned foundation of economic analysis (Foley, 1998) and yields limited empirical insights through rational choice theory (Green & Shapiro, 1996). A key concern is that economic rationality often becomes a reductive shorthand for understanding human behaviour, imposed on complex social and political dynamics that cannot be fully explained through the lens of self-interested calculation alone.

Homo ecologicus provides a different model of human behaviour, in particular, by focusing on a human being's relationship (i) with itself, (ii) the community and (iii) nature (Becker, 2006). *Homo ecologicus* turns out to be inescapably social, unlike *homo economicus* (Dryzek, 1996). In philosopher Henryk Skolimowski's ecological humanism, the concept is used to emphasise an equal position between humans and nature (Fios & Arivia, 2018). As such, the concept of *homo ecologicus* aligns with criticisms of anthropocentrism by modern philosophers such as Bruno Latour, Henk Manschot and Harry Kunneman, who argue for a new relationship with the Earth and other living beings. The decentring of the individual as the political agent and central focus of economic, social and political thinking will have profound implications for "representing" nature in economic and governance systems.

5.2.3   From Individual to Collective Values

The development of a societal vision for a sustainable and just future should revolve around the discussion and identification of guiding principles and a joint value base. Research has acknowledged the important role values might play in driving transformative change (Horlings, 2015). However, different conceptions of values abound in different scholarly communities and beyond, which is why the term "value" is used quite differently in different contexts (Tadaki et al., 2017).

In psychological research, values are understood as principles or standards that guide people's interpretations, judgements, choices and actions (Schwartz, 2012). Values are seen as relatively stable cognitive structures embodying desirable goals that transcend specific situations and actions (Steg, 2016). With respect to environmental behaviour, four types of values have been shown to be most relevant (Steg & De Groot, 2012):

- *Biospheric values* (such as respecting the earth) that focus on the quality of nature and the environment for its own sake.
- *Altruistic values* (such as social justice) that focus on ways to benefit others.
- *Egoistic values* (such as power or status) that focus on furthering one's own resources such as wealth and power.
- *Hedonistic values* (such as gratification for oneself) that focus on what makes individuals feel good and minimises effort.

Research (Steg & De Groot, 2012; Rinscheid et al., 2025) shows that people endorse altruistic and biospheric value orientations much more than egoistic and hedonistic value orientations. However, this does not translate into behavioural change, due to an economic system that is modelled around an incomplete model of human behaviour, i.e. *homo economicus* (Huntjens, 2021). So how to change to a system that is built around a more "natural" model of human behaviour, one that views the human being as relational, embedded in and responsible for the web of life, that is *homo ecologicus*? Research from Bouman et al. (2020) shows that behavioural change can be supported when individuals perceive their group to prioritise biospheric values more strongly. In such a case, their own pro-environmental behaviour also becomes stronger. At the same time, we know that social change often begins with a minority that goes against the established order (Otto et al., 2020). That minority must become large enough so the rest can no longer see it as an exception. Research suggests that for a group to adopt a new behaviour, only 25% of the group need to change to influence and convert the rest (Centola et al., 2018; Otto et al., 2020). From such a critical mass, more and more people "jump on the bandwagon", and the change accelerates itself.

Though there is broad agreement on the potential role of values in driving intentional transformative change towards sustainability, the development of value-based transformation approaches remains relatively new, particularly in the context of multi-level decision-making (Horcea-Milcu et al., 2023). While values have historically underpinned major international frameworks—such as the post-war order built on principles of development, peace and human rights (e.g. the Universal Declaration of Human Rights, UN Charter)—these frameworks did not fundamentally challenge the extractive and colonial foundations of the modern system. Given their profound influence on societal transformations, integrating values into transformative change approaches is essential (IPBES, 2022, 2024a, 2024b).

Although personal, relational and systemic values are widely acknowledged as critical in shaping attitudes, decisions and actions, their operationalisation in transformative change approaches remains inconsistent and underdeveloped (Horlings, 2015; Huntjens & Kemp, 2022; Bouman & Steg, 2022; Huntjens & Kemp, 2025; IPBES, 2022, 2024a; Sagiv & Schwartz, 2022). Effective transformative approaches must not only reflect diverse societal values but also address potential value conflicts and, when necessary, foster shifts in value constellations to enable sustainability transitions. Chapter 16 of this book introduces a values-based transformative change approach for a society-wide transformation towards an eco-social contract.

Eco-social contracts require a shift from anthropocentric to ecocentric visions of life where people are part of an ecosystem; from human-centred individualism to humans as part of a social-ecological system; and from using, exploiting and managing nature to serving the needs of the Earth as a whole. Anthropocentrism seeks to reduce nature to a function of humanity, while ecocentrism considers humanity as part of nature (Scott, 2003). Here, nature has its own status, not under humans but rather beside humanity, the two working together in a dynamic process of interaction and mutual development. And humans are also regarded as *participants* in nature, not just biologically but in having a sense of belonging, connectedness and being related to it (de Groot & van den Born, 2007).

5.3   Shifting Paradigms in Resource Governance

How are worldviews and governance models evolving to shape more ecocentric approaches to policy and decision-making? The transition towards eco-social contracts requires a fundamental shift from hierarchical, top-down resource governance to polycentric, adaptive and participatory models that acknowledge the interconnectedness of humans and nature. Current governance structures, especially in sectors like water management and agriculture, often impose rigid frameworks that seek to categorise and control natural resources. However, in response to growing environmental complexities and uncertainties, there is an increasing need for transformative, reflexive and deliberative governance approaches (Ostrom, 2009; Dryzek & Pickering, 2017; Huntjens, 2021).

Polycentric governance, as emphasised by Huntjens (2021) and Ostrom (2009), highlights the need for multi-level decision-making, where local, regional, national and transnational actors collectively manage resources. This approach fosters resilience by ensuring governance structures remain flexible, decentralised and responsive to changing socio-ecological dynamics (Folke et al., 2005; Pahl-Wostl, 2009). Such an approach is particularly effective in addressing global challenges, as demonstrated by water governance models that integrate ecological feedback loops and community engagement (Huntjens et al., 2011; Pahl-Wostl et al., 2010).

Adaptive governance complements this model by prioritising learning-based, inclusive and iterative decision-making that evolves alongside dynamic ecological and social systems (Berkes et al., 2002; Folke et al., 2005). This requires cross-sectoral collaboration and the incorporation of ecocentric principles that recognise the intrinsic value of ecosystems, moving beyond a purely anthropocentric worldview (Dryzek, 2022; Latour, 2018).

Building on adaptive governance, Transformative Social-Ecological Innovation (TSEI) (Huntjens, 2021) provides a framework for institutional and behavioural shifts necessary for sustainability transitions. TSEI acknowledges that addressing complex governance challenges requires not only technological advancements but also institutional, cultural and economic innovations (Meadowcroft, 2009; Loorbach, 2010). This perspective aligns with governance models that promote multiple value creation, balancing ecological sustainability, economic well-being and social equity (Huntjens, 2021: 107; Schot & Geels, 2013).

While resource governance has traditionally focused on water and agriculture, emerging governance models emphasise the need for integrated and nexus approaches across multiple domains. For instance, urban sustainability transitions (Huntjens, 2021: 163) illustrate how climate-resilient and regenerative city planning requires coordination between local governments, businesses and communities (Avelino et al., 2016). Similarly, circular and regenerative economies (Huntjens, 2021: 165) offer examples of governance that fosters sustainable production and consumption patterns (Raworth, 2018).

A critical dimension of governance transitions is addressing power asymmetries and institutional constraints. As Huntjens (2021) and others argue, governance reforms should incorporate power and network analysis to ensure equitable access to decision-making processes and avoid reinforcing existing socio-economic inequalities (Avelino & Wittmayer, 2016). Institutional design principles for governing the commons (Ostrom, 1990; Huntjens, 2021) provide valuable insights into creating governance structures that balance participation, efficiency and fairness.

An example of adaptive water governance can be seen in the Netherlands and Vietnam, where a shift from a "fight against water" paradigm to a "living with water" approach has transformed river basin management (Huntjens et al., 2011; Huntjens & Zhang, 2016). By reconnecting rivers with wetlands and reversing land reclamation efforts, these countries have enhanced flood resilience and created models for integrated water governance (Folke et al., 2005). The establishment of transboundary institutions such as the Rhine and

Mekong River Commissions further demonstrates how governance can be collaborative, adaptive and ecologically grounded (Pahl-Wostl, 2009).

The transition to adaptive, polycentric and eco-social governance is essential for managing the complexities of social-ecological systems. By moving beyond rigid, hierarchical structures and embracing multi-level governance, transformative innovation and institutional design principles, societies can foster governance systems that are flexible, inclusive and resilient (Meadowcroft, 2009; Loorbach, 2010). Integrating these principles into policy and practice will be crucial to navigate contemporary environmental and social challenges while ensuring sustainable and equitable resource management for future generations.

5.4 Economies at the Service of Life

The new economy movement is another area of inquiry and practice that shows how new economic paradigms, underscored by a shift from anthropocentric to ecocentric values, are envisioning ways of living in reciprocity and partnership with nature. The anthropocentric worldview largely sees nature in relation to its benefit and utility for (some of) humanity and has been employed in a way that has not only exploited, polluted and destroyed nature but has left many behind. Remarkable progress in welfare and life expectancy has been achieved but not for all. Income and wealth inequalities are on the increase in many countries (Qureshi, 2023), and nature loss is in freefall. Yet mainstream economists regard environmental damage caused by economic production as negative externalities, i.e. when prices do not fully capture costs and the social—that is, total—costs of production are larger than the private costs (Helbling, 2010). So, in practice, nature remains a blind spot in economic governance (Dasgupta, 2021; Lewsey, 2024). And moreover, on the whole, people do not have a say in the economic decisions being made in their name (Mohamed, 2023).

Prevailing economic systems and the thinking behind them have contributed to the ecological divide (Scharmer, 2013), as discussed earlier in the chapter. Outdated paradigms of economic thought are among the root causes of the ecological divide and the intertwined social and ecological crisis (Scull, 2017). The anthropocentric foundation of dominant economic models also marginalises beliefs, narratives and values that advance a relational view of the human-nature relationship, producing economic paradigms that are driving the ecological divide we witness today (Bogert et al., 2022). This includes marginalising economic paradigms that view nature beyond its benefit and utility for humans and that seek to structure this material relationship humans have with nature, not on domination and extraction but on reciprocity and care (Hickel, 2020).

Eco-social contracts need a different economy. This economy should enable all people to enjoy prosperity and promote equity within and between generations and across different countries, an economy guided by integrated, accountable and resilient institutions that restores and invests in nature. But first we need to recognise that our economies are embedded within nature and are not external to it. This is the central message of the *Dasgupta Review*, an independent global examination of the economics of biodiversity (Dasgupta, 2021). We therefore need economic paradigms that support prosperity within the ecological limits of the planet and allow nature and people to thrive together.

Alternative economic visions that integrate nature range from Indigenous and regenerative to well-being and degrowth economics. These economic visions offer development pathways to eco-social futures (Table 5.1). They question the purpose and goals of economies and the measurement of economic progress. They adopt a critical view of economic theories that focus on growth above all else and contain a fatal flaw in its "systemic discounting of the environment" (Boehnert, 2018: 355). Indigenous, ecological, post-growth, well-being, sufficiency, regenerative and eco-feminist economics, to name a few, all incorporate one fundamental premise: the economic system is not separate from, but rather embedded within and dependent upon, nature. Further, many of these heterodox economic theories assert that economies should not only function as serving humans but should be in the service of all life, leading to the stewarding of natural resources for the well-being of all living entities.

Post-growth economics is one approach gaining credence amid the evolving landscape of alternative economic models. Rooted in a "worldview that sees society operating better without the demand of constant economic growth", it proposes widespread economic justice, social well-being and ecological regeneration (Post Growth Institute, 2024). It offers not only a broader view of progress but a range of policies that includes better measures of progress. It builds on the recognition that gross domestic product (GDP) is a deeply flawed measure of societal well-being and proposes more holistic measures of progress, such as the Genuine Progress Indicator, well-being indicators and National Happiness Index.

Post-growth theories are gaining ground globally, notably in Europe, where a movement made up of governments, academics and civil society (Widuto et al., 2023; Luciani, 2024) is exploring what post-growth economics means

Table 5.1 Five highlighted alternative economic models

Alternative economic models	Description
Circular economy	It is designed for restorative, regenerative use of raw materials and manufactured goods. It emphasises reuse, extending products' service life through repair, remanufacture, restoration, upgrades and retrofits; turning old goods into as-new resources by recycling materials (upcycling)
Degrowth and steady-state economy	This approach emphasises sustainable and equitable reduction (and stabilisation) of societies' material and energy extraction, processing and transportation. It also recognises that degrowth and/or changing growth patterns in the Global North open the possibility of more autonomous pathways to Well-being in the Global South
Well-being economy	Central to this model is a broad swath of perspectives centred on the idea of a "safe and just space" for humans living in harmony with the natural environment while achieving the social foundation for human well-being. It also entails living with values oriented towards securing human well-being and "sufficiency" in resource consumption to balance nature's capacities and human needs
Economics of biodiversity	This approach emphasises valuing nature or recognising "natural capital", defined as the value of natural resources—in economic models, along with human capital (human knowledge) to deal with ecosystem and biodiversity losses. It emphasises that humans are embedded in nature, including the entire biosphere, and human ingenuity should be focused on achieving the common good without transgressing biosphere boundaries
Eco-social contract for economic transformations	This model postulates that the nature of the societal, environmental and economic problems facing humanity today could be improved by a new social contract, coined as a natural social contract or eco-social contract. It recognises the need for co-evolutionary processes and new social-ecological contracts between humans and nature to create societies that are more humane and fairer and regenerative instead of destructive to nature and people. It regards societies as social-ecological systems focused on people as members of a community and as part of natural ecosystems

Source: Authors (adapted from IPBES, 2024a)

in practice. City-level application in Europe reveals a "patchwork of post-growth concepts and frameworks" and an intent to move beyond small-scale piloting and experimentation to post-growth economies at scale (LSE, 2023). Examples of national adoption of post-growth economic frameworks, such as the Wellbeing Economy Governments (WEGo), a partnership of governments with shared ambitions to promote well-being, exist. While these reflect the potential of national-level adoption of post-growth models, translation into policy, governance and budgets that promote this vision remains limited (Hayden, 2024). This illustrates the challenge of breaking the addiction to unfettered growth that underlies current economic practice.

Other alternative economic approaches embrace ecocentrism wholly. Regenerative and Indigenous economics aim to "heal the story of separation" between humans and nature. Indigenous economics (Swiderska, 2021) offer holistic models that do not privilege "human economic goals" and seek to achieve the well-being of both humans and nature through "sufficiency rather than infinite growth, and equity and redistribution of wealth rather than accumulation". These visions signal a way to reimagine the organisation of production and consumption in the service of all life, offering a pathway that breaks with anthropocentrism. As Jessica Stago, from the Navajo people, says:

> Our economies are currently built on principles that are foreign to our ways of thinking and the lives that we want to live. We need to build our economies based on our own ways of knowing our world. This translates into how we utilise resources and how we conserve those resources for future generations of not just five-fingered beings but other species as well—the forests, the plants, the elk, the fish. These living beings are not assets on a balance sheet, nor are they only spiritual relatives; they are also key stakeholders and contributors to truly Indigenous economies. (Stago, 2021)

While alternative economic models challenge the narrative around growth and environmental limits, and connect social well-being and ecological regeneration, many still privilege human well-being, reflecting an anthropocentric view. Recognition and greater adoption of economic models and approaches rooted in ecocentric worldviews, such as Indigenous or regenerative economics, would address one of the primary failings of mainstream Western economics: anthropocentrism. Economic thinking that can challenge and "change the narrative at the heart of modern economics, which insists that nature is external to our economy, rather than the foundational source of all value" (Stories for Life, 2024), is needed to reimagine eco-social contracts for economic transformation. This requires greater recognition and engagement with knowledges, ideas and practices that can shift the narratives that underpin the design of the economy, from narratives of separation to narratives of connectedness.

5.5   Conclusion

The fundamental paradigm shifts underlying eco-social contracts require deep systemic changes that challenge current power structures, economic models and social norms. Dominant social and economic paradigms that have eroded nature and society, which have emerged largely from the Global North and are now circulating everywhere due to colonial histories and legacy, continue to shape the world we live in. "New ways of imagining the future are critical to shift people's relationships with nature. One way to achieve such changes are stronger imaginative efforts across different partners and stakeholder groups…which propose societal agreements that serve all of life and reflect an understanding that humans are part of and fully interdependent with nature for all they have, do, consume, wear and inhabit" (IPBES, 2024b: 29).

Eco-social contracts exemplify the reimagining of a just and sustainable world. They embody a holistic approach where *human rights and the Rights of Nature* and *ecological integrity and social equity* are mutually reinforcing. An eco-social contract envisions a world where economic systems are designed to serve the planet and people, not the other way around. This requires an eco-centric instead of an anthropocentric worldview. This shift is closely linked to a shift in our view of humanity. A different model of human behaviour is required, including a transition from ego-awareness to eco-awareness, from the incomplete model of *homo economicus* to *homo ecologicus*. This model focuses on a human being's relationship with itself, with community and with nature (Becker, 2006). As described by Huntjens (2025):

> Humans are not only rational and self-interested, as the incomplete human image of *homo economicus* suggests. Humans are essentially eco-social and for our survival we have developed qualities such as solidarity, reciprocity, co-creation and cooperation. We have an innate capacity to share and care for each other and for nature, and an innate aversion to seeing others suffer (including plants and animals). These behaviours are natural to us, they give us energy, meaning, purpose and compassion.

Envisioning society from a different worldview and model of human behaviour has ramifications for economic development and resource management. These paradigm shifts are reflected in alternative economic models ranging from Indigenous and regenerative to well-being and degrowth economics, as well as resource management regimes that respect ecological integrity. They adopt adaptive and integrated approaches to deal with complexity, uncertainty and the need to live within the boundaries of life-supporting ecosystems.

An eco-social contract "embraces the reality that humans are an integral part of the whole living community…and that, in order to flourish, we must govern ourselves in ways that accord with the laws of that community" (Cullinan, 2014: 3). New social contracts, if they are to offer a framework to advance a just and sustainable world, must break with the anthropocentric values and principles of old social contracts. Ecological social contracts (Jennings, 2016), green social contracts (Tagliepetra, 2021) and natural social contracts (Huntjens, 2021) demonstrate that this existential shift in the relationship between humans and nature is already underway in reformulations of the social contract.

The future of nature, and humanity's future, is the focus of the Kunming-Montreal Global Biodiversity Framework (CBD, 2022), an action plan for achieving the vision of living in harmony with nature. To accomplish this we need to heal the ecological divide. We need to rediscover and revive (old and new) stories and narratives of human interconnection with nature, to design an economy and society that prioritises the health and well-being of people and planet. This requires transformative changes in ecological worldviews, from separation to connection with nature, from master over to partner with nature and from social contracts of consumption to eco-social contracts. This re-envisioned pact with nature, rooted in reciprocity, care, and interdependence, captures the transformative spirit of eco-social contracts.

References

Avelino, F., & Wittmayer, J. M. (2016). Shifting power relations in sustainability transitions: A multi-actor perspective. *Journal of Environmental Policy & Planning, 18*(5), 628–649.

Avelino, F., Grin, J., Pel, B., & Jhagroe, S. (2016). The politics of sustainability transitions. *Journal of Environmental Policy & Planning, 18*(5), 557–567.

Becker, C. (2006). The human actor in ecological economics: Philosophical approach and research perspectives. *Ecological Economics, 60*(1), 17–23.

Berkes, F., Colding, J., & Folke, C. (Eds.). (2002). *Navigating social-ecological systems: Building resilience for complexity and change*. Cambridge University Press.

Boehnert, J. (2018). Anthropocene economics and design: Heterodox economics for design transitions. *The Journal of Design, Economics and Innovation, 4*(4), 355–374.

Bogert, J., Ellers, J., Lewandowsky, S., Balgopal, M., &, Harvey, J. (2022). Reviewing the relationship between neoliberal societies and nature: Implications of the industrialized dominant social paradigm for a sustainable future. *Ecology and Society* *27*(2), 7. Accessed September 21, 2024, from https://doi.org/10.5751/ES-13134-270207.

Bosselmann, K. (2004). In search of global law: The significance of the earth charter. *Worldviews: Global Religions, Culture, and Ecology, 8*(1), 62–75.

Bouman, T., & Steg, L. (2022). A spiral of (in)action: Empowering people to translate their values in climate action. *One Earth, 5*(9), 975–978. https://doi.org/10.1016/j.oneear.2022.08.009

Bouman, T., Steg, L., & Zawadzki, S. J. (2020). The value of what others value: When perceived biospheric group values influence individuals' pro-environmental engagement. *Journal of Environmental Psychology, 71*(6), 101470.

Bourdieu, P. (2005). *The social structures of the economy.* Polity.

Brondizio, E., Díaz, S., Settele, J., & Ngo, H. (2019). *Global assessment report on biodiversity and ecosystem services of the Intergovernmental Science-Policy Platform on Biodiversity and Ecosystem Services (IPBES).* IPBES Secretariat. https://doi.org/10.5281/zenodo.3831673

Cecchi, C. (2013). Sostenibilità e Decrescita: dall'Homo OEconomicus all'Homo Ecologicus. (Sustainability and de-growth: From homo economicus to homo ecologicus). In E. Basile, G. Lunghini, & F. Volpi (Eds.), *Pensare il Capitalismo. Nuove Prospettive per l'Economia Politica (Thinking about capitalism. New perspectives for political economy)* (pp. 168–184). Francoangeli.

Centola, D., Becker, J., Brackbill, D., & Baronchelli, A. (2018). Experimental evidence for tipping points in social convention. *Science, 360*(6393), 1116–1119.

Christian Aid and partners. (2024). *New feminist and anti-racist social contracts for people and the planet.* Accessed September 21, 2024, from https://www.christianaid.org.uk/our-work/policy/new-feminist-and-anti-racist-social-contracts

Conference of the Parties to the Convention on Biological Diversity (CBD). (2022). *Decision adopted by the conference of the parties to the convention on biological diversity 15/4.* Kunming-Montreal Global Biodiversity Framework. Accessed September 22, 2024, from https://www.cbd.int/doc/decisions/cop-15/cop-15-dec-04-en.pdf

Cullinan, C. (2014). Governing people as members of the Earth community. In *State of the world 2014. State of the world.* Island Press. https://doi.org/10.5822/978-1-61091-542-7_7

Dasgupta, P. (2021). *The economics of biodiversity: The Dasgupta review. Abridged version.* HM Treasury.

de Groot, M., & van den Born, R. (2007). Humans, nature and god: Exploring images of their interrelationship in Victoria, Canada. *Worldviews, 11*, 324–351.

Desai, M. (2022). Going beyond the social: Communitarian imaginaries as inspirations for rethinking the eco-social contract? In *United Nations Research Institute for Social Development (UNRISD) issue brief 12.* UNRISD. https://cdn.unrisd.org/assets/library/briefs/pdf-files/ib12-communitarian-imaginaries.pdf

Dryzek, J. S. (1996). Foundations for environmental political economy: The search for homo ecologicus? *New Political Economy, 1*(1), 27–40.

Dryzek, J. S. (2022). *The politics of the earth: Environmental discourses*. Oxford University Press.

Dryzek, J. S., & Pickering, J. (2017). Deliberation as a catalyst for reflexive environmental governance. *Ecological Economics, 131*, 353–360.

Extinction Rebellion. (2024). *Declaration of rebellion*. Accessed September 21, 2024, from https://extinctionrebellion.uk/declaration/#:~:text=We%20hereby%20declare%20the%20bonds,citizen%20to%20rise%20with%20us

Fios, F., & Arivia, G. (2018). The concept of homo ecologicus spiritual-ethical (an ethical reflection on the ecological humanism concept of Henryk Skolimowski). In M. Budianta, M. Budiman, A. Kusno, & M. Moriyama (Eds.), *Cultural dynamics in a globalized world*. Routledge.

Foley, D. K. (1998). Introduction (Chapter 1). In P. S. Albin (Ed.), *Barriers and bounds to rationality: Essays on economic complexity and dynamics in interactive systems*. Princeton University Press.

Folke, C., Hahn, T., Olsson, P., & Norberg, J. (2005). Adaptive governance of social-ecological systems. *Annual Review of Environment and Resources, 30*, 441–473.

Green, D., & Shapiro, I. (1996). *Pathologies of rational choice theory: A critique of applications in political science*. Yale University Press.

Hayden, A. (2024). The wellbeing economy in practice: Sustainable and inclusive growth? Or a post-growth breakthrough? *Humanities and Social Sciences Communications, 11*. https://doi.org/10.1057/s41599-024-03385-8

Helbling, T. (2010). *What are externalities? What happens when prices do not fully capture costs. Finance and development. Back to basics*. Accessed October 29, 2024, from https://www.imf.org/external/pubs/ft/fandd/2010/12/pdf/basics.pdf

Hickel, J. (2020). *Less is more. How degrowth will save the world*. Windmill Books.

Horcea-Milcu, I. A., Koessler, A.-K., Martin, A., Rode, J., & Soares, T. (2023). Modes of mobilizing values for sustainability transformation. *Current Opinion in Environmental Sustainability, 64*(46), 101357.

Horlings, L. (2015). The inner dimension of sustainability: Personal and cultural values. *Current Opinion in Environmental Sustainability, 14*, 163–169.

Huntjens, P. (2025). Op weg naar een Eco-Sociaal Contract: een wenkend en realistisch perspectief. Hoofdstuk 10 in Brabander, Richard, Erik Janssen, Maja Rocak. (Eds. 2025) Sociale Professionals en de Klimaatcrisis. Van Gorcum Uitgeverij, Assen, Nederland.

Huntjens, P., & Zhang, T. (2016). *Climate justice: Equitable and inclusive governance of climate action*. The Hague Institute for Global Justice. Working Paper 16.

Huntjens, P., & Kemp, R. (2025). The transformation flower approach for eco-social contracting: Comparative insights from eight case studies in the global south and north. In P. Huntjens, N. Mohamed, K. Hujo, & M. Desai (Eds.), *Eco-social contracts for sustainable and just futures* (Chapter 16, pp. 283–312). Springer Nature.

Huntjens, P. (2019). Sociale innovatie voor een duurzame samenleving: Op weg naar een natuurlijk sociaal contract. In *Lectorale boek, IMPACT Lectoraat Sociale Innovatie in het Groene Domein*. Hogeschool Inholland. https://doi.org/10.48544/90f2b3be-ab4d-4acf-a9c3-13effea07bcf

Huntjens, P. (2021). *Towards a natural social contract: Transformative social-ecological innovations for a sustainable, healthy and just society*. Springer. https://doi.org/10.1007/978-3-030-67130-3

Huntjens, P., & Kemp, R. (2022). The importance of a natural social contract and co-evolutionary governance for sustainability transitions. *Sustainability, 14*(5), 2976. Accessed April 2, 2025, from https://www.mdpi.com/2071-1050/14/5/2976

Huntjens, P., Pahl-Wostl, C., Rihoux, B., Schlüter, M., Flachner, Z., Neto, S., Koskova, R., et al. (2011). Adaptive water management and policy learning in a changing climate: A formal comparative analysis of eight water management regimes in Europe, Africa and Asia. *Environmental Policy and Governance, 21*(3), 145–163.

International Trade Union Confederation (ITUC). (2021). *New social contract: Five workers' demands for recovery and resilience*. Accessed September 21, 2024, from https://www.ituc-csi.org/new-social-contract-five-demands

IPBES. (2022). *Methodological assessment report on the diverse values and valuation of nature of the intergovernmental science-policy platform on biodiversity and ecosystem services*. Balvanera, P., Pascual, U., Christie, M., Baptiste, B., and González-Jiménez, D. (Eds.). IPBES Secretariat. https://doi.org/10.5281/zenodo.6522522

IPBES. (2024a). *Thematic assessment report on the underlying causes of biodiversity loss and the determinants of transformative change and options for achieving the 2050 vision for biodiversity of the intergovernmental science-policy platform on biodiversity and ecosystem services*. O'Brien, K., Garibaldi, L., and Agrawal, A. (Eds.). IPBES Secretariat. doi:https://doi.org/10.5281/zenodo.11382215.

IPBES. (2024b). *Summary for policymakers of the thematic assessment report on the underlying causes of biodiversity loss and the determinants of transformative change and options for achieving the 2050 vision for biodiversity of the intergovernmental science-policy platform on biodiversity and ecosystem services*. O'Brien, K., Garibaldi, L., Agrawal, A., Bennett, E., Biggs, O., Calderón Contreras, R., Carr, E., Frantzeskaki, N., Gosnell, H., Gurung, J., Lambertucci, S., Leventon, J., Liao, C., Reyes García, V., Shannon, L., Villasante, S., Wickson, F., Zinngrebe, Y., and Perianin, L. (Eds.). IPBES Secretariat. 10.5281/zenodo.11382230.

ITUC. (2022). *A new social contract for a gender-transformative agenda. Workers and trade unions major group sectoral position paper to the HLPF (high level political forum HLPF)*. ITUC.

Jennings, B. (2016). *Ecological governance: Toward a new social contract with the Earth*. West Virginia University Press.

Kimmerer, R. W. (2013). *Braiding sweetgrass: Indigenous wisdom, scientific knowledge and the teachings of plants*. Milkweed editions.

Latour, B. (2018). *Down to earth: Politics in the new climatic regime.* Wiley.

Lewsey, F. (2024). *Dasgupta review: Nature's value must be at the heart of economics.* University of Cambridge. Accessed September 21, 2024, from https://www.cam. ac.uk/stories/dasguptareview

London School of Economics (LSE). (2023, April 4). *Implementing post-growth agendas in European cities. Roundtable hosted by LSE Cities' European Cities Programme.* Accessed September 21, 2024, from https://www.lse.ac.uk/Cities/events/ECP-Roundtables/Implementing-post-growth

Loorbach, D. (2010). Transition management for sustainable development: A prescriptive, complexity-based governance framework. *Governance, 23*(1), 161–183.

Luciani, L. (2024). A wave of beyond growth economics rolls across Europe. *Friends of the Earth Europe.* Accessed October 27, 2024, from https://friendsoftheearth. eu/news/a-wave-of-beyond-growth-economics-rolls-across-europe/

Meadowcroft, J. (2009). What about the politics? Sustainable development, transition management, and long term energy transitions. *Policy Sciences, 42*(2009), 323–340.

Messner, D. (2015). A social contract for low carbon and sustainable development. *Technological Forecasting and Social Change, 98*, 260–270.

Mohamed, N. (2023). *Building new social contracts: An overview of participatory mechanisms for economic governance.* Green Economy Coalition.

Mohamed, N., & Huntjens, P. (2023). *Dismantling the ecological divide: Toward a new eco-social contract.* United Nations Research Institute for Social Development (UNRISD) Issue Brief 15. UNRISD. https://cdn.unrisd.org/assets/library/briefs/ pdf-files/2023/ib15-a-new-contract-with-nature.pdf

O'Brien, K., Hayward, B., & Berkes, F. (2009). Rethinking social contracts: Building resilience in a changing climate. *Ecology and Society, 14*(2).

Ostrom, E. (1990). *Governing the commons: The evolution of institutions for collective action.* Cambridge University Press.

Ostrom, E. (2009). A general framework for analysing sustainability of social-ecological systems. *Science, 325*(5939), 419–422.

Otto, I., Donges, J., Cremades, R., Bhowmik, A., Hewitt, R., Lucht, W., Rockström, J., et al. (2020). Social tipping dynamics for stabilizing earth's climate by 2050. *Proceedings of the National Academy of Sciences, 117*(5), 2354–2365.

Pahl-Wostl, C. (2009). A conceptual framework for analysing adaptive capacity and multi-level learning processes in resource governance regimes. *Global Environmental Change, 19*(3), 354–365.

Pahl-Wostl, C., Holtz, G., Kastens, B., & Knieper, C. (2010). Analysing complex water governance regimes: The management and transition framework. *Environmental Science & Policy, 13*(7), 571–581.

Patel, R., & Moore, J. W. (2020). *A history of the world in seven cheap things: A guide to capitalism, nature, and the future of the planet.* Verso.

Plumwood, V. (2005). *Environmental culture: The ecological crisis of reason.* Routledge.

Post Growth Institute. (2024). *What is post-growth economics and why is it necessary?* Accessed September 21, 2024, from https://postgrowth.org/post-growth-economics/

Qureshi, Z. (2023, May 16). Rising inequality: A major issue of our time. *Brookings*. Accessed March 2, 2025, from https://www.brookings.edu/articles/rising-inequality-a-major-issue-of-our-time/

Raworth, K. (2018). *Doughnut economics: Seven ways to think like a 21st century economist*. Chelsea Green Publishing.

Richardson, K., Steffen, W., Lucht, W., Bendtsen, J., Cornell, S. E., Donges, J. F., Drüke, M., et al. (2023). Earth beyond six of nine planetary boundaries. *Science Advances, 9*(37). Accessed October 29, 2024, from https://www.science.org/doi/epdf/10.1126/sciadv.adh2458

Rinscheid, A., Huntjens, P., & Aarts, N. (2025). Do stakeholders' values support transformative change in the food system? Evidence from the Netherlands. *Sustainability: Science, Practice, and Policy, 21*(1). https://doi.org/10.1080/15487733.2025.2549160

Sagiv, L., & Schwartz, S. H. (2022). Personal values across cultures. *Annual Review of Psychology, 73*(1), 517–546. https://doi.org/10.1146/annurev-psych-020821-125100

Scharmer, O. (2013). From ego-system to eco-system economies. *Open Democracy*. Accessed October 27, 2024, from https://www.opendemocracy.net/en/transformation/from-ego-system-to-eco-system-economies/

Schot, J., & Geels, F. W. (2013). Strategic niche management and sustainable innovation journeys: Theory, findings, research agenda, and policy. *The Dynamics of Sustainable Innovation Journeys, 2013*, 17–34.

Schwartz, S. (2012). An overview of the Schwartz theory of basic values. *Online Readings in Psychology and Culture, 2*(1), 1–20.

Scott, P. (2003). *A political theology of nature*. Cambridge University Press.

Scull, J. (2017). The separation from more-than-human nature. *Seeds for Thought: A Collective ICE blog*. Accessed September 21, 2024, from https://www.ecopsychology.org/the-separation-from-more-than-human-nature/

Serres, M. (1995). *The natural contract*. University of Michigan Press.

Stago, J. (2021). This is what indigenous economies look like. *Grand Canyon Trust Advocate Magazine, Fall-Winter*. Accessed September 21, 2024, from https://www.grandcanyontrust.org/advocatemag/fall-winter-2021/Indigenous-Economies-Just-Transition

Steg, L. (2016). Values, norms, and intrinsic motivation to act pro-environmentally. *Annual Review of Environment and Resources, 41*(1), 277–292.

Steg, L., & De Groot, J. (2012). Environmental values. In S. Clayton (Ed.), *The Oxford handbook of environmental and conservation psychology* (pp. 81–92). Oxford University Press.

Stockholm Resilience Centre. (2024). *Planetary boundaries*. Accessed September 21, 2024, from https://www.stockholmresilience.org/research/planetary-boundaries.html

Stories for Life. (2024). *How might our stories help design an economy in service to life?* Accessed September 21, 2024, from https://stories.life/

Swiderska, K. (2021). *Here's why indigenous economics is the key to saving nature.* Green Economy Coalition. Accessed September 21, 2024, from https://www.greeneconomycoalition.org/news-and-resources/nature-based-inclusive-development-why-indigenous-peoples-economic-systems-are-key

Tadaki, M., Sinner, J., & Chanai, K. M. A. (2017). Making sense of environmental values: A typology of concepts. *Ecology and Society, 22*(1).

Tagliepetra, S. (2021). It's time for a green social contract. *Breugel.* Accessed October 27, 2024, from https://www.bruegel.org/opinion-piece/its-time-green-social-contract

United Nations Environment Programme (UNEP). (2021). *Making peace with nature: A scientific blueprint to tackle the climate, biodiversity and pollution emergencies.* United Nations Environment Programme (UNEP).

United Nations Environment Programme and International Science Council. (2024). *Navigating New Horizons: A global foresight report on planetary health and human wellbeing.* Accessed September 21, 2024, from https://wedocs.unep.org/20.500.11822/45890

UNRISD (United Nations Research Institute for Social Development). (2016). Policy innovations for transformative change: Implementing the 2030 Agenda for Sustainable Development. UNRISD.

UNRISD. (2021). *A new eco-social contract vital to deliver the 2030 agenda for sustainable development.* United Nations research Institute for Social Development (UNRISD) issue brief 11. UNRISD.

Widuto, A., Evroux, C. & Spinaci, S. (2023). *European Parliament Briefing: From growth to 'beyond growth': Concepts and challenges.* European Parliamentary Research Service. Accessed September 21, from https://www.europarl.europa.eu/RegData/etudes/BRIE/2023/747107/EPRS_BRI(2023)747107_EN.pdf

Willis, R. (2020). A social contract for the climate crisis. *IPPR Progressive Review, 27*(2), 156–164.

6

Beyond Sustainable Development to Sustainable Societies: Insights from Feminist and Indigenous Theories and Praxis

Manisha Desai

6.1 Introduction

On the release in March 2023 of the report of the Intergovernmental Panel for Climate Change (IPCC), the responses of the United Nations (UN) officials ranged from "the damage is done" to "now or never." Similarly, the 2023 Global Sustainable Development Report (UN Department of Economic and Social Affairs, 2023) focuses on transformation as key to achieve sustainable development goals. Both reports recognise the unjust reality of climate change, namely, those who contribute the least to the problem will bear the greatest brunt of the problem. They also address how the most vulnerable, including women and the poor, will be the worst impacted given the historical and contemporary inequalities of the current, global system. Both, however, rely on sustainable development as the path forward.

Sustainable development entered the international discourse and policy framework in 1987 via the Brundtland Commission's report (World Commission

This chapter was first published in Sociology of Development, pp. 1–11, electronic ISSN: 2374-538X © 2024 by the Regents of the University of California. DOI: https://doi.org/10.1525/sod.2023.0049. Published here with permission. Includes minor edits to reflect the theme of this edited volume.

M. Desai (✉)
Stony Brook University, Stonybrook, NY, USA
e-mail: manisha.desai@stonybrook.edu

© The Author(s) 2025
P. Huntjens et al. (eds.), *Eco-Social Contracts for Sustainable and Just Futures*,
https://doi.org/10.1007/978-3-031-99109-7_6

on Environment and Development, 1987) which defined it as "development that meets the needs of the present without compromising the ability of future generations to meet their own needs." But it does not say who defines those needs or how they are to be met. The role of national and global elite, multinational corporations, and multilateral bodies such as the International Monetary Fund (IMF) and World Bank that dictate one-size-fits-all economic policies that benefit some at the cost of most people and ecosystems, is left unproblematised. Even as the UN Agenda 2030 for Sustainable Development (UN, 2015) focuses on equitable and inclusive growth in harmony with nature, there is no explicit critique of growth-based and market-oriented development nor a rethinking of "nature" that has contributed to the intersecting crises of inequalities, climate change, loss of biodiversity, and species extinction, among others. Rather, in the 40 years since sustainable development entered the international policy arena, there has been even more aggressive spread of neoliberal extractive and exploitative globalisation.

Drawing upon feminist, indigenous, and abolitionist thinkers, I suggest that what we need is a fundamental transformation in how and what we produce, consume, and redistribute and how we relate to each other and to all living and nonliving entities on this earth. This means going beyond sustainable development, a one-size-fits-all model based on neoliberal, market-based growth, to conceptualise place and culture specific, sustainable societies. I define sustainable societies along three dimensions: economic, epistemic, and political.

Economically, drawing on feminist scholarship (Mies et al., 1988; Mies & Shiva, 2014; Shiva, 1988), sustainable societies are organised away from extractive and exploitative profit-oriented growth towards a focus on care and subsistence. Epistemically, as many scientists and Indigenous communities worldwide have understood (Altmann, 2020; Boyd, 2017; Ito, 2019; Kauffman & Martin, 2021; Solon, 2019), we need to shift away from the "modern" understanding of "nature" as separate from "culture," and therefore a resource to be exploited for unending human consumption and growth, to focusing on "ecosystems" as a source of life that need to be regenerated for all living beings in the present and the future. Finally, drawing on Women of Color and abolitionist scholarship (Davis, 2005; Gilmore, 2017; Hull et al., 1982; Moraga & Anzaldua, 1983), sustainable societies are built via coalition politics that ensure liberation from hierarchy and oppression for all.

Most importantly, sustainable societies are varied and multiple in response to place, history, and culture specific needs, desires, and imaginations. As in my previous work (Desai, 2020), I move away from unproductive binaries of, e.g., North/South and academic/activists, to show how in the past several decades, due to multiple and intersecting crises, greater scientific understanding, and social movements all around the world, actors from the Global North and South in dialogues and contestations with each other have contributed to envisioning a paradigmatic shift that I am calling sustainable societies.

This shift is also evident in the framing of a new eco-social contract (UNRISD, 2021). Like sustainable societies, the new eco-social contract also identifies the urgency of moving beyond the anthropocentric view of nature as a resource for fulfilling human needs and desires to an ecocentric understanding of our reciprocal relationship, as part of nature, with all other living entities. Similarly, it envisions economic transformations that move beyond the extractive, exploitative growth-based models of development to ones that ensure equity and sustainability and social transformations that address historical and ongoing injustices to deliver rights for all.

6.2   Care-Centred Economic Restructuring

Beginning in the 1970s, ecofeminists around the world (Mies et al., 1988; Mies & Shiva, 2014; Shiva, 1988) were connecting the ways in which modern, capitalist patriarchies, first via colonialism and later via development, were exploiting both nature and women through a gendered discourse of control and domination that was destructive and unsustainable. Mies and Shiva (2014) critiqued the ways in which, in the name of poverty reduction and development, marginalised communities around the world were being propelled into the global market as cheap labour. Based on these communities' preference to regain control over their livelihoods, Mies and Shiva (2014) proposed a return to subsistence as an alternative to the destructive capitalist-colonialist-patriarchal model of global development. Subsistence, they argued, both produces and preserves life and, while invisibilised in the capitalist economy, is nonetheless central even to its survival. More importantly, subsistence provides a blueprint to move beyond the destruction of capitalist-colonialist-patriarchy to sustainability based on "community self-determination, self-reliance, self-provisioning, food self-sufficiency, regionality, participatory democracy, social equity, and cooperation" (Mies & Shiva, 2014 (1993): 298).

At the same time, thinkers like Meadows et al. (1972) were also articulating a critique of growth. Yet since the beginning of international development in the 1950s, subsistence was assumed to be premodern/"traditional" and hence to be undermined in favour of a "modern," i.e., capitalist industrial future, in which the poor of the world would be incorporated as cheap labour on a global assembly line. Even as climate change has revealed the unsustainability of this approach, the search for alternatives has focused on market-based solutions of a global, green economy. While green economies seek alternatives to fossil fuels, for the most part, they leave intact their capitalist extractive, exploitative, and global underpinnings. Alternatives such as degrowth or zero growth (D'Alisa et al., 2015; Schmelzer et al., 2022) have begun to address the need for a restructuring of the economy away from growth to ecological and social justice and the need for institutional and legal transformations.

But it is feminists who centre care in their renewed call for "subsistence for all" (Di Chiro, 2019), which re-embeds the economy in just and ethical social and ecological relationships based on care rather than growth. They argue that:

> We need a climate policy vision that is grounded not in overcoming subsistence to attain modern (carbon-free) progress, but in promoting a new definition of a subsistence way of living, one that embodies interdependent, socially just, and earth-caring and climate-caring human–nature relationships. Subsistence, so defined, means we're all in; subsistence is how we all 'live well together' within limits on a finite earth (Di Chiro, 2019: 308–9).

In addition to rescuing subsistence from the colonial binaries of tradition/modern, feminists also link it to social reproduction and care work that is central to all economies. In neoliberal global capitalism, care work is commodified as low wage and gendered and becomes part of what Hochschild (2014) has called the global care chain, performed primarily by immigrant women from the Global South for those in the Global North. Beyond such commodification, care work is still primarily unrecognised, unpaid, and performed primarily by women in private in the Global North and South. Hence, the new subsistence feminists claim:

> Care work, or the work of everyday subsistence, will always need to be accomplished (even in the green economy's post-carbon world, and by both humans or non-humans alike), and it cannot remain invisible, privatised, and done for free by women, people of colour, immigrants, or other marginalised groups. In other words, caring for climate, caring for earth, and caring for people should be at the centre of economic value, not at the margins. (Di Chiro, 2019: 308)

Similarly, Tandon (2013: 22) observes that rather than focusing on reducing the carbon footprint, the metric discussed in most mainstream climate change discussions, we should be working to extend the "care footprint," for an "earth friendly caring democracy" that cares for "communities, for future generations in a finite world, and for nature" (Tandon, 2013: 22).

Centring care also opens other possibilities. For example, as Indigenous scholars note: "Material life is only one aspect and cannot be reduced to accumulation of things and objects. We have to learn to eat well, dance well, sleep well, drink well, to practice one's beliefs, work for the community, take care of nature, appreciate elders, respect whatever surrounds us, and learn as well how to die—because death is an integral part of the cycle of life" (Solon, 2019: 257).

Unlike sustainable development's one-size-fits-all economy, care-based economies will unfold differently around the world based on specific material, cultural, and historical contexts. Such a place-based and culturally specific perspective is also evident in the ways many communities around the world engage the nonhuman world.

6.3 From Nature to Life-Sustaining Ecosystems

The concept of humans as separate from nature which they must domesticate for their own ends may be biblical in origin, but the solidification of such a binary was facilitated by colonialism and later the rise of industrial capitalism in the Global North (Boyd, 2017; Ito, 2019; Tanasescu, 2022). Boyd (2017) notes that the disconnect between legal systems in the Global North and the ways that even scientists everywhere understand nature as a complex, fluid ecosystem of interdependent parts, including human societies, are because of anthropocentrism, where everything in nature is property for human domination and the pursuit of endless economic growth, goals specific to modern society. Even scholars such as Marx who critiqued private property saw nature as separate from humans and perpetuated the nature/culture binary. But anthropologists have long noted that nature itself is a cultural construct. Thus, nature appears self-evident, but what constitutes it and who defines it is neither simple nor universal and varies across the world. Yet the Western, modern conception of nature as a totality removed from humanity and as a resource has diffused worldwide first via colonialism and then through international development. Yet for most communities around the world, "nature" is a specific place, to which they relate genealogically and beyond ownership. Today, even scientists in the Global North focus on specific ecosystems rather than an abstract concept of "nature."

It is only in the last few decades that as climate change has become undeniable that environmental movements around the world have begun to merge with Indigenous understandings of our relationship to nonhumans. Among the most well-known are the Andean indigenous discourses of *Sumak Kawsay*, better known by its Spanish translation of *buen vivir* or *vivir bien* (living well) and more recent feminist articulations of commons and commoning.

Sumak Kawsay or *buen vivir* has been appropriated by environmental justice movements globally as an ecological concept, but its origin is in the decolonial struggles of the Andean people (Altmann, 2017, 2020). As such, although well intentioned, its unmooring from its territorial, historical, and political contexts is a cause of concern for many Indigenous communities in the Andes and elsewhere. For example, Whyte (2018) observes that for Indigenous people, climate justice and environmental movements are "about stopping sexual and state violence against Indigenous peoples, reclaiming ethical self-determination across diverse urban and rural ecosystems, empowering gender justice and gender fluidity, transforming lawmaking to be consensual, healing intergenerational traumas, and calling out all practices that erase Indigenous histories, cultures, and experiences" (cited in Di Chiro, 2019:307–9).

Thus, *Sumak Kawsay* is a complex cosmology whose main elements are wholeness, coexisting in multipolarity, equilibrium, complementarity of diverse subjects, and decolonisation (Solon, 2019). Wholeness refers to *Pacha*, the earth, in its connection to the heavens above and the ancestors below along the time space continuum of the past, present, and future. Such an understanding of time is evident in the Aymara proverb "to walk forward, one must have eyes on the past." Beyond totality is the multipolarity of thought in which binaries such as good/evil are not oppositional but coexist, just as individual and community also coexist and are cared for by collaboration and community labour. The concept of equilibrium refers to balance, a contentment rather than a pursuit of endless growth or happiness. But this equilibrium is not static or permanent but needs to be reestablished as new circumstances bring new challenges that must be addressed. In this context, human beings are seen "not as producers, conquerors, transformers, but care takers, cultivators, mediators" (Solon, 2019: 259). Such an equilibrium of a whole made up of oppositions or binaries can be maintained only by complementarity, which "means seeing the differences as part of a whole." This leads to new dynamics as people work together learning from each other. Within this worldview, there is no single alternative for all societies but rather multiple or pluriversal formations informed by specific historical and cultural contexts. What unites them, however, is the focus on decolonising territories and beings "to recover the past to redeem the future." For *Sumak Kawsay* to flourish, it needs to be in relations of complementarity with similar, holistic alternatives worldwide. Elsewhere (Desai, 2022), I have discussed other alternatives such as *Ubuntu* in Southern Africa and *Eco-Swaraj* in India.

Yet such complex and dynamic Indigenous cosmovisions have often been disciplined within a liberal, modern discourse and jurisprudence of "Rights of Nature." Thus, the constitution of Ecuador grants rights to *Pacha Mama* as does the law of "mother earth" in Bolivia. But in both Bolivia and Ecuador global, extractive mining by multinational industries continues, making a mockery of such legal recognition. Similarly, legal personhood has been granted to rivers in Aotearoa, India, and Bangladesh, among other countries. Since 2006, there are more than 409 initiatives in 39 countries to grant some recognition to natural entities as having intrinsic value beyond human interests (Putzer et al., 2021). Additionally, there is the UN's Harmony with Nature and the Global Alliance for the Rights of Nature's (RoN) call for a Universal Declaration of the Rights of Mother Earth (GARN, 2025). But as Kauffman and Martin (2021) note, the scope and strength of the RoN initiatives are context specific and determine to what extent they work towards a transformative process. Similarly, Tanasescu (2022) notes that the neoliberal expansion of rights, including for nature, has occurred alongside the expansion of global exploitation of nature and labour. Therefore, rights may not always be the tool of emancipation of either humans or nature. "…personhood and rights are not fundamentally threatening to dominant modes of organizing social, political, and economic life. Thinking genealogically with landscapes that make the person look insignificant—now that is something truly revolutionary" (Tanasescu, 2022:45).

Therefore, even in the Global North, thinkers like Berry (2001), Ito (2019), and others have called for Earth Jurisprudence. It is different from the Right of Nature discourse and jurisprudence which have, since the 1970s, focused on legal personhood for ecosystems such as rivers, forests, and the Earth, as in the case of Bolivia and Ecuador mentioned above. "Earth Jurisprudence asserts that there is a lawful order to the universe that maintains a web of life. All elements of Nature, including humans, are inextricably connected into this order and linked to one another through interdependent relationships. Consequently, human well-being is dependent on the well-being of the ecosystems that sustain all life" (cited in Kauffman & Martin, 2021: 4). Thus, Earth Jurisprudence goes beyond legal change to economic and political change.

For Earth Jurisprudence the focus is the responsibility to live in reciprocal relationship with the "biotic communities" in which humans live. Like *Sumak Kawsay*, it prioritises balance and dynamic equilibrium of all living entities, not just humans who are part of the various ecosystems. Rather than extending human rights to ecosystems, thinkers like Berry (2001) even suggest each part of the ecosystem needs substantive rights of their own, such as tree rights for trees and river rights for rivers, each according to their unique substantive nature.

Similarly, the postcapitalist feminist political ecological approaches, e.g., Clement et al. (2019) and Sato and Alarcon (2019), reframe nature as commons produced through community, expressing gendered relationships in society that are not separate from relations to nature. Like Indigenous cosmologies and ecological sciences, their understanding of commons goes beyond natural resources to social relations and knowledge forms. Commons are seen as "… a practice, or a knowledge that is shared by a community" (Gibson-Graham et al., 2013: 130). Such commons must be ensured through collective action and transformative politics. Moreover, for feminists, commoning is multispecies and multiscalar. As Harcourt and Escobar (2005) and Harcourt (2019) suggest, humans and nonhumans all reappropriate, reconstruct, and reinvent in relation to each other and other living and nonliving entities.

Thus, many movements and scientists are thinking systemically in terms of ecosystems rather than nature. Such new conceptions will also require coalition politics evident in the work of Women of Color feminists and other social justice movements.

6.4   Coalition Politics of Freedom

To reorient economies from growth to care and to think holistically of ecosystems and our place in them and responsibility towards them will require a radical coalitional politics of freedom from all oppressions. Among the earliest formulators of such politics were Women of Color in the United States, from Sojourner Truth in the nineteenth century to Frances Beal (1970), Shirley Chisolm (1972), and to the now classic Combahee River Collective (1977), Hull et al. (1982), and Moraga and Anzaldua (1983). With her "Ain't I a Woman," speech, Sojourner Truth called out race and class privilege of white women. Nearly a century later, Frances Beal in her "double jeopardy" similarly called out the myopic women's liberation movement for foregrounding a sisterhood based on gender and failing to understand that without fighting racism and capitalism there could be no liberation even for white women. The Combahee River Collective's statement not only articulated an understanding of the simultaneous and intersecting oppressions of queer Women of Color but also a coalition politics of working on their own terms with other radical movements that sought freedom for all. It was an anti-racist, anti-heteropatriarchal, and anti-capitalist radical politics. This radical, political work was not about organising around a single identity nor about finding common ground but a long-term process that recognises that to transform societies to be free of all oppressions will involve collective action with others across differences.

Anzaldua (1990: 23–24) identified four strategies for such coalition work. First, serving as a bridge, i.e., being a mediator between "yourself and your community and white people." Knowing that a bridge gets used and abused, her second strategy is to "drawbridge," i.e., decide to "be down" and engage with white people or temporarily "be up" and withdraw from white people "in order to regroup, recharge our energies, and nourish ourselves before wading back into the frontlines" (Anzaldua, 1990: 223). Third is to "island," i.e., withdraw completely. "Yet an island cannot be a way of life—there are no life-long islands because no one is totally self-sufficient." Fourth is "sandbar," a sedimentation that is either connected or disconnected to the shore based on tides. This means "getting a breather from being a perpetual bridge without having to withdraw completely" (Anzaldua, 1990: 224). Anzaldua suggests that each group of women will have to decide which strategy to engage depending on where they are in their struggle and might involve moving back and forth between those strategies.

Bernice Reagon (1983) similarly emphasised political commitment over identity and like Anzaldua noted that it is a struggle that can be both risky and dangerous. She further emphasised the need to ensure that we do not reproduce oppressive structures in our activist spaces and to be open to existential transformation. Lugones and Keating (2023) similarly emphasise that radical coalition politics is not just about struggling against oppressions but also the possibilities of self and collective transformations in how we understand and relate to each other. I would add to this how we relate to all other living and nonliving entities as well. Contemporary, intersectional social justice feminists in the United States (e.g., Taylor, 2017) have returned to these earlier and, unfortunately, still necessary articulations of coalition politics given the rise of white supremacist and authoritarian forces. They emphasise the need for political commitment informed by collective identities but not reducible to those identities, for as Keeanga-Yamahtta Taylor, a founder of the Black Lives Matter movement, notes: "Lived experience of oppression does not, alone, constitute radical politics. ... Experiencing oppression is not nearly the same as knowing what to do about it. That comes from history, politics, and, ultimately, organizing" (cited in Taylor, 2022: 231).

Such a radical, coalition politics is just as relevant for the project of sustainable societies as it enables us to see that climate justice is not just about climate but about all the intersecting inequalities perpetuated by the global neoliberal political economy. As Barbara Smith noted in the wake of The Women's March to Washington DC in 2016, such coalitional political commitment requires being "Complicit in liberation for all and accountable to each other … as we cannot remove one roadblock to freedom without addressing all" (cited in Taylor, 2022: 229).

Given that histories of patriarchies, colonialism, and contemporary neoliberal global capitalism vary across the globe, coalitional politics will look different in different places. But its logic will be the same everywhere: freedom from all oppressions and self and collective transformation in how we relate to each other and our ecosystems. For example, in the United States, activists have articulated abolition ecologies to address how for communities of colour, particularly African Americans, climate justice is not just about climate but is linked to historical and ongoing injustices, and therefore abolition climate justice must be an intersectional, political struggle rooted in the homes, neighbourhoods, workplaces, and in an ethics of care (Ranganathan & Bratman, 2021). For Gilmore (2017: 228, cited in Ranganathan & Bratman, 2021: 116), abolition is the "unfinished liberation … [from] processes of hierarchy, dispossession, and exclusion that congeal in and as group-differentiated vulnerability to premature death." Furthermore, it is about "rehumanizing those dehumanized by the Eurocentric Man." Gilroy (2017, cited in Ranganathan & Bratman, 2021: 116) calls this political work "reparative humanism."

Abolition ecologies (Heynen & Ybarra, 2021; Ranganathan & Bratman, 2021) therefore move away from the language common in much climate change and sustainable development discourses of resilience. Resilience language, they argue, ignores what creates the need to be resilient in the first place. Furthermore, as Leitner et al. (2018, cited in Ranganathan & Bratman, 2021) observe, we now have a "global resilience complex," of contractors, architects, designers, and corporations in the wake of various climate change crises ready to profit from these new crises. Therefore, abolition ecologies focus on resourcefulness as a counter approach, which is rooted in the lived experiences of frontline communities' experience, knowledge, and approaches to dealing with the structural violence of daily injustices such as low wages, food insecurity, and healthcare of frontline communities.

Similarly, for Heynen and Ybarra (2021: 29) abolition ecologies define communities, not by the oppressions they face but rather how a land-based coalition politics "makes freedom a place." It is a "political ecology of liberated life-ways through radical place-making." Thus, struggles for land as resource and identity are shared across spaces of those who are free and unfree. "Abolition is a theory of change, a theory of social life, of making things. Place making as prefiguration for liberation."

Similarly, feminists have defined a coalition politics based on reimagining "commons" as an active process of "communing." For example, Clements et al. (2019) see feminist commoning as "the process of making and remaking the commons" based on an emphasis on the everyday practices, social relations, and spaces of creativity and social reproduction. Thus, the "commons" is not a disembodied abstract notion. It acknowledges power in everyday experiences and practices at various scales but seeks to transform those unequal relations via collective action.

In India, grassroots movements have articulated coalition politics as *Eco-Swaraj* (Kothari et al., 2014). It brings together Mahatma Gandhi's concept of *Swaraj* or self-rule—forged in the fight against British colonial rule—with the current intersecting crises of caste, class, gender, religion, and climate to develop an alternative paradigm of economics and politics based on local, self-determination. "Ecological Swaraj is an evolving worldview, not a blueprint set in stone. In its very process of democratic grassroots evolution, it forms an alternative to top-down ideologies and formulations…This is the basis of its transformative potential" (Kothari et al., 2014: 368). Sustainable societies, I believe, are the key to such global transformations.

6.5 Towards Sustainable Societies

The need for sustainable societies has been made urgent by the current climate crises. But as I have discussed above, feminists, among others, have long critiqued the unsustainable colonial, patriarchal, and capitalist mode of production for its social, economic, ecological, and political crises. Therefore, sustainable societies are about more than just climate justice or sustainable development. They are about social relations within and between groups, with nonhuman species and nonliving entities across time-space. They are about the nature of those relationships, free from hierarchies, and oppositional binaries, based on reciprocity and collective practices of freedom at all levels.

I have defined sustainable societies along three dimensions: economic, epistemic, and political. Economically, there is an urgent need to go beyond growth and green, techno fixes and reducing our carbon footprints. As feminists have been arguing for decades, we need to centre care and subsistence in ways that are culturally specific rather than to be governed by a one-size-fits-all neoliberal, global economy that is extractive and exploitative of people and the planet. Epistemically, we need to think beyond the liberal, modern understanding of nature in opposition to culture and solely as a resource for endless human consumption. As indigenous communities everywhere have long understood and scientists are now reinforcing, humans are part of the ecosystems which make all life possible, and therefore we have a responsibility to live in ways that regenerate those systems for all life. Finally, to do so we will need a radical coalition politics that ensures freedom from all oppressions for all and is built on collective action and transformation.

Like sustainable societies, eco-social contracts are another conception of how we can move beyond the polycrisis in which we find ourselves today. As the chapters in this edited book demonstrate, eco-social contracts can be in conversation with alternative knowledge and aesthetic systems; influence governance practices from the global to the local levels; and are being enacted in various ways around the world through the state and social movements.

Whether such efforts will suffice to address contemporary injustices and build sustainable societies or will humans—in the words of Geoffrey Hinton the father of AI—be just a passing phase in the evolution of intelligence (quoted in Mearian, 2023), only time will tell. But as scholars such as Gibson-Graham (1996) and Wright (2010) remind us, across the world people are already practicing forms of what I call sustainable societies, what the United Nations Research Institute for Social Development (UNRISD) calls eco-social contracts, and what Wright has called "real utopias." We need both to recognise and support such experiments so they become ubiquitous not through scaling up but through exchange and dialogue, in response to specific histories, cultures, and ecologies.

References

Altmann, P. (2017). *Sumak Kawsay* as an element of local decolonization in Ecuador. *Latin American Research Review, 52*(5), 749–759.

Altmann, P. (2020). The commons as colonisation – The well-intentioned appropriation of *Buen Vivir. Bulletin of Latin American Research, 39*(1), 83–97.

Anzaldúa, G. (Ed.). (1990). *Making face, making soul/haciendo caras; Creative and critical perspectives by feminist of color.* Aunt Lute Books.

Beal, F. M. (1970). Double jeopardy: To be black and female. In Third World Women's Alliance (Ed.), *Black woman's manifesto* (pp. 19–34). Third World Women's Alliance.

Berry, T. (2001, January 11). *The origin, differentiation and role of rights.* Institute for Educational Studies. Retrieved March 14, 2025, from http://www.ties-edu.org/wp-content/uploads/2018/09/Thomas-Berry-rights.pdf

Boyd, D. (2017). *The rights of nature: A legal revolution that could save the planet.* ECW.

Chisholm, S. (1972). The politics of coalition. *Black Scholar, 4*(1), 30–32.

Clement, F., Harcourt, W., Joshi, D., & Sato, C. (2019). Feminist political economy of the commons and communing. *International Journal of the Commons, 13*(1), 1–15.

Combahee River Collective. (1977). The Combahee River collective statement. *Black Past*. Retrieved March 14, 2025, from https://www.blackpast.org/african-american-history/combahee-river-collective-statement-1977/

D'Alisa, G., Demaria, F. & Kallis, G. (Eds.). (2015). *Degrowth: A vocabulary for a new era.* Routledge.

Davis, A. (2005). *Abolition democracy: Beyond empire, prison, and torture.* Seven Stories Press.

Desai, M. (2020). Troubling the Southern turn in feminisms. *European Journal of Women's Studies, 27*(4), 381–393.

Desai, M. (2022, April). *Going beyond the social: Communitarian imaginaries as inspiration for rethinking the eco-social contract.* Issue Brief 12. UNRISD.

Di Chiro, G. (2019). Care not growth: Imagining a subsistence economy for all. *British Journal of Politics and International Relations, 21*(2), 303–311.

GARN. (2025). *Universal Declaration for the Rights of Mother Earth.* https://www.garn.org/

Gibson-Graham, J. K., Cameron, J., & Healy, S. (2013). *Take Back the economy: An ethical guide for transforming our communities.* University of Minnesota Press.

Gilmore, R. W. (2017). Abolition geography and the problem of innocence. In G. T. Johnson & A. Lubin (Eds.), *Futures of black radicalism* (pp. 225–240). Verso.

Harcourt, W. (2019). Feminist political ecology practices of worlding: Art, communing and the politics of hope in the classroom. *International Journal of the Commons, 13*(1), 153–174.

Harcourt, W., & Escobar, A. (Eds.). (2005). *Women and the politics of place.* Kumarian Press.

Heynen, N., & Ybarra, M. (2021). On abolition ecologies and making "Freedom as a Place". *Antipode, 53*(1), 21–35.

Hochschild, A. (2014). *Global care chains and emotional surplus value.* Routledge.

Hull, G. T., Scott, P. B., & Smith, B. (Eds.). (1982). *All the women are white, all the blacks are men, but some of us are brave: Black women's studies.* Feminist Press.

Ito, M. (2019). Nature's rights: Why the European Union needs a paradigm shift in law to achieve its 2050 vision. In C. La Follete & C. Maser (Eds.), *Sustainability and the rights of nature in practice* (pp. 311–330). CRC.

Kauffman, C., & Martin, P. (2021). *The politics of rights of nature: Strategies for building a more sustainable future.* MIT Press.

Kothari, A., Demaria, F., & Acosta, A. (2014). Buen Vivir, degrowth and ecological Swaraj: Alternatives to sustainable development and green economy. *Development, 57*(3–4), 362–375.

Lugones, M., & Keating, C. (2023). *Educating for coalition: Popular education and contemporary political praxis.* SUNY Press.

Meadows, D., Randers, J., & Behrens, W. (Eds). (1972). *The limits to growth: A report of the Club of Rome's project on the predicament of mankind*. Universe Books. https://www.library.dartmouth.edu/digital/digital-collections/limits-growth

Mearian, L. (2023, May 4). Q&A: Google's Geoffrey Hinton: Humanity just a 'passing phase' in the evolution of intelligence. *Computerworld*. https://www.computerworld.com/article/1625138/qa-googles-geoffrey-hinton-humanity-just-a-passing-phase-in-the-evolution-of-intelligence.html

Mies, M., & Shiva, V. (2014 [1993]). *Ecofeminism*. Zed Books.

Mies, M., Bennholdt-Thompson, V., & Von Werholf, C. (1988). *Women: The last colony*. Zed Books.

Moraga, C., & Anzaldua, G. (Eds.). (1983). *This bridge called my back: Writings by radical women of color*. Kitchen Table/Women of Color Press.

Putzer, A., Lambooy, T., Jeurissen, R., & Kim, E. (2021). Putting the rights of nature on the map: A quantitative analysis of rights of nature initiatives across the world. *Journal of Maps, 18*(1), 89–96.

Ranganathan, M., & Bratman, E. (2021). From urban resilience to abolitionist climate justice in Washington D.C. *Antipode, 53*(1), 115–137.

Reagon, B. J. (1983). Coalition politics: Turning the century. In B. Smith (Ed.), *Home girls: A black feminist anthology* (pp. 343–356). Rutgers University Press.

Sato, C., & Alarcon, J. (2019). Towards a postcapitalist feminist political ecology's approach to the commons and communing. *International Journal of the Commons, 13*(1), 36–61.

Schmelzer, M., Vetter, A., & Vansintjan, A. (Eds.). (2022). *The future is degrowth: A guide to a world beyond capitalism*. Verso.

Shiva, V. (1988). *Staying alive: Women, ecology, and development*. Zed Books.

Solon, P. (2019). Is vivir bien possible? Candid thoughts about alternatives. In K. Bhavanani, J. Foran, P. Kurian, & D. Munshi (Eds.), *Climate futures* (pp. 253–262). Zed Press.

Tanasescu, M. (2022). *Understanding the rights of nature: A critical introduction*. Transcript Verlag.

Tandon, N. (2013). Harmonising our footprints: Reducing the ecological while accumulating the care. In Genanet (Ed.), *Sustainable economy and green growth: Who cares?* (pp. 22–25). http://www.genanet.de/fileadmin/user_upload/dokumente/Care_Gender_Green_Economy/Int_WS_Sustainable_Economy_Green_Growth_who_cares_EN.pdf

Taylor, K.-Y. (Ed.). (2017). *How we get free: Black feminism and the Combahee River collective*. Haymarket.

Taylor, L. (2022). *Feminism in coalition: Thinking with US women of color feminism*. Duke University Press.

UN (United Nations). (2015). *Transforming our world: The 2030 agenda for sustainable development*. UN Doc. No. A/RES/70/1. 21 October.

UN Department of Economic and Social Affairs. (2023). *Global sustainable development report*. https://sdgs.un.org/gsdr/gsdr2023

UNRISD. (2021). *A new eco-social contract. Vital to Deliver the 2030 Agenda for sustainable development.* UNRISD Issue Brief 11.

Whyte, K. P. (2018, April 3). White allies, let's be honest about decolonization. *YES! Magazine.* www.yesmagazine.org/issues/decolonize/white-allies-lets-be-honest-about-decolonization-20180403

World Commission on Environment and Development. (1987). *Our common future.* Oxford University Press.

Wright, E. O. (2010). *Envisioning real utopias.* Verso.

Part II

Policies and Transformative Change Agendas at Different Governance Levels

7

Governance that Inspires Collective Action for a Just and Sustainable World

Mary Robinson

The rapidly worsening climate, nature and inequality crises confront us with an existential challenge. Unless we reverse greenhouse gas emissions and restore nature, temperatures will rise and ecosystems degrade to a point where large parts of the planet will become unliveable. This will not be felt equally: and it will exacerbate other forms of injustice and inequalities in countries of all incomes, for example through its disproportionate impact on women, Indigenous communities and the poor. The urgency of the interconnected existential threats we face requires changes in how we govern our societies and economies. We share one solitary, fragile planet. We will all suffer the consequences of inaction. Yet humanity is struggling to put aside individual interests and act and govern in our common interest.

The climate, nature, and inequality crises are governance challenges at multiple scales, requiring not only national governments but also regional authorities, cities, businesses, civil society, and grassroots movements to act collectively. Collective action needs transparent and accountable leadership that is responsive to the challenges that people face. It involves taking a long-term view that moves beyond short-term political cycles and recognises the interests and voice of young people and future generations. While pressure on leaders to take a whole-of-society approach is increasing, change is also being driven by bottom-up movements, including citizen assemblies, corporate

M. Robinson (✉)
The Elders, London, UK
e-mail: Mary.Robinson@tcd.ie

P. Huntjens et al. (eds.), *Eco-Social Contracts for Sustainable and Just Futures*,
https://doi.org/10.1007/978-3-031-99109-7_7

sustainability initiatives, and local governance innovations. Cities, businesses, and Indigenous communities are leading in areas where national governments lag, showing that governance transformation is possible even in the absence of strong political leadership. Polycentric governance, which integrates decision-making across these levels, provides a more resilient and adaptive approach to systemic transformation.

For instance, robust multilateral institutions are essential, but so are regional agreements, city-led coalitions, and transnational networks that drive action even when global governance structures stall. Yet despite this urgent need for collective action, the world is increasingly facing a crisis of multilateralism and worsening trust deficit, with powerful states prioritising short-term national agendas over long-term global stability, while practising double standards in their application of international law and breaking their promises. Forums to meaningfully integrate peoples' priorities and aspirations in multilateralism are needed. Transformative change agendas at different governance levels, including reforms to the multilateral system, finance architecture and the UN's peacebuilding/security structures, must be embraced.

This will require that we integrate climate and nature into people-centred policy and governance at all levels. Climate and nature are not distinct policy areas; we need to actively identify the links between them and other challenges facing communities, countries and the planet. Reductions in carbon emissions and the protection of nature can support all the Sustainable Development Goals, including through the creation of jobs, the reduction of conflict and migration, and the provision of healthy air, clean water and food systems that support health and well-being. Recognising these wider impacts can also help build broader political coalitions and social movements for more urgent collective action. The technology and finance are largely available, but governance structures and economic incentives must be reformed to enable transformative change. This includes restructuring global finance to align investments with sustainability goals, introducing legal accountability measures such as carbon pricing and climate litigation, and ensuring economic policies internalise environmental costs. What we need is not just more political will, but institutional frameworks and accountability mechanisms that embed long-term climate and nature commitments into governance structures. This includes legally binding climate policies, financial regulations that redirect investments towards sustainability, and participatory governance models that ensure decisions are inclusive and accountable.

8

Advancing Earth System Governance: Key Achievements and Propositions for Meaningful Progress Towards a Global Eco-Social Contract

Maja Groff, Georgios Kostakos, and Patrick Huntjens

8.1 Introduction: Eco-social Contracting for Effective Earth System Governance

In this chapter we define "eco-social contracts" in an ideal way as "collective agreements (at multiple governance levels) among members of a society" aimed at "deal[ing] with the polycrisis of the twenty-first century. This entails cooperation and the recognition of norms targeted at sustainability, equity and justice, with the associated rights and duties of care for the environment and the well-being of others, including future generations and all life on this planet" (Huntjens & Kemp, 2025, pp. 284: Chap. 16). These contracts are

M. Groff (✉)
Climate Governance Commission, The Hague, The Netherlands
e-mail: m.groff@climategc.org

G. Kostakos
Foundation for Global Governance and Sustainability (FOGGS), Brussels, Belgium
e-mail: georgios.kostakos@foggs.org

P. Huntjens
Inholland University of Applied Sciences, Delft, The Netherlands
e-mail: patrick.huntjens@inholland.nl

© The Author(s) 2025
P. Huntjens et al. (eds.), *Eco-Social Contracts for Sustainable and Just Futures*,
https://doi.org/10.1007/978-3-031-99109-7_8

not static but tend to reflect their time and context (Norton & Greenfield, 2023; UNRISD, 2022: Chap. 4). They constitute the modern incarnation of the old social contracts (e.g., by Rousseau, 1762), adapted to the greater awareness of and knowledge about the interaction between nature and humanity, the latter having taken the upper hand in our era of the Anthropocene (Crutzen & Stoermer, 2000). Such contracts are different at each governance level and within each cultural or temporal context but essentially comprise the web of relationships that bind together diverse citizens, communities, institutions, and governments into a just and sustainable society (Mohamed & Huntjens, 2023).

Twentieth-century social contracting efforts in the West have tended to focus on reforms towards "welfare capitalism". Although they have (unevenly) sought to expand social and workers' rights and increased social protection and social services coverage, these contracts have not sufficiently tackled inequalities along gender lines, have neglected ecological issues, and have only glossed over the Global North-South divide (UNRISD, 2022). Moreover, the predominance of the neoliberal economic model that followed the end of the Cold War undermined even the European post-World War II (WWII) welfare state model of strongly regulated capitalism. In addition to reversing a range of hard-fought gains in social inclusion, this "unfettered market" primacy, extractivism, and wasteful and noncircular material production that ignores the Earth's finite resources has led to the "triple planetary crisis" of climate change, biodiversity loss, and pollution that defines the state of the world today (United Nations, 2020a; UNEP, 2022; UNFCCC, 2022). Recent science likewise shows that six out of nine planetary boundaries crucial to Earth's stability have been breached, risking irreversible damage to planetary systems (Richardson et al., 2023; Caesar et al., 2024). Advocates across the globe and at all governance levels have been calling for an "ecosocial" contract that integrates environmental sustainability, social justice, and global accountability (Huntjens, 2019, 2021; Huntjens & Kemp, 2022; UNRISD, 2022; Gough, 2022; Bogert et al., 2022; Kempf & Hujo, 2022; Krause et al., 2022; UNEP and International Science Council, 2024). This approach would balance planetary boundaries with social needs, creating a "safe and just operating space for humanity" (Raworth, 2018; Rockström et al., 2023).

8.2   Earth System Governance: Concept and Progress

Earth (system) governance has been defined as the interlinked processes, mechanisms, and institutions by which societies collectively address environmental challenges on a planetary scale, aiming to ensure sustainable development, maintain ecological integrity, and foster global cooperation (Biermann et al., 2010). This concept has gained traction as a field within global governance, emphasising the need for coordinated international action on issues like climate change, biodiversity loss, and sustainable development. It can be considered to subsume related concepts launched earlier, like "global environmental governance" that focuses on the institutions and processes managing "the human-environment relationship" on a global scale (Young, 2002) and the environmental aspects of "sustainable development governance" introduced by the Brundtland Commission's iconic report *Our Common Future* (World Commission on Environment and Development (WCED), 1987). The latter's focus on managing economic growth and environmental stewardship together to ensure intergenerational equity has become foundational for policies targeting Earth governance, as it ties together the environment with social and economic dimensions of human well-being.

More recently, Rockström et al. (2009) introduced the planetary boundaries framework to define safe limits for human impact on Earth systems, which Fernández and Malwé (2019) note has gained political recognition but not yet legal integration. Ostrom (2009) advocates for polycentric governance, while Raworth (2018) highlights a safe operating space between planetary and social boundaries, focusing on human dignity and well-being.

Despite such strong and widely accepted conceptual advances, it remains true that international processes associated with sustainable development and Earth governance "have not led to an international legally binding framework that adequately addresses the challenges we face" (cf. Rühs & Jones, 2016, 174). Nevertheless, important elements of such a framework are already in place, including, to name a few, the Sustainable Development Goals (SDGs) (United Nations, 2015), the Paris Agreement on Climate Change (UNFCCC, 2015), and the Sendai Framework for Disaster Risk Reduction (UNDRR, 2015). Despite their different legal status—of the above, only the Paris Agreement is of legally binding character—they represent serious attempts at advanced international tools for a new "eco-social contract" or proactive understandings in that direction. All three, however, operate on the basis of consensus, with voluntary and nationally determined targets, thus lacking adequate implementation and enforcement mechanisms (Huntjens, 2021).

We should not discount the importance of other pieces of sustainable development and Earth governance frameworks that are already in place and which have had growing prominence in policymaking and in public consciousness, such as the Convention on Biological Diversity (CBD) and its Kunming-Montreal Global Biodiversity Framework (GBF) adopted in 2022. These and hundreds of other multilateral environmental agreements support the achievement of the SDGs and set out an ambitious pathway towards reaching the global vision of a people living in harmony with nature by 2050 (CBD, 2022).

As the GBF example illustrates, the relationship between humans and nature is becoming more prominent and recognised at the international level in various ways (Huntjens, 2021). Articulation of this relationship and attendant values has been under development by various groups in the international community for decades, with an official milestone reached more than 50 years ago in 1972, when the United Nations Conference on the Human Environment was held in Stockholm. Other milestones have been less official but still quite influential, like the collaboratively developed Earth Charter. The Earth Charter is an ethical framework for sustainable development published in 2000, which involved the leadership of former politicians such as Maurice Strong (Canada), Joaquín Almunia (EU, Spain), Mikhail Gorbachev (Soviet Union), Fernando Cardoso (Brazil), as well as environmentalists and thought leaders from various fields. It reserves a central place for environmental protection, human rights, equitable human development, and peace and argues that these values are interdependent and indivisible (Earth Charter, 2000). Similarly, the seminal Universal Declaration of the Rights of Mother Earth was proclaimed at the World People's Conference on Climate Change and the Rights of Mother Earth, a citizens' initiative convened in Cochabamba, Bolivia, in 2010 (International Rights of Nature Tribunal, 2010).

Overall, multiple examples exist showing a growing awareness of the sets of values underpinning humanity's more conscientious relationship with the Earth and its ecosystems. This is evidenced by diverse initiatives across regions, alongside the significant development of international environmental law within international legal practice (Dupuy & Viñuales, 2018), which may mark a turning point after the destructive exuberance of the industrialisation, modernisation, and urbanisation period covering the last two centuries. In quantitative terms at least, articulated commitments to respecting the natural environment now abound; according to the International Environmental

Agreements Database (IEADB), there are now over 3000 multilateral and bilateral environmental agreements in place (Mitchell et al., 2020).

8.3 Nine Propositions for Effective Earth System Governance

The question remains how to advance this agenda. In this section we highlight nine propositions, which, while non-exhaustive, could lead to meaningful progress towards effective Earth system governance and a global eco-social contract, responding to the unprecedented crisis conditions scientists have alerted us to and to which populations around the world are increasingly exposed. Incorporating even some of these proposals would represent a significant advance towards a more equitable, sustainable, and resilient global governance framework for the future of humanity and planet Earth.

We divide the nine propositions into four clusters that correspond to proposed key elements of advancing Earth system governance, namely:

1. Building consensus on the moral imperative for effective Earth system governance.
2. Establishing strengthened legal frameworks for implementation to regulate effective Earth system governance.
3. Implementing an institutional framework for accountable Earth system governance.
4. Redirecting resources to invest in and finance structural economic transformations.

8.3.1 Building Consensus on the Moral Imperative for Effective Earth System Governance

Proposition 1: Catalysing Urgency of Action in the International System – UNGA Declaration of Planetary Emergency and Creation of a Planetary Emergency Platform

A UN General Assembly (UNGA) Declaration of Planetary Emergency and the establishment of a Planetary Emergency Platform could significantly enhance global governance for addressing the triple planetary crisis of climate change, biodiversity loss, and pollution. This declaration could reflect scientific findings that humanity has exceeded six of nine planetary boundaries

(Rockström et al., 2009; Steffen et al., 2015) and respond to calls from climate coalitions, scientific communities, and nations already declaring climate emergencies (Climate Emergency Declaration Group, 2024). Beyond symbolism, it could provide or catalyse a framework for coordinated multilevel action and establish a baseline for addressing global governance gaps (Climate Governance Commission, 2023).

A Planetary Emergency Platform, convened by the UN Secretary-General or otherwise in the international system, elaborating on the proposal from the UN Secretary-General (Climate Governance Commission, 2023), and inspired by such initiatives/statements as the Global Pact for the Environment (UNEP, 2017) and the Pact for the Future (United Nations, 2024; see Action 54), could assist in coordinating and integrating efforts across intergovernmental and non-state actors. It could develop a Planetary Emergency Plan (Club of Rome, 2019) to align targets from existing agreements (e.g., Paris Agreement, CBD, SDGs), address implementation gaps, and catalyse key initiatives such as a global decarbonisation framework. Additionally, the platform could collaborate with alliances like the Breakthrough Agenda and Race to Zero to accelerate clean energy solutions and foster knowledge exchange. This coordinated approach would strengthen the overall global response to the planetary emergency (Climate Governance Commission, 2023).

8.3.2 Establishing Strengthened Legal Frameworks for Implementation

Proposition 2: Establish an International Court for the Environment (ICE) Addressing the urgent challenges of the Anthropocene requires robust accountability mechanisms and legal forums for effective global environmental governance. Current venues such as the International Court of Justice (ICJ), the International Tribunal for the Law of the Sea (ITLOS), and the Permanent Court of Arbitration (PCA) occasionally deal with global environmental issues, with regional bodies like the Inter-American Commission of Human Rights (IACHR) and the European Court of Human Rights (ECHR) also handling climate-related claims (Sabin Center for Climate Change Law, 2019; ECHR, 2024). Recent developments include advisory opinions from ITLOS on states' obligations to protect marine environments (ITLOS, 2024) and ongoing cases at the ICJ and IACHR addressing climate responsibilities (ICJ, 2024; IACHR, 2024).

Despite progress, these institutions face limitations, including fragmented jurisdiction and limited scientific capacity (Koivurova, 2007; Bruce, 2016). In response, proposals for an International Court for the Environment (ICE) envision a central forum to resolve disputes, provide authoritative advisory opinions, and establish coherent global environmental law.

An ICE could accept cases from non-state actors, including individuals, communities, and NGOs, granting citizens worldwide standing to seek enforcement of existing environmental obligations and redress for environmental harms (ICE Coalition, 2011). Furthermore, the ICE could operate across multiple legal frameworks, integrating international, local, and hybrid laws to suit specific cases (ICE Coalition, 2011).

The court could address the fragmentation of international environmental law, which currently spans several hundred multilateral agreements, by assisting in ensuring that these form a coherent legal framework (Bruce, 2016). The ICE could issue binding decisions and remedies, including compensation, restoration, fines, injunctive relief, and interim measures to prevent environmental harm (Hockman, 2010; Bruce, 2016). To ensure technical rigour, the ICE could include judges with specialised expertise and scientific experts to provide technical assessments (Hockman, 2010). It could operate as a voluntary tribunal, a permanent stand-alone body, or a judicial wing of a broader global environmental agency, which would represent a strengthened UNEP, the leading global authority on the environment (Bruce, 2016; Karlsson-Vinkhuyzen & Dahl, 2021).

By internalising future environmental risks into legal processes, the ICE could adapt international law to the realities of the Anthropocene and hold actors accountable for addressing the triple planetary crisis. Such an innovative legal institution could fill critical gaps in global environmental governance.

Proposition 3: The Incorporation of the Rights of Nature into International Law

Recognising the Rights of Nature in international law could transform environmental governance by granting ecosystems legal rights similar to human rights.[1] This shift would address the anthropocentric bias of current legal frameworks, offering more robust protection for biodiversity and ecosystems and fostering a balanced relationship between humanity and nature. While

---

[1] For an in-depth discussion of the legal, philosophical, and institutional dimensions of the Rights of Nature as a foundation for eco-social contracts, see Chap. 11: *The Role of the Rights of Nature in Establishing Eco-Social Contracts* in this volume.

human rights and environmental protection have historically been linked, they often treat nature as a resource for human benefit (Borràs, 2016). Embedding nature's intrinsic rights into international law could enhance existing treaties, as well as reflect diverse ecological knowledges and values, including Indigenous worldviews, which often conceive of nature as a community member with rights (Gilbert, 2023).

Since Ecuador recognised Nature, or *Pacha Mama*, as a legal subject in 2008, eco-jurisprudence initiatives have grown globally, with over 510 initiatives across 40 jurisdictions tracked by the Eco-Jurisprudence Monitor (2022). Ecuador, Bolivia, and New Zealand have pioneered legal frameworks granting ecosystems enforceable rights, demonstrating the feasibility of possible legal developments at the international level.

Incorporating the Rights of Nature requires a paradigm shift to recognise nature's intrinsic rights, international institutional capacity to uphold these rights, and integration of Indigenous knowledge into legal frameworks. Operational challenges would still need to be overcome, but a global convention could provide consistency and strengthen international efforts to protect ecosystems and biodiversity (Borràs, 2016; Gilbert, 2023).

Proposition 4: Update the UN Charter to Include Obligations for Environmental Protection and Trusteeship

The United Nations Charter, largely unchanged since its drafting 80 years ago, needs some updating, not least because of the existential threats posed by the triple planetary crisis (Global Governance Forum, 2024). Proposals from organisations (e.g., IGEP (2021), UNEP (2017), UNDESA (2021)) and scholars (e.g., Chapin et al., 2011; Biermann, 2012; Biniaz, 2019; Gilbert et al., 2023) suggest several possible updates to the Charter text:

- *Environmental Protection as a Core UN Pillar:* One option is to add environmental protection alongside peace and security, sustainable development, and human rights as fundamental UN pillars. Member states would be obligated to safeguard the environment and sustainably manage Earth's resources. This update could empower the UN to pass binding decisions to protect the global environment or Earth system, in the fundamental interests of all peoples, supported by enforcement and dispute resolution mechanisms.
- *Reviving the Trusteeship Council:* The dormant UN Trusteeship Council could be given a new mandate focused on environmental stewardship and

protection of global commons, including the atmosphere, oceans, and biodiversity.

- *The Earth Trusteeship Initiative*: This concept recognises humans as trustees of the planet, responsible for its care and sustainable management. Embedding this principle in the Charter, in connection with the two above measures, would create obligations for member states to act as stewards of the Earth.

Amending the UN Charter would require approval by a two-thirds majority of the UN General Assembly and ratification by all five permanent UN Security Council members (UNSC). This process demands extensive diplomatic negotiations and consensus building to address contemporary challenges effectively. However, these proposed Charter amendments would powerfully align the UN's governance framework with modern environmental realities, ensuring global collaboration to protect the planet for present and future generations.

8.3.3 Implementing an Institutional Framework for Accountable Earth System Governance

Proposition 5: Establish an International Panel on Planetary Boundaries
Maintaining human development within planetary boundaries is vital for international policymaking, given the consequences of exceeding these limits on health, well-being, the economy, and global society, and the current grave Earth system risks and international environmental deterioration we are already experiencing (Climate Governance Commission, 2023). To guide decision-makers, an enhanced scientific advisory panel or panels are needed to monitor full Earth system risks and functions, inform policy, and improve public awareness. Such a facility would build on the work of the Intergovernmental Panel on Climate Change (IPCC), Intergovernmental Science-Policy Platform on Biodiversity and Ecosystem Service (IPBES), and similar bodies to create a unified, comprehensive Earth system assessment and monitoring framework.

Such enhanced scientific monitoring capacity could take the form of an International Panel on Planetary Boundaries. Such a panel would consolidate the latest planetary science, ensure policy coherence across the monitoring and analysis of Earth system risks, and develop common assessment and reporting frameworks (Obura, 2023). It could also establish nimble, semi-autonomous task forces to address science-policy gaps and emerging challenges requiring technical expertise. These groups would enhance the implementation of responses to Earth system hazards and crises. By reducing fragmentation in policymaking and scientific efforts, these measures would integrate the human and social dimensions of Earth system science, ensuring effective and equitable responses to the triple planetary crisis. Such frameworks are critical for mobilising the data and coherent policies necessary to achieve sustainable global development.

Academia and civil society have already initiated efforts to bridge science and policy gaps. The Earth Commission,[2] for example, defines "safe and just" planetary boundaries to ensure ecological stability and social equity. By setting science-based targets, it aligns policy and private sector efforts with the Sustainable Development Goals (SDGs) to address climate change, biodiversity loss, and resource degradation (Rockström et al., 2009; Raworth, 2018). Supported by Future Earth[3] and the Global Commons Alliance,[4] the Earth Commission synthesises interdisciplinary research, emphasising open data, cross-sector partnerships, and actionable policies to balance human needs and ecological limits (Steffen et al., 2015). The Earth Commission could, for example, form the basis of or feed into an official international entity, integrating its work into global governance frameworks.

Proposition 6: A Global Resilience Council for Human and Planetary Security, Responding to Eco-social Crises

The concept of human security, emphasising freedom from fear and want, was the focus of the 1994 Human Development Report and has since evolved to include resilience building, addressing poverty, and sustainable development (UNDP, 1994; United Nations, 2012). Climate change intensifies the link between human and environmental security by exacerbating resource scarcity, vulnerability, and conflict risks. Policies that treat climate change as a security issue focus on adaptation, resilience, and conflict prevention, highlighting the interconnected impacts on humans and ecosystems (Barnett & Adger, 2007; Huntjens & Nachbar, 2015; Huntjens & Zhang, 2016).

---

[2] Earth Commission provides science-based targets to safeguard a stable and just planet (https://earthcommission.org).

[3] Future Earth is a global research network advancing science for sustainability and planetary health (https://futureearth.org).

[4] Global Commons Alliance is a coalition driving collective action to protect Earth's life-support systems (https://globalcommonsalliance.org).

Integrating human security with planetary security is essential for an eco-social contract that safeguards both human well-being and ecosystem stability. Earth system governance frameworks advocate for policies addressing social-ecological interactions, requiring governance that transcends national borders (Biermann, 2014). The UNDP's 2022 Special Report on Human Security underscores systemic risks—climate change, biodiversity loss, and pandemics—that fragment global efforts and threaten progress towards the SDGs (UNDP, 2022).

A shift in priorities is needed to redirect resources from armaments and exploitation to tackling systemic risks that undermine eco-social balance and disproportionately harm the most vulnerable, endangering SDG achievement. FOGGS, ISI, and IGD (2024) propose the creation of a Global Resilience Council (GRC) to address nonmilitary threats to human and planetary security.

A GRC could be established as an intergovernmental body under ECOSOC, with regional and functional representation and no veto powers. It could coordinate with the UN system, regional organisations, and non-state actors, such as scientists, youth groups, Indigenous communities, and businesses. The GRC could eventually hold compulsory jurisdiction on some issues, offering a comprehensive governance mechanism to oversee implementation of eco-social agreements on climate change, biodiversity loss, pandemics, etc., thus helping to effectively tackle global, interconnected threats (FOGGS et al., 2024).

This integrated approach aligns with human security principles and the transformative potential of the 2030 Agenda, fostering resilience and sustainability for current and future generations.

8.3.4 Redirecting Resources to Invest in and Finance Structural Economic Transformations

Proposition 7: Transitioning to a Green Economy Through Large-Scale Global Investment

Transitioning to a green economy requires significant global investment, with estimates ranging from billions to trillions of dollars, depending on climate scenarios and other factors (McCollum et al., 2018; IRENA, 2020; IPCC, 2022; WRI, 2022). Low- and middle-income countries face particular challenges due to limited fiscal space, cost of capital, underdeveloped financial markets, and high debt burdens, which increase investment costs and deter private financing (Ameli et al., 2021; Buhr et al., 2018; Calcaterra et al., 2024). These constraints make the transition more expensive and delay progress.

For instance, accelerated climate action is central to transitions, yet many developing countries' Nationally Determined Contributions (NDCs) are conditional on financial support from developed nations (Pauw et al., 2020; WRI, 2022). To meet the Paris Agreement targets of limiting temperature increases to 2 or 1.5 °C, accessible, equitable, and sufficient financing is essential. At COP29, developed countries committed to mobilising USD 300 billion annually by 2035 for developing nations, with aspirations to scale this to USD 1.3 trillion. However, critics argue that these commitments fall short of addressing the needs of developing countries (Masood, 2024).

Global financing mechanisms, such as the Global Environment Facility (GEF), Green Climate Fund (GCF), and others, aim to support climate action in developing nations (UNFCCC, 2024). Yet these mechanisms face challenges, including fragmented structures and complex application processes, which demand significant institutional capacity that smaller countries often lack. The GCF, in particular, has been criticised for its slow project approval process, delaying critical initiatives (Darby, 2017; GCF, 2021).

A new eco-social contract must prioritise sufficient public finance, including through novel and collaborative measures to raise new sources of (public) international financing (e.g., the proposed minimum global tax on high-net-worth individuals, proposed windfall taxes on fossil fuel companies, etc.), complemented by debt relief instruments like carefully designed debt-for-nature swaps, where debt repayments genuinely fund conservation efforts (Hujo & Fuentes-Nieva, 2024). Also, simplified application processes are vital for timely access to funds. Additionally, private capital must be redirected from unsustainable practices to green investments; well-regulated (to protect against misuse) efforts to de-risk private investments and the offering of concessional loans or grants can further incentivise private sector participation, ensuring a just and timely transition to a sustainable economy (UNEP, 2023).

Proposition 8: Implementing a Global Carbon Pricing Mechanism

A universal carbon pricing mechanism could significantly accelerate the reduction of greenhouse gas emissions by internalising the costs of carbon and incentivising emission reductions, as well as potentially raising additional funds for the energy transitions needed (see Proposition 7). Carbon pricing encompasses tools like carbon taxes and emissions trading systems (ETS), requiring a globally coordinated approach that accommodates the Global South's concerns while enabling national-level policies within a global framework. Such a system, potentially managed by the World Bank, IMF, or another facility, could redistribute funds towards Sustainable Development Goals (SDGs) in low-income nations (Thisted & Thisted, 2020).

Well-designed carbon pricing policies have proven effective in reducing emissions and fostering clean technology innovation and investment (Aldy & Stavins, 2012; Fabian et al., 2023; Akindote et al., 2023). While some countries have implemented these policies at various levels, challenges remain, including concerns about equity, competitiveness, and political acceptability. Addressing distributional impacts on vulnerable populations and energy-intensive industries is crucial for ensuring fairness and effectiveness (Feindt et al., 2021).

To maximise impact, dynamic and adaptable mechanisms are essential. Lessons from implemented policies underscore the importance of optimising ETS designs, adapting to local contexts, and ensuring robust policy support (Emeka-Okoli et al., 2024; Narassimhan et al., 2018). A clear yet flexible and equitable global framework could enhance carbon pricing's potential as a key tool in global climate mitigation efforts.

Proposition 9: Corporate Accountability Mechanisms and Corporations as Systemic Solutions Providers

Corporations account for a substantial share of global greenhouse gas emissions, with investor-owned companies responsible for an estimated 31% of emissions from 1854 to 2022 and investor- and state-owned companies combined contributing 64% (Carbon Majors, 2024). However, corporations can also drive economic and societal change towards eco-social objectives, as demonstrated by sustainability leaders like Patagonia, B Corps, and the Science-Based Targets initiative (SBTi).

As one example, American retail company Patagonia has implemented measures to reduce its environmental footprint, such as transitioning to 100% renewable energy for its operations, using organically grown cotton since 1996, and participating in sustainability initiatives like the Zero Emission Maritime Buyers Alliance (ZEMBA). The company also donates 1% of its sales to environmental restoration (Patagonia, 2024a, 2024b). Similarly, B Corps certify companies that meet high social and environmental performance standards, promoting a shift from maximising shareholder value to shared value (Honeyman & Jana, 2014; B Corp, 2024). Meanwhile, the Science-Based Targets initiative (SBTi) provides guidance and validation for companies to set science-based emission reduction targets, supporting corporate climate action (SBTi, 2024). Despite these efforts, voluntary initiatives are insufficient on their own. Loopholes in standards often allow greenwashing where companies present themselves as greener than they are (Day et al., 2022). Romito et al. (2024) found that participation in initiatives like SBTi helps companies reduce emissions, but offsetting strategies often lack credibility, and direct emissions remain largely unaffected (Day et al., 2022; Coen et al., 2023).

Mandatory accountability mechanisms and increased transparency are necessary to complement voluntary efforts. Addressing the discrepancies between direct and indirect emissions impacts is critical for ensuring that corporate sustainability actions are both credible and effective. By integrating voluntary initiatives with robust, mandatory policies, corporations can transition from contributors to climate change to systemic solution providers.

8.4 Conclusions and Outlook

Earth system governance necessitates transformative legal, institutional, and financial frameworks that prioritise environmental sustainability and social equity in a mutually reinforcing way. This chapter reviewed a range of achievements to date and potential further steps in eco-social contracting for global Earth governance (Huntjens & Kemp, 2025; Norton & Greenfield, 2023). The twentieth century's social contracts, which emphasised growth over environmental or social considerations, have led to a perilous overshooting of planetary boundaries (Rockström et al., 2009; Caesar et al., 2024). Recognising this, recent frameworks like the UN 2030 Agenda for Sustainable Development and the Kunming-Montreal Global Biodiversity Framework signal progress, albeit without binding enforcement (UNEP, 2023; CBD, n.d.). Propositions for further progress put forward in this chapter emphasise legal, institutional, and financial innovations, including an International Court for the Environment, a near-term Planetary Emergency Platform, and a permanent Global Resilience Council, as well as global carbon pricing. If implemented, these can be core elements of the much-needed new eco-social contract for the twenty-first century, ensuring equitable and effective Earth governance while promoting social justice and human flourishing.

Unfortunately, a significant opportunity for a fresh departure in this direction was missed, through the Summit of the Future and its outcome document, the Pact for the Future. Out of 56 "Actions," to which world leaders committed through this document, only 6 refer to the governance of planetary systems, in a general, declaratory way, without resource allocation or an implementation timeline.[5] There are no explicit references to any of the propositions presented in this chapter, although a number had been put on the table, at least by global civil society. Their absence from the Pact may well be attributed to the continuing disconnect between developed and developing countries regarding prioritisation between environment and broader sustainable development, the bad faith created by the non-provision of climate and other (development) finance, the increasing preoccupation of even developed country citizens with jobs, inflation, and other such life and livelihood issues, and the overall negative geopolitical climate, including the resurgence of major military confrontations. In such a context, environmental concerns are again relegated to "luxury items" of secondary importance.

Nevertheless, there are ways to build on the modest Pact provisions and continue to advance this agenda, enriching the emerging global eco-social contract with much-needed elements, making the necessary adjustments that would take into consideration the geopolitical context and the continuing North-South divide, as indicated above. Going forward, strengthening accountability and cross-sector collaboration at every governance level will be essential to securing a resilient, just future for all of humanity and a healthy planet.

References

Akindote, O., Egieya, Z., Ewuga, S., Omotosho, A., & Adegbite, A. (2023). A review of data-driven business optimization strategies in the US economy. *International Journal of Management & Entrepreneurship Research, 5*(12), 1124–1138.

Aldy, J. E., & Stavins, R. N. (2012). The promise and problems of pricing carbon: Theory and experience. *The Journal of Environment & Development, 21*(2), 152–180.

---

[5] Action 9. We will strengthen our actions to address climate change;

Action 10. We will accelerate our efforts to restore, protect, conserve, and sustainably use the environment;

Action 12. We will plan for the future and strengthen our collective efforts to turbocharge the full implementation of the 2030 Agenda for Sustainable Development by 2030 and beyond;

Action 52. We will accelerate the reform of the international financial architecture so that it can meet the urgent challenge of climate change;

Action 53. We will develop a framework on measures of progress on sustainable development to complement and go beyond gross domestic product;

Action 54. We will strengthen the international response to complex global shocks.

Ameli, N., Dessens, O., Winning, M., Cronin, J., Chenet, H., Drummond, P., Calzadilla, A., Anandarajah, G., & Grubb, M. (2021). Higher cost of finance exacerbates a climate investment trap in developing economies. *Nature Communications, 12*(1), 1–12.

Barnett, J., & Adger, N. W. (2007). Climate change, human security and violent conflict. *Political Geography, 26*(6), 639–655.

B Corp. (2024). *Building the movement.* Accessed October 7, 2024, from https://www.bcorporation.net/en-us/movement/

Biermann, F., Betsill, M. M., Gupta, J., Kanie, N., Lebel, L., Liverman, D., Schroeder, H., Siebenhüner, B., & Zondervan, R. (2010). Earth system governance: A research framework. *International Environmental Agreements: Politics, Law and Economics, 10*(2010), 277–298.

Biermann, F. (2012). Greening the United Nations charter: World politics in the Anthropocene. *Environment: Science and Policy for Sustainable Development, 54*(3), 6–17.

Biermann, F. (2014). *Earth system governance: World politics in the anthropocene.* MIT Press.

Biermann, F., Abbott, K., Andresen, S., Bäckstrand, K., Bernstein, S., Betsill, M. M., Bulkeley, H., Cashore, B., Clapp, J., Folke, C., Gupta, A., Gupta, J., Haas, P. M., Jordan, A., Kanie, N., Kluvánková-Oravská, T., Lebel, L., Liverman, D., Meadowcroft, J., Mitchell, R. B., Newell, P., Oberthür, S., Olsson, L., Pattberg, P., Sánchez-Rodríguez, R., Schroeder, H., Underdal, A., Vieira, S. C., Vogel, C., Young, O. R., Brock, A., & Zondervan, R. (2012). Transforming governance and institutions for global sustainability: Key insights from the Earth system governance project. *Current Opinion in Environmental Sustainability, 4*(1), 51–60.

Biniaz, S. (2019). The UNGA resolution on a 'global pact for the environment': A chance to put the horse before the cart. *Review of European, Comparative & International Environmental Law, 28*(1), 33–39.

Bogert, J. M., Ellers, J., Lewandowsky, S., Balgopal, M. M., & Harvey, J. A. (2022). Reviewing the relationship between neoliberal societies and nature: Implications of the industrialized dominant social paradigm for a sustainable future. *Ecology and Society: A Journal of Integrative Science for Resilience and Sustainability, 27*(2), 7.

Borràs, S. (2016). New transitions from human rights to the environment to the rights of nature. *Transnational Environmental Law, 5*(1), 113–143.

Buhr, B., Donovan, C., Kling, G., Lo, Y., Murinde, V., Pullin, N., & Volz, U. (2018). *Climate change and the cost of capital in developing countries.* SOAS.

Bruce, S. (2016). The project for an international environmental Court. In: C. Tomuschat, R. Mazzeschi, & D. Thürer (Eds.), *Conciliation in international law: The OSCE Court of Conciliation and Arbitration* (pp. 133–170). Brill.

Caesar, L., Sakschewski, B., Andersen, L., Beringer, T., Braun, J., Donovan, D., Gerten, D., Hellemann, A., Kaiser, J., Kitzmann, N., Lorlani, S., Lucht, W., Ludescher, J., Martin, M. A., Mathesius, S., Paolucci, A., te Wierik, S., &

Rockström, J. (2024). *Planetary health check report 2024*. Potsdam Institute for Climate Impact Research.

Calcaterra, M., Reis, L. A., Fragkos, P., Briera, T., de Boer, H. S., Egli, F., Emmerling, J., et al. (2024). Reducing the cost of capital to finance the energy transition in developing countries. *Nature Energy, 9*(10), 1241–1251.

Carbon Majors. (2024). *The carbon majors database: Launch report*. Carbon Majors.

CBD. (n.d.). *List of parties*. Accessed November 13, 2024, from https://www.cbd.int/information/parties.shtml

CBD. (2011). *Convention on biological diversity: Texts and annexes*. Secretariat of the Convention on Biological Diversity.

CBD. (2022). *Kunming-Montreal Global Biodiversity Framework* (19 December 2022) Document CBD/COP/DEC/15/4.

Chapin, S., Power, M. E., Pickett, S., Freitag, A., Reynolds, J. A., Jackson, R. B., Lodge, D. M., Duke, C., Collins, S. L., Power, A. G., & Bartuska, A. (2011). Earth stewardship: Science for action to sustain the human-Earth system. *Ecosphere, 2*(8), 1–20.

Club of Rome. (2019). *Planetary emergency plan*. Accessed November 14, 2024, from https://www.clubofrome.org/publication/the-planetary-emergency-plan/

Coen, D., Kreienkamp, J., & Pegram, T. (2020). *Global climate governance*. Cambridge University Press.

Coen, D., Herman, K., & Pegram, T. (2023). Market Masquerades? Corporate climate initiative effects on firm-level climate performance. *Global Environmental Politics, 23*(4), 141–169.

Climate Emergency Declaration. (2024). *Climate emergency declarations in 2,356 jurisdictions and local governments cover 1 billion citizens*. Accessed November 14, 2024, from https://climateemergencydeclaration.org/climate-emergency-declarations-cover-15-million-citizens/

Climate Governance Commission. (2023). *Governing our planetary emergency*. Climate Governance Commission.

Crutzen, P. J., & Stoermer, E. F. (2000). The anthropocene. *IGBP Global Change Newsletter, 41*, 17–18.

Darby, M. (2017). *Green Climate fund 'a laughing stock', say poor countries*. Accessed November 14, 2024, from https://www.climatechangenews.com/2017/04/06/green-climate-fund-laughing-stock-ethiopia-bid-left-limbo/

Day, T., Mooldijk, S., Smit, S., Posada, E., Hans, F., Fearnehough, H., Kachi, A., Warnecke, C., Kuramochi, T., & Höhne, N. (2022). *Corporate Climate Responsibility Monitor 2022*. NewClimate Institute.

Desai, B. (2022). Regulating global climate change: From common concern to planetary concern. *Environmental Policy and Law, 52*(5–6), 331–347.

Dupuy, P.-M., & Viñuales, J. E. (2018). *International environmental law*. Cambridge University Press.

Earth Charter. (2000). *The Earth Charter*. Accessed November 14, 2024, from https://earthcharter.org/wp-content/uploads/2020/03/echarter_english.pdf

ECHR. (2024). *Grand chamber rulings in the climate change cases*. Accessed November 14, 2024, from https://www.echr.coe.int/w/grand-chamber-rulings-in-the-climate-change-cases

Emeka-Okoli, S., Otonnah, C. A., Nwankwo, T. C., & Nwankwo, E. E. (2024). Review of carbon pricing mechanisms: Effectiveness and policy implications. *International Journal of Applied Research in Social Sciences, 6*(3), 337–347.

Engel, J. R. (2019). Can the Earth Charter movement be renewed? The covenantal promise of the Earth Charter movement. In *The crisis in global ethics and the future of global governance* (pp. 20–30). Edward Elgar Publishing.

European Commission. (n.d.). *Corporate sustainability reporting*. Accessed November 13, 2024, from https://finance.ec.europa.eu/capital-markets-union-and-financial-markets/company-reporting-and-auditing/company-reporting/corporate-sustainability-reporting_en

Fabian, A. A., Uchechukwu, E. S., Okoye, C. C., & Okeke, N. M. (2023). Corporate outsourcing and organizational performance in Nigerian investment banks. *Scholars Journal of Economic, Business and Management, 10*(3), 46–57.

Feindt, S., Kornek, U., Labeaga, J. M., Sterner, T., & Ward, H. (2021). Understanding regressivity: Challenges and opportunities of European carbon pricing. *Energy Economics, 103*(2021), 105550.

Falkner, R. (2020). Global environmental responsibility in international society. *The Rise of Responsibility in World Politics, 1*, 21–23.

Fernández, E. F., & Malwé, C. (2019). The emergence of the 'Planetary Boundaries' concept in International Environmental Law: A proposal for a framework convention. *Review of European, Comparative & International Environmental Law, 28*(1), 48–56.

FOGGS, ISI and IGD. (2024). *Global resilience council and a rebalancing of the global governance system*. Accessed November 24, 2024, from https://www.foggs.org/wp-content/uploads/2024/09/GRC-concept-FOGGSIGDISI_16Aug2024.pdf

GCF. (2021). *Independent evaluation of the adaptation portfolio and approach of the Green Climate Fund*. Report No. 9, February 2021. Green Climate Fund.

Gilbert, J. (2023). Creating synergies between international law and rights of nature. *Transnational Environmental Law, 12*(3), 671–692.

Gilbert, J., Macpherson, E., Jones, E., Dehm, J., Jong, D. D.-d., & Amtenbrink, F. (2023). The rights of nature as a legal response to the global environmental crisis? A critical review of international law's 'greening' agenda. *Netherlands Yearbook of International Law, 52*, 47–74.

Global Governance Forum. (2024). *A second United Nations Charter: Modernizing the UN for a new generation*. Retrieved November 22, 2024, from https://global-governanceforum.org/

Gough, I. (2022). Two scenarios for sustainable welfare: A framework for an eco-social contract. *Social Policy and Society, 21*(3), 460–472.

Grantham Research Institute. (2018). *The role and influence of the UK's Committee on Climate Change*. Accessed November 14, 2024, from https://www.lse.ac.uk/

granthaminstitute/wp-content/uploads/2018/10/The-role-and-influence-of-the-UKs-Committee-on-Climate-Change_policy-brief.pdf

Gupta, J., & Prodani, K. (2022). Sustainable development, climate change and planetary justice: Governance challenges. In W. Hout & J. Hutchinson (Eds.), *Handbook on governance and development* (pp. 212–229). Edward Elgar Publishing.

Hockman, S. (2010). The case for an international court for the environment. *Journal of Court Innovation, 3*(1), 215–230.

Honeyman, R., & Jana, T. (2014). *The B Corp handbook: How to use business as a force for good.* Berret-Koehler Publishers.

Hujo, K. (2021). *A new eco-social contract.* UNRISD.

Hujo, K., & Fuentes-Nieva, R. (2024). *System change for economic transformation: Toward fair fiscal contracts.* UNRISD Working Paper, No. 2024-01, United Nations Research Institute for Social Development (UNRISD).

Huntjens, P. (2019). *Sociale Innovatie voor een Duurzame Samenleving: Op Weg naar een Natuurlijk Sociaal Contract.* Lectorale boek, Lectoraat Sociale Innovatie in het Groene Domein, Hogeschool Inholland, Delft. https://doi.org/10.48544/90f2b3be-ab4d-4acf-a9c3-13effea07bcf

Huntjens, P. (2021). *Towards a natural social contract: Transformative social-ecological innovation for a sustainable, healthy and just society.* Springer Nature. https://www.springer.com/gp/book/9783030671297

Huntjens, P., & Kemp, R. (2022). The importance of a natural social contract and co-evolutionary governance for sustainability transitions. *Sustainability, 14*(5), 2976. https://www.mdpi.com/2071-1050/14/5/2976

Huntjens, P., & Nachbar, K. (2015). *Climate change as a threat multiplier for human disaster and conflict.* The Hague Institute for Global Justice. Working Paper No. 9: 1-24.

Huntjens, P., & Kemp, R. (2025). The transformation flower approach for eco-social contracting: Comparative insights from eight case studies in the globalsouth and north. In P. Huntjens, N. Mohamed, K. Hujo, & M. Desai (Eds.), *Eco-social contracts for sustainable and just futures* (Chapter 16, pp. 283–312). Springer Nature.

Huntjens, P., & Zhang, T. (2016). *Climate justice: Equitable and inclusive governance of climate action.* The Hague Institute, Working Paper 16 (2016).

IACHR. (2024). *Solicitud de Opinión Consultiva presentada por la República de Chile y la República de Colombia.* Accessed October 8, 2024, from https://www.corteidh.or.cr/observaciones_oc_new.cfm?nId_oc=2634

ICE Coalition. (2011). *Draft Protocol for an International Court for the Environment.* Accessed October 8, 2024, from https://static1.squarespace.com/static/658826c47399bd578d3c8822/t/65cb2e3a3f74df597b3b18e6/1707814458395/Draft%2BProtocol%2Bfor%2Ban%2BICE.pdf

ICJ. (2024). *Obligations of states in respect of climate change.* Accessed October 8, 2024, from https://www.icj-cij.org/case/187.

IGEP (Independent Group of Experts on the UN Charter). (2021). *A global pact for the future: Modernizing the UN Charter for People and Planet*. Global Governance Forum. https://globalgovernanceforum.org

International Rights of Nature Tribunal. (2010). *Universal Declaration of the Rights of Mother Earth*. Accessed November 14, 2024, from https://www.rightsofnaturetribunal.org/wp-content/uploads/2018/04/ENG-Universal-Declaration-of-the-Rights-of-Mother-Earth.pdf

IPCC. (2022). *Climate Change 2022: Mitigation of Climate Change. Contribution of Working Group III to the Sixth Assessment Report of the Intergovernmental Panel on Climate Change*. Cambridge University Press.

IRENA. (2020). *Global renewables outlook: Energy transformation 2050*. International Renewable Energy Agency.

ITLOS. (2024). *Request for an Advisory Opinion Submitted by the Commission of Small Island States on Climate Change and International Law (Request for Advisory Opinion submitted to the Tribunal)*. Accessed October 8, 2024, from https://www.itlos.org/en/main/cases/list-of-cases/request-for-an-advisory-opinion-submitted-by-the-commission-of-small-island-states-on-climate-change-and-international-law-request-for-advisory-opinion-submitted-to-the-tribunal/

Karlsson-Vinkhuyzen, S., & Dahl, A. (2021). *Towards a global environment agency: Effective governance for shared ecological risks*. Global Challenges Foundation.

Kempf, I., & Hujo, K. (2022). Why recent crises and SDG implementation demand a new eco-social contract. In A. Antoniades, A. Antonarakis, & I. Kempf (Eds.), *Financial crises, poverty and environmental sustainability: Challenges in the context of the SDGs and Covid-19 recovery* (pp. 171–186). Springer Nature.

Koivurova, T. (2007). International legal avenues to address the plight of victims of climate change: Problems and prospects. *Journal of Environmental Law and Litigation, 22*, 267–299.

Krause, D., Stevis, D., Hujo, K., & Morena, E. (2022). Just transitions for a new eco-social contract: Analysing the relations between welfare regimes and transition pathways. *Transfer: European Review of Labour and Research, 28*(3), 367–382.

Masood, E. (2024). *Is COP29 climate deal a historic breakthrough or let-down? Researchers react?* Accessed October 26, 2024, from https://www.nature.com/articles/d41586-024-03875-4

Mohamed, N., & Huntjens, P. (2023). *Dismantling the ecological divide: Toward a new eco-social contract*. UNRISD Issue Brief 15. https://cdn.unrisd.org/assets/library/briefs/pdf-files/2023/ib15-a-new-contract-with-nature.pdf

McCollum, D., Zhou, W., Bertram, C., de Boer, H.-S., Bosetti, V., Busch, S., Després, J., Drouet, L., Emmerling, J., Fay, M., Fricko, O., Fujimori, S., Gidden, M., Harmsen, M., Huppmann, D., Iyer, G., Krey, V., Kriegler, E., Nicolas, C., Pachauri, S., Parkinson, S., Poblete-Cazenave, M., Rafaj, P., Rao, N., Rozenberg, J., Schmitz, A., Schoepp, W., van Vuuren, D., & Riahi, K. (2018). Energy investment needs for fulfilling the Paris Agreement and achieving the sustainable development goals. *Nature Energy, 3*, 589–599.

Mitchell, Ronald B., Liliana B. Andonova, Mark Axelrod, Jörg Balsiger, Thomas Bernauer, Jessica F. Green, James Hollway, Rakhyun E. Kim, Jean-Frédéric Morin. (2020). What we know (and could know) about international environmental agreements. *Global Environmental Politics 2020, 20*(1), 103–121. https://doi.org/10.1162/glep_a_00544

Narassimhan, E., Gallagher, K. S., Koester, S., & Alejo, J. R. (2018). Carbon pricing in practice: A review of existing emissions trading systems. *Climate Policy, 18*(8), 967–991.

Norton, A., & Greenfield, O. (2023). *Eco-social contracts for the polycrisis: Participatory mechanisms, green deals and a new architecture for just economic transformation.* Green Economy Coalition.

Obura, D. (2023). *Responding to Earth system risk in the global multilateral system.* Global Challenges Foundation.

Ostrom, E. (2009). *A polycentric approach for coping with climate change.* World Bank Policy Research. Working Paper No. 5095: 97–134

Patagonia. (2024a). *Our environmental responsibility programs.* Accessed October 7, 2024, from https://www.patagonia.com/our-responsibility-programs.html

Patagonia. (2024b). *Patagonia action works.* Accessed October 7, 2024, from https://www.patagonia.com/actionworks/about/

Pauw, P., Castro, P., Pickering, J., & Bhasin, S. (2020). Conditional nationally determined contributions in the Paris Agreement: Foothold for equity or Achilles heel? *Climate Policy, 20*(4), 468–484.

Raworth, K. (2018). *Doughnut economics: Seven ways to think like a 21st-century economist.* Chelsea Green Publishing.

Richardson, K., Steffen, W., Lucht, W., Bendtsen, J., Cornell, S. E., Donges, J. F., Drüke, M., Fetzer, I., Bala, G., & von Bloh, W. (2023). Earth beyond six of nine planetary boundaries. *Science Advances, 9*(37), 1–16.

Rockström, J., Gupta, J., Qin, D., et al. (2023). Safe and just Earth system boundaries. *Nature, 619*, 102–111. https://doi.org/10.1038/s41586-023-06083-8

Rockström, J., Steffen, W., Noone, K., Persson, Å., Chapin, S., Lambin, E., Lenton, T. M., Scheffer, M., Folke, C., Schellnhuber, H. J., Nykvist, B., de Wit, C. A., Hughes, T., van der Leeuw, S., Rodhe, H., Sörlin, S., Snyder, P. K., Costanza, R., Svedin, U., Falkenmark, M., Karlberg, L., Corell, R. W., Fabry, V. J., Hansen, J., Walker, B., Liverman, D., Richardson, K., Crutzen, P., & Foley, J. (2009). Planetary boundaries: Exploring the safe operating space for humanity. *Ecology and Society, 14*(2), 32.

Romito, S., Vurro, C., & Pogutz, S. (2024). Joining multi-stakeholder initiatives to fight climate change: The environmental impact of corporate participation in the science based targets initiative. *Business Strategy and the Environment, 33*(4), 2817–2831.

Rousseau, J.-J. (1762). *Du contrat social; ou, Principes du droit politique / On the Social Contract; or, Principles of Political Right.*

Rühs, N., & Jones, A. (2016). The implementation of Earth jurisprudence through substantive constitutional rights of nature. *Sustainability, 8*(2), 174.

Sabin Center for Climate Change Law. (2019). *Hearing on climate change before the Inter-American Commission on Human Rights.* Accessed November 14, 2024, from https://climatecasechart.com/non-us-case/hearing-on-climate-change-before-the-inter-american-commission-on-human-rights/

Sachs, J. D. (2015). *The age of sustainable development.* Columbia University Press.

SBTi. (2024). *SBTi Monitoring Report 2023.* Accessed October 7, 2024, from https://sciencebasedtargets.org/resources/files/SBTiMonitoringReport2023.pdf

Sobion, J. (2023). *Earth trusteeship: A framework for a more effective approach to international environmental law and governance.* Doctoral dissertation, The University of Auckland.

Steffen, W., Richardson, K., Rockström, J., Cornell, S. E., Fetzer, I., Bennett, E. M., Biggs, R., Carpenter, S. R., de Vries, W., de Wit, C. A., Folke, C., Gerten, D., Heinke, J., Mace, G. M., Persson, L. M., Ramanathan, V., Reyers, B., & Sörlin, S. (2015). Planetary boundaries: Guiding human development on a changing planet. *Science, 347*(6223), 736.

Thisted, E., & Thisted, R. V. (2020). The diffusion of carbon taxes and emission trading schemes: The emerging norm of carbon pricing. *Environmental Politics, 29*(5), 804–824.

United Nations. (1982). *World Charter for Nature.* (28 October 1982) Document A/RES/37/7.

United Nations. (2012). *Follow-up to Paragraph 143 on Human Security of the 2005 World Summit Outcome.* (25 October 2012) Document A/RES/66/290.

United Nations. (2015). *Transforming Our World: The 2030 Agenda for Sustainable Development.* (25 September 2015) Document A/RES/70/1.

United Nations. (2019a). *Protests Around the World: Politicians Must Address 'Growing Deficit of Trust'.* Urges Guterres. Accessed November 13, 2024, from https://news.un.org/en/story/2019/10/1050031

United Nations. (2019b). *Opening remarks at press encounter.* Accessed November 13, 2024, from https://www.un.org/sg/en/content/sg/speeches/2019-10-25/remarks-press-encounter

United Nations. (2020a). *Alongside pandemic, world faces 'Triple Planetary Emergency', Secretary-General Tells World Forum for Democracy, Citing Climate, Nature, Pollution Crises.* Accessed November 12, 2024, from https://press.un.org/en/2020/sgsm20422.doc.htm

United Nations. (2020b). *Tackling inequality: A new social contract for a new era.* Accessed November 13, 2024, from https://www.un.org/en/coronavirus/tackling-inequality-new-social-contract-new-era

United Nations. (2021). *Our common agenda - Report by the Secretary-General.* United Nations Publications.

United Nations. (2024). *The pact for the future*. UN General Assembly Resolution A/RES/79/1 of 22 September 2024. Accessed November 24, 2024, from https://documents.un.org/doc/undoc/gen/n24/272/22/pdf/n2427222.pdf

UNDESA (United Nations Department of Economic and Social Affairs). (2021). *Our Common Agenda – Summary of Consultations and Recommendations*. United Nations. Retrieved from https://www.un.org/en/common-agenda

UNDP. (1994). *Human development report 1994*. Oxford University Press.

UNDP. (2022). *New threats to human security in the anthropocene: Demanding greater solidarity*. United Nations Development Programme.

UNEP. (2017). *Environmental rule of law: First global report*. United Nations Environment Programme. Retrieved from https://www.unep.org/resources/environmental-rule-law-first-global-report

UNEP. (2022). *Scoping the Seventh Edition of the Global Environment Outlook: Action for a healthy planet*. UNEP.

UNEP. (2023). *Adaptation Gap Report 2023*. UNEP.

UNEP and International Science Council. (2024). *Navigating New Horizons: A global foresight report on planetary health and human wellbeing*. Kenya.

UNFCCC. (2015). *Paris Agreement*. (12 December 2015) Document 16-1104

UNFCCC. (2022). *What is the triple planetary crisis?*. Accessed November 12, 2024, from https://unfccc.int/news/what-is-the-triple-planetary-crisis

UNFCCC. (2024). *Funds and financial entities*. Accessed October 8, 2024, from https://unfccc.int/process-and-meetings/bodies/funds-and-financial-entities

UNDRR. See: https://www.undrr.org/publication/sendai-framework-disaster-risk-reduction-2015-2030#:~:text=Sendai%20Framework%20for%20Disaster%20Risk,2030%20PDF%2C%201%20MB%20English

UNTFHS. (2016). *Human security handbook*. Accessed November 14, 2024, from https://www.un.org/humansecurity/wp-content/uploads/2017/10/h2.pdf.

UNRISD. (2022). *Overcoming inequalities: Towards a new eco-social contract*. United Nations.

White House. (2024). *FACT SHEET: Biden-Harris Administration Announces New Principles for High-Integrity Voluntary Carbon Markets*. Accessed November 13, 2024, from https://www.whitehouse.gov/briefing-room/statements-releases/2024/05/28/fact-sheet-biden-harris-administration-announces-new-principles-for-high-integrity-voluntary-carbon-markets/

World Commission on Environment and Development. (1987). *Our common future*. University Press.

WRI. (2022). *The state of nationally determined contributions: 2022*. World Resources Institute.

Young, O. (2002). *The institutional dimensions of environmental change: Fit, interplay, and scale*. MIT Press.

Young, O. (2023). *Addressing the grand challenges of planetary governance: The future of the global political order*. Cambridge University Press.

Young, O., Underdal, A., Kanie, N., Kim, R., & Biermann, F. (2017). *Goal setting in the Anthropocene: The ultimate challenge of planetary stewardship*. MIT Press.

9

The European Green Deal: An Eco-Social Contract for Europe?

Davide Sofia and Patrick Huntjens

9.1 Introduction

The European Union's Green Deal (EGD), unveiled in December 2019, marks a transformative step towards addressing the intertwined challenges of ecological sustainability, economic growth, and social equity. Hailed as Europe's "man on the moon moment" (European Commission, 2019), the EGD reflects a growing recognition of the urgent need to reimagine governance frameworks in the face of mounting global crises. It proposes a bold, integrated strategy to achieve climate neutrality—an economy with net-zero greenhouse gas emissions (GHG) by 2050—with a roadmap that aligns environmental sustainability with the principles of fairness, competitiveness, and welfare (European Parliament, 2019). This ambition situates the EGD as a nascent eco-social contract, redefining the relationship between society, the economy, and the environment in Europe (Huntjens, 2021).

D. Sofia (✉)
Friends of Europe, Brussels, Belgium
e-mail: davide.sofia@friendsofeurope.org

P. Huntjens
Inholland University of Applied Sciences, Delft, The Netherlands

© The Author(s) 2025
P. Huntjens et al. (eds.), *Eco-Social Contracts for Sustainable and Just Futures*,
https://doi.org/10.1007/978-3-031-99109-7_9

The concept of an eco-social contract, central to this chapter, builds on historical theories of the social contract that emphasise mutual obligations between the state and its citizens (Rhodes, 2016). However, the EGD extends these principles to include ecological responsibility, acknowledging the interconnectedness of human and natural systems. This expanded framework is essential in addressing the inadequacies of Europe's post-World War II welfare state model, which prioritised economic growth but neglected ecological boundaries, leading to unsustainable resource consumption and growing social inequities (Steffen et al., 2015; OXFAM, 2023).

In this context, the EGD seeks to bridge the gaps in governance by embedding sustainability into EU policies across sectors, including energy, agriculture, and trade. It introduces mechanisms like the Just Transition Mechanism to ensure equitable outcomes for communities most affected by the green transition (Filipović et al., 2022; Sandmann et al., 2024). However, as this chapter highlights, the EGD's implementation faces systemic barriers such as policy fragmentation, insufficient funding, and resistance from entrenched political and economic interests (Leonard et al., 2021; Bocquillon, 2024).

This chapter examines the political origins and governance structures underpinning the EGD. It explores how it addresses social equity and ecological sustainability, drawing on interdisciplinary perspectives and theoretical models of eco-social contracts (Dobson & Eckersley, 2006; Huntjens, 2021; Huntjens & Kemp, 2022; Kempf & Hujo, 2022). Ultimately, the chapter argues that the EGD must integrate stronger just transition policies and adopt a holistic approach that aligns with the principle of common but differentiated responsibilities (CBDR), first enshrined in Principle 7 of the Rio Declaration at the first Earth Summit in 1992 and redefined in the Paris Agreement, to realise its potential as an eco-social contract (IPCC, 2018; Huntjens, 2021). This chapter contributes to the discourse on transformative governance in the Anthropocene by critically analysing the EGD's framework, challenges, and pathways for improvement.

9.2   The European Green Deal: A Step Towards a New Eco-social Contract

In December 2019, Ursula von der Leyen, President of the European Commission, unveiled the European Green Deal (EGD) (European Commission, 2019). This announcement followed the European Parliament's declaration of a climate and environmental emergency, underscoring the urgency of global warming threats. Backed by scientific evidence from the Intergovernmental Panel on Climate Change (IPCC), the resolution called for comprehensive action to reduce carbon emissions while preserving Europe's social model—fairness, competitiveness, and high welfare standards (European Parliament, 2019). From the onset, the EGD was indeed presented not simply as an environmental plan but as a "new growth strategy [...] to transform the EU into a fair and prosperous society, with a modern, resource-efficient and competitive economy" that "must put people first, and pay attention to the regions, industries and workers who will face the greatest challenges" (European Commission, 2019).

A crucial aspect of the EGD is the systematic evaluation of environmental impacts in EU laws and budget proposals. This integrated approach has reshaped policymaking across sectors such as agriculture, energy, and infrastructure, aligning investment policies to limit global warming to 1.5 °C above pre-industrial levels (IPCC, 2018). However, recent reports from the European Environment Agency (European Environment Agency, 2024a) and United Nations Environment Programme (UNEP, 2024a) warn that current efforts to achieve the 1.5 °C target are insufficient, urging countries to accelerate their commitments to avoid dangerous climate tipping points.

The EGD pioneers a governance framework that combines social and economic progress with ecological sustainability. Scholars and social movements see it as a step towards an "eco-social contract", addressing gaps in contemporary governance. This concept integrates human well-being, economic systems, and ecological interdependence into a cohesive framework for the twenty-first century (Dobson & Eckersley, 2006; O'Brien et al., 2009; Huntjens, 2021).

9.2.1 A Post-WWII Legacy and Evolving Social Contracts

The classical social contract theory evolved significantly during the Enlightenment. This period saw the development of the dominant social contract frameworks that view governance as an implicit agreement where rulers provide protection and good governance in exchange for citizen allegiance. These Enlightenment formulations are not the only conceptual approaches to social contracts. Eco-social contracts, for instance, integrate environmental and social dimensions into governance structures.

Post-World War II Europe built on the classical social contract framework, creating welfare states to balance labour and capital interests, guided by Keynesian economics and principles of solidarity to develop social rights like healthcare, education, and job security (Rhodes, 2016). During the mid-twentieth century, there has been a drastic increase in resource use for growth and substantial changes in the features of the planet's ecosystem. This "great acceleration" thus took a toll on the planet's biocapacity, disproportionately impacting lower-income groups (Steffen et al., 2015; OXFAM, 2023; Global Footprint Network, 2024).

Today, mounting crises—climate change, inequality, and societal disillusionment—highlight the limitations of this classical social contract framework that still depends heavily on the extraction of raw materials, necessitating the integration of environmental and social dimensions into new social contract frameworks.

9.2.2 Symptoms of a Broken Social Contract in Europe?

Discontent with inadequate climate action has led to the rise of various movements advocating for stronger environmental policies. Fridays for Future (FFF), a youth-led initiative, has called on governments to align policies with scientific recommendations to mitigate climate change, emphasising the need for urgent and evidence-based action (UN News, 2019). Resistance to fossil fuel projects has emerged from different sectors, including local and Indigenous communities, highlighting concerns over land rights, economic justice, and environmental degradation (Temper et al., 2020). These parallel but distinct movements reflect growing demands for accountability and systemic change in climate governance, illustrating a crisis of trust in existing policy frameworks.

Environmental sustainability features alongside migration, security, and geopolitical concerns as a top priority in the latest Eurobarometer surveys (European Commission, 2024). Most Europeans view climate adaptation as a national priority, and many agree that a sustainable, carbon-neutral transition must also address inequalities (Bocquillon, 2024). Survey and focus group research by the citizen engagement platform, Debating Europe, reinforce this view; in one of the focus groups conducted in Poland, a participant stressed the importance of framing the sustainability transition:

By framing the fight against climate change in terms of introducing modern solutions, investments and educating the public, we are just giving the impression that this is progressive, modern and that it is a class problem: the rich solve, the poor don't get it. (Friends of Europe, 2024a)

Environmental sustainability remains a multifaceted topic; the criticism towards current climate policy relates, for instance, to the lack of appropriate political will and courage, the dangers of greenwashing, the high cost of sustainable products, social inequalities, public transport and car dependency, and localised environmental issues that affect citizens' health (Friends of Europe, 2024a).

9.3   Towards an Eco-social Contract

The convergence of global crises—financial instability, climate change, pandemics, and inflation—necessitates a shift from anthropocentrism, a viewpoint containing elements of human supremacy which places humankind at the centre of existence, at the expense of other species (Kopnina et al., 2018), to an eco-social framework, where society is acknowledged as a social-ecological system. In such a framework, humankind is placed in the larger terrestrial ecosystem and pursues self-preservation and well-being by rejecting the concepts of unlimited economic growth, overconsumption, and over-individualisation (Huntjens, 2021). Providing a roadmap are theoretical models like eco-welfare states, where environmental protection and citizen well-being (for instance, through social protection and investments) are defined as interconnected objectives, and eco-social contracts, which link environmental sustainability with equity and institutional accountability (UNRISD, 2016; Huntjens, 2019, 2021; Kempf & Hujo, 2022; Norton & Greenfield, 2023; Hasanaj, 2023), including recent approaches like the Transformation Flower Approach (Huntjens & Kemp, 2025), which offer practical tools for aligning governance innovation with "multiple value creation".

9.3.1 Taking Stock

In Europe, "a good and healthy life in 2050 within our planet's ecological boundary" is a core component of environmental policy (Decision (EU) 1386/2013, 2013). The EGD has presented an all-encompassing strategy to introduce a large body of legislation, focusing on halting climate change, transforming the energy system by shifting to low-carbon sources, improving circularity and resource efficiency, defining an industrial strategy for growth and innovation, strengthening the agri-food sector, protecting biodiversity, and restoring ecosystems.

At the time of its announcement, the European Commission was facing criticism for its lack of capacity to meet the objectives of the Seventh Environment Action Programme. Civil society organisations such as the European Environmental Bureau (EEB) and Climate Action Network Europe (CAN Europe) and multi-stakeholder platforms of policy experts such as Think2030 were all demanding the European Union to step up its commitment to foster integration across its environmental, economic, and security policy (Baldock & Charveriat, 2018; Charveriat, 2023). In 2018, the European Environment Agency released a report to highlight that the three priority action areas of the seventh Environment Action Programme[1] were far away from being met, especially regarding the Union's natural capital and the reduction of environment-related pressures and risks to health and well-being (European Environment Agency, 2018).

The emergence of the EGD also stemmed from international dynamics, such as the United States (US) announcement of withdrawal from the Paris Agreement and concerns about the international competitiveness of goods and services aligned with environmental objectives. With the EGD, the Commission strived to become a global standard setter (von der Leyen, 2019b) and make a significant step forward compared to previous institutional cycles to address the substantiated climate emergency and the Paris Agreement commitments. Coinciding with the outbreak and consequences of the COVID-19 pandemic, the Commission took the opportunity to implement a new public EU-level debt programme to link the green transition with the economic recovery from the crisis (Charveriat, 2023; Eckert, 2021; Christie et al., 2021; Dorrucci & Freier, 2023).

The broader agenda of the EGD has managed to drive progress on air, water, and soil pollution. Legislation on chemicals has been modernised, and the use of many toxic or hazardous substances has been restricted. Thanks to its zero-pollution ambition, the EU can boast high standards in water quality compared to the rest of the world. It has also progressed on air pollution and industrial emissions, being on track to reduce premature deaths due to fine

---

[1] The three priority action areas of the seventh Environment Action Programme were (1) protecting, conserving, and enhancing fertile soil, productive land and seas, fresh water and clean air, as well as the biodiversity that supports this natural capital; (2) transforming the EU into a resource-efficient, low-carbon economy; and (3) addressing the challenges to human health and well-being, such as air and water pollution, excessive noise, and chemicals (Decision (EU) 1386/2013, 2013).

particulate matter by 55% by 2030 (Decision (EU), 2022/591 2022). Thanks to Natura 2000 sites and other complementary designations for the protection of nature, 26.1% of the EU's land is a protected area (European Environment Agency, 2024b). In addition, The European Climate Law sets binding targets for the European member states to reach the level of net zero greenhouse gas (GHG) emissions by 2050, meaning countries will need to drastically reduce their emissions and implement carbon capture, utilisation, and storage (CCUS) methods to absorb remaining carbon dioxide from the atmosphere.

However, many problems remain, such as water scarcity affecting almost 33% of the EU's population, the pressure on biodiversity caused by land degradation and destruction, and disproportionate exposure to climate change based on socio-economic status (Decision (EU), 2022/591 2022).

These issues must be tackled in a structured way, improving the efforts and instruments for EU-wide convergence. To solve these problems and achieve the goals set in environmental policy, the EU will need to make far-reaching changes in its production and consumption systems. Even after adopting a circular economy action plan in 2020, which focuses on material flows, resource productivity, and waste reduction and management, indicators from the European Environment Agency show that the EU must increase its efforts in using recycled material, decreasing its unsustainable material footprint. While scoring an improved landfill rate, much is still to be achieved in waste reduction (European Environment Agency, 2024c).

9.3.2 Financial and Technical Challenges for the EGD

The EGD is facing considerable challenges in its implementation and enforcement, which threaten to undermine its objectives. A key issue lies in legislative delays and inconsistencies. Several central proposals have encountered significant delays due to prolonged political negotiations. Moreover, the implementation of EGD measures is uneven across member states, as differing economic capacities and political will hinder consistent enforcement. This variation in adoption risks creating a fragmented approach to climate action across the EU.

Another pressing challenge is the insufficiency of funding and resources to address regional asymmetries. Although the Just Transition Mechanism provides financial support of EUR 100 billion, targeting specifically the regions where economies are heavily carbon intensive or dependent on fossil fuels, the available funding falls short of the substantial resources required to achieve the fairness element integrated into the EGD's climate and sustainability goals. Local governments and smaller stakeholders, in particular, often lack the necessary technical expertise and financial capacity to implement the required measures effectively. This resource disparity further exacerbates the uneven implementation of EGD initiatives.

Climate investments have also been an integral part of NextGenerationEU (NGEU), the revolutionary recovery package introduced by the Commission to respond to the economic consequences of the COVID-19 pandemic. One of the substantial risks that could hamper the NGEU's target to boost the EU's GDP by 1.5% and accelerate the green transition is related to the absorption rate, which is the percentage of the funds committed by the EU that are actually paid to member states. While the Recovery and Resilience Facility (RRF), the key funding instrument of the NGEU, is supposed to reach the full 100% absorption rate by 2026, the latest scoreboard shows that as of February 2025, only 55% of the grants (EUR 197.5 billion) and 37% of the loans (EUR 108.7 billion) available have been disbursed, and 28% of the milestones and targets have been achieved. In addition, asymmetry of absorption across member states remains a challenge (Loi et al., 2025).

Connected to insufficient funding and resources, private finance also plays a significant role in reaching the investments required to obtain climate neutrality targets. Indicators from the European Central Bank show that currently sustainable finance is heavily concentrated in a few Central and Western European countries (European Central Bank, 2024). The fact that a significant number of countries in the European Union have only recently started to access the market for green financial instruments or have yet to access it represents another asymmetry that the European Union will need to consider in advancing its sustainability agenda.

Monitoring and accountability also pose significant hurdles. Mechanisms designed to track compliance and measure the impact of EGD policies—such as emissions reductions and pesticide usage—are not uniformly robust across member states. Weak enforcement of critical policies, including carbon pricing and energy efficiency regulations, further hampers progress towards the EGD's targets (Gherardi et al., 2021). Without strengthened monitoring frameworks and more rigorous enforcement, the transformative potential of the EGD remains at risk (Villiers, 2022).

9.3.3 Political Challenges Surrounding the EGD

The political context surrounding the EGD reveals a landscape fraught with contestation and challenges, which significantly constrain its transformative potential. Resistance from stakeholders in carbon-intensive industries and member states with fossil-fuel-dependent economies highlights the tension between climate ambitions and entrenched economic interests. This resistance, coupled with lobbying from powerful sectors such as agriculture, energy, and transport, often dilutes the strength of EGD proposals and slows their adoption (Haas & Sander, 2020; Eckert & Kovalevska, 2021).

The calls for industrial competitiveness to be a key priority of the new institutional cycle of the European Commission and increased simplification in sustainability legislation and reporting have captured significant attention (Antwerp Declaration, 2024). Both points are present in the European Commission's programme and mandate, with competitiveness being the cornerstone of a new "European Prosperity Plan" (von der Leyen, 2024) and an "Omnibus simplification package" being one of the first actions of the new Commission (European Commission, 2025). This has raised concerns about deregulation and backtracking on legislation such as the EU Taxonomy Regulation, the Corporate Sustainability Reporting Directive (CSRD), and the Corporate Sustainability Due Diligence Directive (CSDDD) (WWF EU, 2024).

Divergent national interests within the EU further complicate the implementation of the EGD. Economic disparities among member states lead to varying levels of commitment, with wealthier nations more readily embracing ambitious targets while others reliant on fossil fuels or intensive agriculture push back against stringent measures. These divisions reflect broader structural inequalities, where economic vulnerabilities shape differing capacities to transition to sustainable practices and lead to different priorities for national policymaking.

Populist and Eurosceptic movements add another layer of resistance to the EGD. The rise of populism in some member states fosters opposition to centralised EU initiatives, including the EGD, as leaders frame these policies as threats to national sovereignty. Public scepticism and misinformation regarding the economic costs of the green transition increase this resistance, particularly during energy crises or economic downturns when citizens prioritise immediate economic relief over long-term sustainability goals.

An example of the resistance posed by conservative groups and policymakers can be found in the adoption process of the Regulation on nature restoration, which was approved in June 2024 after strenuous negotiations (Aubert & Underwood, 2024). This Regulation, addressing commitments in the EGD and the EU Biodiversity Strategy, sets the objective of restoring at least 20% of land and sea areas in Europe by 2030 and all other ecosystems requiring restoration by 2050. In addition, it requires member states to draft national restoration plans where they identify all the measures necessary to meet their obligations (Regulation (EU) 2024/1991, 2024). During the negotiations, intense criticism from conservative groups and increasing pressure from farmers' protests led to the approval of a much more diluted text. For example, the restoration of 30% of drained peatland for farmers and private landowners has become voluntary, and until 2030, member states should give priority to improving the conditions of habitats already located in Natura 2000 sites (COM/2022/304 final, 2022; Manzanaro, 2024).

The shifting geopolitical landscape is also affecting the progress on the EGD. Russia's war against Ukraine has brought Europe's energy security concerns to the forefront, highlighting the difficulty of ensuring stable energy supplies due to vulnerabilities caused by the import dependence on fossil fuels. Additionally, the EU's reliance on global supply chains for critical raw materials, minerals and technologies necessary for the green transition exposes it to geopolitical risks further straining efforts to achieve its climate ambitions (Hool et al., 2023). The result of the elections in the United States has brought another point of disruption, as President Trump has restarted the process of withdrawing from the Paris Agreement and limited the United States' financial contributions to support other countries' efforts in mitigating and adapting to climate change (Pullins & Knijnenburg, 2025).

Together, these political and geopolitical dynamics underscore the complexity of navigating a unified pathway towards sustainability within the EU, displaying substantial obstacles that hamper the realisation of the EGD's vision and potential as a new eco-social contract for Europe.

Social equity issues represent another critical challenge. Critics argue that the EGD risks exacerbating social inequalities if policies are not designed in a fair and just manner and it does not adequately address the uneven distribution of transition burdens. Vulnerable communities are particularly at risk, as rising energy costs linked to carbon pricing and renewable energy investments disproportionately impact low-income households. An example of public backlash can be found in movements like France's "Gilets Jaunes", incorrectly described as an anti-environmental movement but, in reality, a movement sparked by a flat fuel tax proposed by the French government which would overwhelmingly affect lower and middle classes in rural areas (Martin, 2019). This case underscores the political and social risks of implementing climate or environmental policies and regulations which end up being unfair or inequitable.

9.4 Pathways to Overcome the Challenges for the EGD

Addressing these challenges to implement the EGD as a European eco-social contract requires targeted pathways for reform at multiple levels. Strengthening policy coherence is essential, with efforts to align EGD initiatives with national energy and industrial policies to minimise fragmentation. The holistic nature of the EGD offers a chance to integrate environmental and social justice goals, aligning economic policy with sustainable development objectives (Stevis & Felli, 2020).

Investing in a just transition is equally important, particularly by expanding funding for regions and communities disproportionately affected by the green transition. Enhancing public engagement through participatory policymaking and job creation in green sectors can help build public support and legitimacy. Emphasising democratic decision-making and citizen involvement can further strengthen the EGD's effectiveness (Dryzek & Pickering, 2019).

At the same time, the EGD must also expand its scope beyond the EU, acknowledging the global dimensions of just transitions, particularly for developing economies reliant on resource exports (Raworth, 2018). By positioning itself as a global leader in climate action, the goal of the EU would be that of inspiring similar frameworks internationally and setting a precedent for sustainable development (von der Leyen, 2019a, b). However, it must be acknowledged that even if currently the European Union represents only 6% of the world's GHG emissions (European Commission et al., 2024), this percentage does not take into account the historical contribution of European countries to global GHG emissions nor its high consumption levels affecting GHG emissions outside of its borders.

Following the recent European Parliament elections, and national election results in countries such as France, Germany, and the United States, a new and more environmentally conservative political reality is now taking shape in Europe. Ultimately, the rise of conservative political groups and backlash from various economic sectors, such as agri-food and transportation, has led to the European Commission drafting an agenda that has adapted to the difficulty of "selling sustainability" to more nationalist and conservative business leaders and citizens (Naldzhiev, 2024; Moore & Haertner, 2024). This shift is exemplified by the echo generated among policymakers, business actors, and analysts to the report on the future of

European competitiveness from former Italian prime minister and president of the European Central Bank, Mario Draghi. The topic of industrial
competitiveness has taken centre stage in the new Clean Industrial Deal,
which considers decarbonisation as an opportunity for Europe to define
key drivers for growth, focusing on the competitive advantages that the
continent has on innovative clean technologies, reducing strategic dependencies from other global actors, and creating an equal level playing field
for the business sector (COM(2025) 85 final, 2025; Draghi, 2024). While
the Clean Industrial Deal stresses the EU's dedication to its climate goals,
its framing heightens the risk that decarbonisation remains on a lower priority level due to increasing global competition, security risks, and the
impact of higher energy prices in Europe. What then are the pathways that
could keep the region on track to deliver the EGD?

9.4.1 Multilevel Governance

The EGD exemplifies the EU's capacity to leverage its unique multilevel governance structure to drive transformative change across diverse member states.
This governance framework, characterised by the interplay of supranational,
national, and local authorities, allows for centralised policy direction while
fostering decentralised implementation. The EGD's objectives, such as achieving climate neutrality and shifting to a circular economy, can thus be obtained
through tailored measures thanks to the National Energy and Climate Plans
(European Commission, 2021).

The multilevel governance approach is supposed to balance uniformity
with subsidiarity, addressing the diverse economic capacities and political priorities across the Union. While the just transition has become a critical point
in the public discourse around the green transition, the lack of specification
and common framing of the concept of justice and what makes a transition
"fair" across the EU can sometimes be a double-edged sword. Different interpretations and lack of trust between various actors (policymakers, private sector, civil society organisations, etc.) and governance levels (supranational,
national, local) create frustration and do not always allow the development of
political narratives that connect, for instance, local energy provisions to the
"bigger picture" of the EU-wide energy transition (Sandmann et al., 2024).

Enhanced participatory governance and localised action are critical to
ensuring the EGD delivers cohesive and just outcomes (Dryzek & Pickering,
2019). Taking the energy sector as an example, the role of local governments,
at times neglected in existing studies around the EGD, is essential and can be

more effective than the supranational and national level in facilitating green energy demand and production, energy efficiency, and housing solutions for local private sector actors who in most cases are small and medium enterprises (SMEs) (Pūķis et al., 2023). In addition, local governments can better stimulate green decisions by local private sector actors, a relevant point considering that 99% of European businesses are SMEs, with a strong understanding of local communities, preferences, and networks.

9.4.2 Greater Focus on Just Transition Elements

The EGD highlights the importance of ensuring a just transition as an integral component of its ambitious agenda to achieve climate neutrality by 2050. The EGD incorporates instruments such as the Just Transition Mechanism to provide financial support to regions and industries most affected by the shift away from fossil fuels. However, the lack of a strong political will and appropriate funding from national governments negatively affects the challenges faced by disadvantaged populations, particularly those in economically fragile regions and reliant on carbon-intensive industries.

Central to the just transition is the principle of common but differentiated responsibility, as outlined in the Paris Agreement, which recognises that countries have varying capacities and starting points in their journey towards sustainability (IPCC, 2018). This principle underscores the need for targeted policies and resources to address structural inequalities, both within and among EU member states. For example, enhanced social measures accompanying environmental taxes, such as energy subsidies or income supports, are essential to prevent regressive impacts on low-income regions and households. Friends of Europe's "10 Policy Choices for a Renewed Social Contract" advocates for the redirection of energy taxation revenues towards investments in low-carbon and energy-efficient goods or subsidies through lump sum transfers and supports its integration for the ongoing negotiations to reform the Energy Taxation Directive (Friends of Europe, 2024b).

Job creation in green sectors and investment in reskilling programmes for workers displaced by the transition are pivotal to ensuring widespread benefits from the EGD. Programmes that promote digital and ecological skills, along with green job initiatives, can help build resilience among affected communities while fostering economic innovation (Stevis & Felli, 2020). In addition to protecting local communities, reskilling and upskilling are also relevant to reducing inequalities among the various socio-economic groups. Indeed, women are generally under-represented in green jobs (Causa et al., 2024), which are also usually more frequent in the professions that women are under-represented in, such as science, technology, and engineering innovations that are core to green transitions. Geographical perspectives are equally important as rural areas are characterised by a higher frequency of high-polluting jobs, such as manufacturing, mining, and mineral processing (Causa et al., 2024). This means that just transition policies must be designed to target these existing structural differences to avoid widening current inequalities.

While the EGD provides a framework for just transitions, gaps remain in its implementation. Strengthening the Just Transition Mechanism with increased funding and robust accountability measures will be critical. Furthermore, aligning the EGD's goals with broader international policy and human rights frameworks, such as the Sustainable Development Goals (SDGs), can enhance the global equity dimensions of Europe's green transition (UNRISD, 2016).

In summary, embedding stronger just transition elements within the EGD is essential to achieving an inclusive and equitable transformation. By addressing social and economic disparities, the EU can ensure that its ambitious environmental goals are met without compromising the well-being of its most vulnerable populations and regions.

9.4.3 Specific Policy Areas Relevant to a New Eco-social Contract for Europe

The EGD has presented the opportunity to embed the concept of an eco-social contract within European policymaking, aiming to balance ecological sustainability, economic resilience, and social equity. Central to this vision are policy proposals addressing critical areas such as climate mitigation and adaptation, energy efficiency, sustainable agriculture, and inclusive growth.

A priority for the EGD is accelerating carbon reduction through immediate and sustained measures to meet global warming targets and, considering the current evolving geopolitical landscape, also to reduce vulnerabilities caused by dependence. This includes expanding renewable energy deployment via initiatives like REPowerEU (European Commission, 2022), improving energy efficiency through better building practices and digital technologies, and reducing overall energy consumption (European Commission, 2021). Addressing fiscal barriers, the reform to the EU Energy Taxation Directive (ETD) could incentivise clean energy adoption while incorporating social measures to shield low-income households from the regressive impacts of environmental taxes.

Digitalisation also plays a vital role in supporting the green economy. With the increasing relevance of artificial intelligence (AI), there will be the potential to drastically accelerate innovation and improve modelling for climate change and nature action. At the same time, AI can also be very problematic for the environment due to the high necessity of raw materials, production of electronic waste, and high energy usage (UNEP, 2024b). In addition, the alignment between climate change policy and AI policy will need to be carefully addressed to understand how to design transparency and accountability on GHG emissions and how to avoid exacerbating social inequalities by, for instance, further widening the existing digital divide (Kaack et al., 2022).

The Farm to Fork Strategy was announced as another critical component of the EGD, designed to create sustainable and resilient food systems. This strategy aimed to cut chemical pesticide use by 50%, reduce nutrient losses by 50%, and expand organic farming to 25% of agricultural land by 2030. The end goal of the Farm to Fork Strategy was to create environmental, social, and health benefits. Promoting healthier, sustainable diets and harmonised food labelling supports consumers in making informed choices, while measures to halve food waste and improve food loss management strengthen supply chain efficiency and equity. Supporting farmers through fair prices and fostering cooperation among stakeholders ensures equitable governance in food production (Yébenes, 2024).

While at the onset the strategy managed to complete some of its deliverables, such as the action plan on organic farming, or the recommendations to EU countries for their strategic plans as part of the Common Agricultural Policy (Šajn, 2025), farmers' discontent across the European Union and the urgency to address the effect of climate change on food systems led to the announcement of a Strategic Dialogue on the Future of Agriculture. This multi-stakeholder dialogue was initiated in January 2024 and culminated in the release of a final report in September 2024 (Strohschneider et al., 2024). The result of the dialogue has been seen as a step in the right direction by organisations such as the European Environmental Bureau (EEB), Slow Food and Agroecology Europe, for instance, regarding increased funding for more sustainable farming practices and promoting targeted income support in reviewing the Common Agricultural Policy (CAP) (Ibbott, 2024; Agroecology Europe, 2024). In the new European institutional cycle (2024–2029), the European Commission has followed up on the Strategic Dialogue with the release of the Vision for Agriculture and Food, which provides a roadmap for the upcoming years and, in particular, will be an important element in the upcoming reform of the Common Agricultural Policy for the new 7-year European budget (the Multiannual Financial Framework 2028–2034).

The EGD also emphasises research and innovation to promote sustainable agriculture technologies, alternative proteins, and improved resource efficiency. Aligning trade policies with sustainability goals and fostering global partnerships underscore the EU's commitment to international standards for sustainable food systems. Revising the Common Agricultural Policy (CAP) to support green farming practices and ensuring compliance with sustainability measures solidify the EGD's role as a driver of systemic change.

9.5   Conclusion

From its inception, the EGD has represented a pivotal step towards addressing the complex and interconnected challenges of ecological sustainability, economic resilience, and social equity. The EGD positioned itself as more than a policy package—by proposing such an all-encompassing agenda, it provided a transformative framework for an eco-social contract that redefines the relationships between governments, citizens, industries, and the environment (European Commission, 2019; Huntjens, 2021). However, realising the full potential of this ambitious vision requires overcoming systemic barriers, addressing the entrenched economic interests in highly polluting sectors such as the fossil fuel industry, and integrating a more inclusive and equitable approach to policy and governance processes.

While highly innovative in its transformative potential, the European institutional cycle of the EGD has been impacted by severe economic shocks and other challenges. The combination of competing political priorities, policy fragmentation, resistance from entrenched interests, insufficient funding, top-down policymaking, and widening social inequalities negatively contributed to the public backing of the EGD, as exemplified by the increased support received by conservatives and anti-environmental political groups in recent elections in Europe. To maintain the core ambitions of the EGD, a new narrative for the multifaceted benefits of accelerating a just decarbonisation pathway must be supported to realise a cohesive and holistic governance model that bridges the gap between ambition and action, stressing the costs of inaction on climate change and reinforcing that vulnerable communities will be most affected by them.

The concept of an eco-social contract underscores the need for a change in basic assumptions in governance, one that integrates ecological limits and social justice into the social and economic fabric of societies. An eco-social contract also has to ensure that the negative effect of climate change on human health, environmental degradation, and natural disasters will not further affect vulnerable communities, both within the EU and globally (Klein, 2014; Kempf & Hujo, 2022). Looking forward, the EGD remains an example with great potential to inspire a broader reimagining of regional and global governance. However, this leadership must be demonstrated through actions that can respond to geopolitical challenges, the entrenched economic interests in highly polluting sectors, and supply chain dependencies that threaten to undermine progress (UNEP, 2024a; European Environment Agency, 2024a).

In the Anthropocene, the convergence of environmental, social, and economic crises necessitates innovative governance and economic models that prioritise sustainability and justice. The EGD, as an evolving eco-social contract, has offered pathways to navigate these challenges. By committing to inclusive policies, equitable resource distribution, and collaborative international action, the EU can not only meet its climate goals but also lay the groundwork for a sustainable and just future for all.

References

Agroecology Europe. (2024). *Joint press release: Slow food and agroecology Europe on the strategic dialogue on the future of agriculture and food: For greater ambition.* https://www.agroecology-europe.org/joint-press-release-slow-food-and-agroecology-europe-on-the-strategic-dialogue-on-the-future-of-agriculture-and-food-for-greater-ambition/

Antwerp Declaration for a European Industrial Deal. (2024). Accessed December 15, 2024, from https://cms.antwerp-declaration.eu/uploads/declaration.pdf

Aubert, G., & Underwood, E. (2024). *The nature restoration law—A hard-fought victory for biodiversity and society.* Institute for European Environmental Policy. https://ieep.eu/publications/the-nature-restoration-law-a-hard-fought-victory-for-biodiversity-and-society/

Baldock, D., & Charveriat, C. (2018). *30x30 actions for a sustainable Europe. #Think2030 action plan.* https://think2030.eu/think-timeline/uploads/2022/07/30x30-Actions-for-a-Sustainable-Europe.pdf

Bocquillon, P. (2024). *Climate and energy transitions in times of environmental backlash?: The EU 'Green Deal' from adoption to implementation.* JCMS-Journal of Common Market Studies (2024).

Causa, O., Nguyen, M., & Soldani, E. (2024). *Lost in the green transition? Measurement and stylized facts.* OECD Economics Department Working Papers, No. 1796, OECD Publishing. https://doi.org/10.1787/dce1d5fe-en

Charveriat, C. (2023). *The green deal: Origins and evolution.* https://geopolitique.eu/en/articles/the-green-deal-origins-and-evolution/

Christie, R., Claeys, G., & Weil, P. (2021). *Next generation EU borrowing: A first assessment.* Bruegel. https://www.bruegel.org/policy-brief/next-generation-eu-borrowing-first-assessment

Communication from the Commission to the European Parliament, The Council, The European Economic and Social Committee and the Committee of the Regions. (2025). The Clean Industrial Deal: A joint roadmap for competitiveness and decarbonisation. COM/2025/85 final. https://eur-lex.europa.eu/legal-content/EN/TXT/?uri=CELEX%3A52025DC0085

Decision (EU) 2022/591 of the European Parliament and of the Council of 6 April 2022 on a General Union Environment Action Programme to 2030, OJ L 114, 12.4.2022. pp. 22–36. https://eur-lex.europa.eu/legal-content/EN/TXT/?uri=CELEX:32022D0591

Decision (EU) 1386/2013 of the European Parliament and of the Council of 20 November 2013 on a General Union Environment Action Programme to 2020 'Living well, within the limits of our planet'. https://eur-lex.europa.eu/legal-content/EN/TXT/?uri=celex%3A32013D1386

Dobson, A., & Eckersley, R. (Eds.). (2006). *Political theory and the ecological challenge.* Cambridge University Press.

Dorrucci, E., & Freier, M. (2023, February 15). *The opportunity Europe should not waste.* https://www.ecb.europa.eu/press/blog/date/2023/html/ecb.blog.230215-4aad7004cf.en.html

Draghi, M. (2024). *The future of European competitiveness.* https://commission.europa.eu/topics/eu-competitiveness/draghi-report_en

Dryzek, J. S., & Pickering, J. (2019). *The politics of the Anthropocene.* Oxford University Press.

Eckert, S. (2021). The European green deal and the EU's regulatory power in times of crisis. *Journal of Common Market Studies, 59,* 81–91.

Eckert, E., & Kovalevska, O. (2021). Sustainability in the European Union: Analyzing the discourse of the European green deal. *Journal of Risk and Financial Management, 14*(2), 80.

European Central Bank. (2024). *Indicators on sustainable finance.* Accessed December 15, 2024, from https://www.ecb.europa.eu/stats/all-key-statistics/horizontal-indicators/sustainability-indicators/data/html/ecb.climate_indicators_sustainable_finance.en.html

European Commission. (2019). *Press remarks by president von Der Leyen on the occasion of the adoption of the European Green Deal.* Communication. https://ec.europa.eu/commission/presscorner/detail/en/speech_19_6749

European Commission. (2021). *Fit for 55: Delivering the EU's 2030 climate target on the way to climate neutrality.* https://eur-lex.europa.eu/legal-content/EN/TXT/HTML/?uri=CELEX%3A52021DC0550. COM (2021) 550 final.

European Commission. (2022). *REPowerEU plan.* https://eur-lex.europa.eu/legal-content/EN/TXT/?uri=CELEX%3A52022DC0230. COM/2022/230 final.

European Commission. (2024). *Flash Eurobarometer 550. (2024). EU challenges and priorities. Eurobarometer survey: Public opinion in the European Union.* https://europa.eu/eurobarometer/surveys/detail/3232

European Commission. (2025). Proposal for a Directive of the European Parliament and of the Council amending Directives 2006/43/EC, 2013/34/EU, (EU) 2022/2464 and (EU) 2024/1760 as regards certain corporate sustainability reporting and due diligence requirements. COM (2025) 81 final. https://finance.ec.europa.eu/document/download/161070f0-aca7-4b44-b20a-52bd879575bc_en?filename=proposal-directive-amending-accounting-audit-csrd-csddd-directives_en.pdf

European Commission, Joint Research Centre, Crippa, M., Guizzardi, D., Pagani, F., Banja, M., Muntean, M., Schaaf, E., Monforti-Ferrario, F., Becker, W. E., Quadrelli, R., Risquez Martin, A., Taghavi-Moharamli, P., Köykkä, J., Grassi, G., Rossi, S., Melo, J., Oom, D., Branco, A., San-Miguel, J., Manca, G., Pisoni, E., Vignati, E., & Pekar, F. (2024). *GHG emissions of all world countries.* Publications Office of the European Union. https://data.europa.eu/doi/10.2760/4002897 JRC138862.

European Environment Agency. (2018). *Environmental indicator report 2018 - In support to the monitoring of the 7th Environment Action Programme.* EEA Report 19/2018. https://www.eea.europa.eu/en/analysis/publications/environmental-indicator-report-2018

European Environment Agency. (2024a). *Trends and projections in Europe 2024.* EEA Report 11/2024. https://www.eea.europa.eu/en/analysis/publications/trends-and-projections-in-europe-2024

European Environment Agency. (2024b). *Terrestrial protected areas in Europe.* https://www.eea.europa.eu/en/analysis/indicators/terrestrial-protected-areas-in-europe?activeAccordion=ecdb3bcf-bbe9-4978-b5cf-0b136399d9f8

European Environment Agency. (2024c). *Accelerating the circular economy in Europe.* State and outlook 2024. https://www.eea.europa.eu/en/analysis/publications/accelerating-the-circular-economy

European Parliament. (2019). *Climate and environmental emergency.* https://www.europarl.europa.eu/doceo/document/TA-9-2019-0078_EN.pdf

Filipović, S., Lior, N., & Radovanović, M. (2022). The green deal—just transition and sustainable development goals nexus. *Renewable and Sustainable Energy Reviews, 168*(2022), 112759.

Friends of Europe. (2024a). *2024 voices – Citizens speak up!.* https://debatingeurope.eu/wp-content/uploads/2024/04/2024Voices_report_layout_rev.pdf

Friends of Europe. (2024b). *10 policy choices for a renewed social contract for Europe.* https://www.friendsofeurope.org/wp/wp-content/uploads/2024/03/10-policy-choices-for-a-Renewed-Social-Contract.pdf

Gherardi, L., Linsalata, A. M., Gagliardo, E. D., & Orelli, R. L. (2021). Accountability and reporting for sustainability and public value: Challenges in the public sector. *Sustainability, 13*(3), 1097.

Global Footprint Network. (2024). Accessed December 15, 2024, from www.over-shootday.org

Haas, T., & Sander, H. (2020). Decarbonizing transport in the European Union: Emission performance standards and the perspectives for a European Green Deal. *Sustainability, 12*(20), 8381.

Hasanaj, V. (2023). The shift towards an eco-welfare state: Growing stronger together. *Journal of International and Comparative Social Policy, 39*(1), 42–63.

Hool, A., Helbig, C., & Wierink, G. (2023). Challenges and opportunities of the European critical raw materials act. *Mineral Economics, 37*(3), 661–668.

Huntjens, P. (2019). *Sociale innovatie voor een duurzame samenleving: Op weg naar een Natuurlijk Sociaal Contract.* Lectorale boek. IMPACT Lectoraat Sociale Innovatie in het Groene Domein. Hogeschool Inholland. https://doi.org/10.48544/90f2b3be-ab4d-4acf-a9c3-13effea07bcf

Huntjens, P. (2021). *Towards a natural social contract. Transformative social-ecological innovation for a sustainable, healthy and just society.* Springer Nature. https://doi.org/10.1007/978-3-030-67130-3

Huntjens, P., & Kemp, R. (2022). The importance of a natural social contract and co-evolutionary governance for sustainability transitions. *Sustainability, 14*(5), 2976. https://www.mdpi.com/2071-1050/14/5/2976

Huntjens, P., & Kemp, R. (2025). The transformation flower approach for eco-social contracting: Comparative insights from eight case studies in the Global South and North. In P. Huntjens, N. Mohamed, K. Hujo, & M. Desai (Eds.), *Eco-social contracts for sustainable and just futures* (Chapter 16, pp. 283–312). Springer Nature.

Ibbott, S. (2024). *Strategic dialogue: Farmers and NGOs reach historic consensus calling for a fair and sustainable transition for EU food and farming.* https://eeb.org/strategic-dialogue-farmers-and-ngos-reach-historic-consensus-calling-for-a-fair-and-sustainable-transition-for-eu-food-and-farming/

IPCC. (2018). Annex I: Glossary [Matthews, J.B.R. (ed.)]. In Masson-Delmotte, V., Zhai, P., Pörtner, H.-O., Roberts, D., Skea, J., Shukla, P. R., Pirani, A., Moufouma-Okia, W., Péan, C., Pidcock, R., Connors, S., Matthews, J. B. R., Chen, Y., Zhou, X., Gomis, M.I., Lonnoy, E., Maycock, T., Tignor, M., & Waterfield, T. (Eds.), *Global Warming of 1.5°C. An IPCC Special Report on the impacts of global warming of 1.5°C above pre-industrial levels and related global greenhouse gas emission pathways, in the context of strengthening the global response to the threat of climate change, sustainable development, and efforts to eradicate poverty* (pp. 541–562). Cambridge University Press. https://doi.org/10.1017/9781009157940.008

Kaack, L., et al. (2022). Aligning artificial intelligence with climate change mitigation. *Nature Climate Change, 12*, 518–527. https://doi.org/10.1038/s41558-022-01377-7

Kempf, I., & Hujo, K. (2022). Why recent crises and SDG implementation demand a new eco-social contract. In *Financial crises, poverty and environmental sustainability: Challenges in the context of the SDGs and Covid-19 recovery* (pp. 171–186).

Klein, N. (2014). *This changes everything: Capitalism vs. the climate*. Simon and Schuster.

Kopnina, H. (2018). Anthropocentrism: More than just a misunderstood problem. *Journal of Agricultural and Environmental Ethics, 31*, 109–127.

Leonard, M., Pisani-Ferry, J., Shapiro, J., Tagliapietra, S., & Wolff, G. B. (2021). *The geopolitics of the European green deal* (No. 04/2021). Bruegel policy contribution.

Loi, G., et al. (2025). *Recovery and resilience dialogue with the European Commission*. In-depth analysis. BUDG-ECON Committee meeting on 10 February 2025. https://www.europarl.europa.eu/RegData/etudes/IDAN/2025/764345/ECTI_IDA(2025)764345_EN.pdf

Manzanaro, S. S. (2024). *Nature restoration law, myths and facts for farmers*. Euractiv. https://www.euractiv.com/section/agriculture-food/news/nature-restoration-law-myths-and-facts-for-farmers/

Martin, B. (2019). *Green hearts and gilets jaunes*. Green Economy Coalition. https://www.greeneconomycoalition.org/news-and-resources/green-hearts-and-gilets-jaunes

Moore, B., & Härtner, J. (2024). *Where exactly is the Green Deal in Europe's new agenda?* Accessed February 24, 2025, from https://www.epc.eu/publication/Where-exactly-is-the-Green-Deal-in-Europes-new-agenda/

Naldzhiev, B. (2024). *EU sustainability policy: a choice between spreadsheet stagnation and climate leadership*. Accessed February 24, 2025, from https://www.friendsofeurope.org/insights/critical-thinking-eu-sustainability-policy-a-choice-between-spreadsheet-stagnation-and-climate-leadership/

Norton, A., & Greenfield, O. (2023). *Eco-social contracts for the polycrisis*. Green Economy Coalition. https://www.greeneconomycoalition.org/assets/reports/GEC-Reports/GEC_Eco-Social-Contracts-Polycrisis-FINAL-Nov23.pdf

O'Brien, K., Hayward, B., & Berkes, F. (2009). Rethinking social contracts: Building resilience in a changing climate. *Ecology and Society, 14*(2), 12.

OXFAM. (2023). *Climate equality: A planet for the 99%, Oxfam International*. https://policy-practice.oxfam.org/resources/climate-equality-a-planet-for-the-99-621551

Proposal for a regulation of the European parliament and of the Council on nature restoration. COM/2022/304 final. 2022. https://eur-lex.europa.eu/legal-content/EN/TXT/?uri=celex:52022PC0304

Pūķis, M., Bičevskis, J., Gendelis, S., Karnītis, E., Karnītis, Ģ., Eihmanis, A., & Sarma, U. (2023). Role of local governments in green deal multilevel governance: The energy context. *Energies, 16*(12), 4759.

Pullins, T., & Knijnenburg, S. (2025). *US withdrawal from the Paris Agreement: Impact and next steps*. White & Case. https://www.whitecase.com/insight-alert/us-withdrawal-paris-agreement-impact-and-next-steps

Raworth, K. (2018). *Doughnut economics: Seven ways to think like a 21st-century economist*. Chelsea Green Publishing.

Regulation (EU) 2024/1991 of the European Parliament and of the Council of 24 June 2024 on nature restoration and amending Regulation (EU) 2022/869 (Text with EEA relevance). OJ L, 2024/1991, 29.7.2024. ELI: http://data.europa.eu/eli/reg/2024/1991/oj

Rhodes, M. (2016). *The future of European welfare: A new social contract?* Springer.

Šajn, N. (2025). *Farm to fork strategy on sustainable food system. Legislative train schedule.* https://www.europarl.europa.eu/legislative-train/theme-a-european-green-deal/file-farm-to-fork-strategy

Sandmann, L., Bülbül, E., Castano-Rosa, R., Hanke, F., Großmann, K., Guyet, R., Jiglau, G., Laakso, S., Nuorivaara, E., & Vornicu, A. (2024). The European green deal and its translation into action: Multilevel governance perspectives on just transition. *Energy Research & Social Science, 115,* 103659.

Steffen, W., Broadgate, W., Deutsch, L., Gaffney, O., & Ludwig, C. (2015). The trajectory of the Anthropocene: The great acceleration. *The Anthropocene Review, 2*(1), 81–98.

Stevis, D., & Felli, R. (2020). Planetary just transition? How inclusive and how just? *Earth System Governance, 6,* 100065. https://doi.org/10.1016/j.esg.2020.100065

Strohschneider, P., et al. (2024). *Strategic dialogue on the future of EU agriculture. A shared prospect for farming and food in Europe.* https://agriculture.ec.europa.eu/document/download/171329ff-0f50-4fa5-946f-aea11032172e_en?filename=strategic-dialogue-report-2024_en.pdf

Temper, L., Avila, S., Del Bene, D., Gobby, J., Kosoy, N., Le Billon, P., Martinez-Alier, J., Perkins, P., Roy, B., Scheidel, A., & Walter, M. (2020). Movements shaping climate futures: A systematic mapping of protests against fossil fuel and low-carbon energy projects. *Environmental Research Letters, 15*(12), 123004.

UN News. (2019). *Greta Thunberg tells world leaders 'you are failing us', as nations announce fresh climate action.* https://www.un.org/uk/desa/greta-thunberg-tells-world-leaders-%E2%80%98you-are-failing-us%E2%80%99-nations-announce-fresh

UNEP (United Nations Environment Programme). (2024a). *Emissions gap report 2024: No more hot air ... please! With a massive gap between rhetoric and reality, countries draft new climate commitments.* Nairobi. https://doi.org/10.59117/20.500.11822/46404

UNEP (United Nations Environment Programme). (2024b). *AI has an environmental problem. Here's what the world can do about that.* https://www.unep.org/news-and-stories/story/ai-has-environmental-problem-heres-what-world-can-do-about

UNRISD (United Nations Research Institute for Social Development). (2016). *Policy innovations for transformative change: Implementing the 2030 agenda for sustainable development.* UNRISD.

Villiers, C. (2022). New directions in the European Union's regulatory framework for corporate reporting, due diligence and accountability: The challenge of complexity. *European Journal of Risk Regulation, 13*(4), 548–566.

von der Leyen, U. (2019a). *A Union that Strives for More: My Agenda for Europe.* Political Guidelines for the Next European Commission 2019-2024. European Commission. https://commission.europa.eu/document/download/063d44e9-04ed-4033-acf9-639ecb187e87_en?filename=political-guidelines-next-commission_en.pdf

von der Leyen, U. (2019b). Speech in the European Parliament plenary session. Ursula von der Leyen. President-elect of the European Commission. https://ec.europa.eu/commission/presscorner/detail/en/SPEECH_19_6408

von der Leyen, U. (2024). *Europe's choice.* Political guidelines for the next European Commission 2024–2029. https://commission.europa.eu/document/download/e6cd4328-673c-4e7a-8683-f63ffb2cf648_en?filename=Political%20Guidelines%202024-2029_EN.pdf

WWF EU. (2024). *Smart implementation of EU sustainability reporting standards: Make complying with rules easy.* https://www.wwf.eu/?16294391/A-call-for-smart-implementation-of-EU-sustainability-rules

Yébenes, M. O. (2024). Climate change, ESG criteria and recent regulation: Challenges and opportunities. *Eurasian Economic Review, 14.* https://doi.org/10.1007/s40822-023-00251-x

10

Radical Democracy, Ecology, and Justice in India: Experiences from Four Decades of Activist Research

Shrishtee Bajpai, Ashish Kothari,
and Neema Pathak Broome

10.1 Introduction and Conceptual Framework

This chapter attempts to distil key findings and lessons from a history of engagement of members of Kalpavriksh with various aspects of local community political governance in India. Kalpavriksh is a civil society organisation focusing on environment, development, conservation, livelihoods, and related issues, active since 1979. Its main work has been located in India, but through global networks it also has wider engagements.

The first section of the chapter gives a brief history of political governance in India followed by describing the conceptual framework that the authors use to analyse some of the cases of radical democracy mentioned here. The next section gives an overview of examples where communities are asserting self-governance and control over their territories. The last section highlights key learnings from these experiences and lists some key elements of swaraj (self-governance) embedded in these varied examples. The conclusion briefly weaves the intersectional nature of transformation in the initiatives described.

S. Bajpai (✉) • A. Kothari • N. P. Broome
Kalpavriksh – Environment Action Group, Pune, Maharashtra, India

© The Author(s) 2025
P. Huntjens et al. (eds.), *Eco-Social Contracts for Sustainable and Just Futures*,
https://doi.org/10.1007/978-3-031-99109-7_10

10.1.1 Brief History of Governance in India

Indigenous peoples[1] and other local communities across the world have had a diversity of systems of self-governance and decision-making guided by their cultural-social contexts and the geographies that they inhabit. These systems have operated through a diversity of institutional forms and unwritten or sometimes written codes of conduct and deliberations.

Over the several millennia that the Indian subcontinent has been inhabited, a wide diversity of governance forms has existed, including those with strong elements of direct or participatory 'democracy'.[2] This phenomenon is still relevant in the case of some of India's oldest communities, especially Adivasis, as we describe below. In more recent times, during the reign of princely states before the British Empire colonised India, most Adivasi regions remained on the periphery or as a nominal part of the realm. The rule of the monarchies and princes rarely extended to these regions beyond collecting some payments or using their habitats for aspects like hunting (Bijoy, 2024). They were usually left to govern themselves through their customary and traditional governance systems and institutions (Bijoy, 2024).

Such local systems of governance informed people's interaction with fellow community members as well as the rest of nature. These include village assemblies, selected or elected councils, and groups of village headpersons, who would form the key institutional pillars of self-governance. Broadly, these institutions and individuals were responsible for internal conflict resolution, management of village commons and water and other resource distribution, liaising with government agencies, livelihood activities, religious/spiritual ceremonies, and other cultural relations (Bajpai et al., 2022). They were in turn based on or guided by principles or norms, handed down over generations (Bhaskar et al., 2021).

During the colonial period (approximately mid-eighteenth to midtwentieth century), these communities and their governance systems came in direct conflict with British policies of commercialisation of lands, forests, waters, and other ecosystems and elements of nature. By 1864, a centralised forest department was established, draconian forest laws were enacted, and a large part of India's forests were taken over by the state (Skaria, 1999). These steps were strongly resisted by Adivasis and other communities such as those

---

[1] The term 'indigenous' is not used officially in India, but many communities do identify themselves as such or as adivasis (original inhabitants); the term more commonly used for official purposes is Scheduled Tribes, denoting those peoples who are listed in the relevant list of the Constitution of India.

[2] For glimpses into ancient Indian democratic or republic-like practices, see Muhlberger (1998); on clan assemblies, village assemblies, and gana-sanghas, see Thapar (2002).

in the Himalaya, leading to frequent and brutal suppression or (more rarely) conceding ground such as in the case of Van Panchayats (local forest governance institutions) in the Kumaon hills in northern India (Arnold, 1989).

The colonial state's uniform and centralised systems of governance were continued after India gained independence in 1947 but embedded within modern notions of electoral or liberal democracy. These systems did not take into consideration the already existing customary or local governance mechanisms across India, except where the colonial regime had already made concessions or let local systems remain. Only in some places where the newly established Indian nation-state did not have the reach to impose its power, or where constitutional exemptions were made due to local resistance, such as in north-east India, communities continued their traditional forms of governance in a state of relative autonomy (Bijoy, 2024).

Records of the constituent assembly during the framing of the Indian Constitution indicate that there was a recognition of the importance of local self-government, but this was not initially included as a justiciable part of the Constitution (Bijoy, 2024). Local governance was placed under the purview of state governments, leading to varied implementations across states. The Assembly ultimately agreed on the necessity of incorporating provisions for Panchayats (village councils) in the Directive Principles of State Policy, leading to Article 40 of the Constitution, which directs the state to organise village Panchayats and endow them with necessary powers and authority to function as units of self-government.

The idea of democratic decentralisation had been Mahatma Gandhi's key point of struggle but neglected by policymakers. In 1957, the Balwantray G. Mehta Committee, constituted by the Government of India, recommended 'democratic decentralisation' in the form of a three-tier Panchayati Raj System (PRS): the Gram Panchayat (council) at the village level, the Panchayat Samiti (committee) at the block level, and Zila Parishad (committee) at the district level. This system was adopted by state governments during the 1950s and 1960s (Mehta, 1957), and in 1992, it got constitutional backing, with the 73rd amendment to the Constitution of India. The powers and responsibilities devolved to the Panchayats under this system include preparation of plans for economic development and social justice, along with implementation of 29 subjects listed in the Constitution. Recognising that Adivasis needed something more tuned to their context, this system was extended in modified form to Scheduled V areas (GoI, 1996), where Adivasi populations were predominant, by the 73rd Constitution Amendment in 1996.

Additionally, over time, special provisions have been made for north-east India, to recognise its diverse and unique governance institutions. This includes empowering Autonomous District Councils or other such bodies, 'to make laws in respect of areas under their jurisdiction, which cover the land, forest, cultivation, inheritance, indigenous customs and traditions of tribals, etc. and also to collect land revenues and certain other taxes'. The Constitution's Sixth Schedule, providing relative autonomy, was provided for the states of Assam, Mizoram, Meghalaya, and Tripura (The Constitution of India, Sixth Schedule).

In the case of communities still practising traditional occupations and ways of life (forest based, pastoral, fishing, and/or farming), many traditional systems of governance are still being followed in parallel with the formal governance systems brought in by the state or being reinvented in combination with modern forms of governance (Bajpai et al., 2022). We explore this now in relation to specific sites.

10.1.2 Conceptual Framework

Over four decades of work, Kalpavriksh has tried to interrogate conventional notions and practices of democracy and understand from grounded initiatives how communities conceptualise and practice it. Democracy (demos = people + cracy = rule) is meant to be the rule of, by, and for people. In its original meaning this would imply that all of us, wherever we are, have the power to govern our lives. However, across the world its dominant meaning has been constrained by the form of 'liberal' governance in which representatives elected by people have power at varying degrees of centralisation. It is necessary to understand this crucial difference between direct or radical democracy and representative democracy (Kothari & Das, 2016). In the former, 'ordinary' citizens self-govern for various essential aspects of life, expressing and exercising power, which is felt inherently, rather than 'given' down by a 'higher' authority. In the latter, power is concentrated in representatives (elected or delegated), and typically the higher-level institutions (state and national level) where these representatives exercise their power, forming the state, are far removed from those who have voted or selected them. These two forms of democracy are not necessarily antithetical to each other, and conceivably one can formulate systems of subsidiarity where all decisions that can be taken at the level of local, face-to-face units of direct democracy are taken there. Only those decisions requiring larger-scale coordination are taken by units comprising representatives or delegates. In such a system, or even in those where direct democracy does not exist or is very weak, there can be various processes to ensure that representatives are accountable, transparent, and participatory in their decision-making and that there are methods such as the right to recall, periodic rotation, and so on that reduce unaccountable concentration of power (Bajpai & Kothari, 2020).

This understanding of democracy has been embedded in Kalpavriksh's work towards alternatives more generally, whether on ground or through social and political action. Elements of this are many decades old, for instance, in its action research and advocacy related to community-led conservation. More recently, in 2014, Kalpavriksh initiated a process involving participants from grassroots movements, NGOs, and others engaged in resistance against destructive models of development and working towards alternative ways of being, called Vikalp Sangam (VS) (Vikalp Sangam, 2025) or Confluence of Alternatives. This platform has since then brought together nearly 100 organisations and movements across India.

One of the first questions that the VS process explored was what is an alternative? Through deliberations at multiple gatherings, a collective vision called the 'The Search for Radical Alternatives' was drafted and has been evolving since then till its current (late 2024) seventh version (Vikalp Sangam, 2024). Additionally, as part of this, a self-reflecting, activist-academic tool to help evaluate the holism and intersectionality of transformative alternatives was produced, the Alternative Transformation Framework (ATF) (The Alternatives Framework, 2017). These documents contain a concept of alternatives and holistic transformation depicted by a 'Flower of Transformation', with five intersecting petals: ecological wisdom and resilience, social well-being and justice, direct and delegated political democracy, radical economic democracy, and cultural-knowledge diversity and commons (Kothari, 2021a, b). Also envisioned is the core of this Flower, comprised of multiple values and ethics, such as solidarity, diversity, equality, dignity, rights of humans as also non-human nature, and others. The documents define 'alternatives' as those that challenge the structures of oppression and unsustainability (capitalism, statism, patriarchy, racism, colonialism, anthropocentrism) and provide systemic pathways out of these as also to meet human needs and aspirations without trashing the earth and without creating or furthering inequities and injustices.

10.2 Kalpavriksh's Work on Local Governance

Since the 1980s, one strand of work within Kalpavriksh has been to understand common learnings, experiences, and challenges around local governance in India and the intersectionality of political governance with ecological, sociocultural, and economic dimensions.

One common thread over these decades has been the documentation of, and advocacy related to, *Community Conserved Areas*. Challenging exclusionary, top-down models of protected areas, Kalpavriksh has advocated community-led approaches of conservation by providing evidence of their viability and necessity, both in India and globally (Pathak Broome, 2009). Another common thread has been to document and visibilise traditional governance systems in such CCAs and at other sites, some of which we describe below.

10.2.1 Adivasis in Mendha Lekha

Mendha Lekha village in Gadchiroli district of Maharashtra is home to over 400 people, all belonging to the Gond tribe (indigenous people) or the Koya (human) as the tribe refers to itself. With a traditional cosmology of interconnection of all beings, humans and forests, forests have been synonymous with existence with these communities. For over 200 years such communities across India have been subjected to exploitation by government—both colonial and non-colonial—oppressed by centralised and corrupt bureaucracy, and marginalised by economic and industrial interests. In this context Mendha became the first village in the country in August 2009 to have its legal rights and responsibilities to use, manage, and conserve the 1800 ha of forests falling within its customary boundary recognised under the Scheduled Tribes and Other Traditional Forest Dwellers (Recognition of Forests Rights Act) 2006 or Forest Rights Act (FRA). This was a milestone in the history of forest governance not just in the village but for the country as a whole. This legal change was a consequence of more than two decades of struggle against oppressive forest governance regimes—as part of a larger regional movement called the Jungle Bachao Manav Bachao Andolan (Save Forests Save Humanity Movement).

Mendha's historic struggle and subsequent internal processes of transformation are an iconic and remarkable testimony of what a small tribal village, politically weak and voiceless, could do to empower itself to become a formidable force, through principles of consensus decision-making, commons, and non-violence. A village where until a few decades back the villagers would run into the forest and hide at the sight of an outsider to a village today where no activities even by the highest political or administrative functionaries can be carried out without their free prior informed consent.

Through processes of internal reflection over many years, the village realised that what weakened them in the face of external pressures was a fractured and non-inclusive governance system. This led to the revival of the customary practice of village assembly meetings for all decisions in the village, ensuring that at least one woman and one man from each family participate, ensuring adequate information and in-depth and transparent dialogues before decisions (through abhyas gats or study circles) and consensus decision-making. For decisions beyond the village, the Gram Sabha nominates delegates. Through its resistance and self-strengthening, the village has consistently questioned, challenged, and eventually catalysed a change in laws and policies or dominant social, political, and economic norms. In 2013 it also converted all private lands into the village commons, a decision taken after a decade of discussions in the Gram Sabha. Within Gadchiroli, Mendha is now part of a collective of over 30 Gram Sabhas in the region to work as a political and economic pressure group. It continues to face challenges from the state and corporations that eye the land and forests from a commercial mindset, as also the fragility that comes from being within a territory with a long history of conflict between the state and 'Maoist'[3] groups (Pathak Broome, 2018).

10.2.2 Adivasi Assemblies in Korchi

Korchi is an administrative subdivision or taluka of Gadchiroli in Maharashtra, inhabited by 133 Gram Sabhas (village assemblies), which were traditionally divided into three Ilakas (territories). Although officially the taluka is administered by Gadchiroli District Administration, informally and independently the Ilakas continue to have their traditional village level to supra-village-level self-governance structures.

Communities in the region have been resisting mining and exclusionary forest laws for several decades. These, and the struggle to gain greater control over jal, jungal, aur jameen (water, forest, and land), have significantly impacted economic, political, cultural, and social life as a whole.

Political decentralisation led to moving towards an engendered direct democracy aiming to achieve greater autonomy for the local Gram Sabhas (village assemblies) and greater accountability of the state institutions, particularly the local and administrative institutions. These processes were catalysed by the Forest Rights Act (FRA) and Maharashtra State Rules under the Panchayat Extension to Scheduled Areas Act (PESA). Localised, equitable, and transparent economy through assertion of rights over traditional forests and forest produce, particularly from the extraction and sale of forest produce

---

[3] For more detailed analysis, see series of articles in https://www.epw.in/journal/2006/29

or its use for domestic food and livelihood security, has led to bringing back as well as reinventing notions of well-being of people and forests.

Simultaneously, a need to strengthen the traditional tribe level governance systems such as the Jat Panchayats led to making them more inclusive of women and youth, through initiatives taken by women's collectives. As key activist Kumaritai says, 'the lives of the tribal women & the forest are intricately woven with each other. Our relation with forests is much older than the Forest Rights Act that was enacted in 2006'. Kumaritai and several other Adivasi women are articulating the need to visibilise and strengthen their role in decision-making.

To deal with newer challenges from the extractive industry as well as support for economic activities such as managing equitable collection, trade, and benefit sharing of and from forest produce, a federation of Gram Sabhas called the Maha Gramsabha emerged. The Maha Gramsabha and its executive body both have 50% women decision-makers along with representation from the youth, elders, non-tribal, and physically challenged persons from within the community. As in the case of Mendha Lekha, the context of India's neoliberal economy and top-down state remain ongoing challenges (Pathak Broome et al., 2022).

10.2.3 Communities in Ladakh and Spiti

The northernmost parts of India, adjoining Tibet, are culturally very different from most other parts of the country. They comprise of high-altitude ecosystems, Buddhist-Islamic cultures, and livelihoods based on pastoralism, farming, and trade. In Ladakh, the traditional governance system of yulpa (village assembly) and goba (village headman) appears to be very old, possibly from well before the time Ladakh came under the Kashmiri Dogra rule in 1846 (Bajpai et al., 2022). The goba acts as a representative of the village with social and cultural-ritual responsibilities. Traditionally, a well-respected person with a good comprehension of local history, communication skills, and good relationships with people was usually considered for selection to the post of the goba. In some places it was a hereditary position.

Key functions of the goba (currently carried out in varying degrees and combinations in different villages) include calling for all village-level meetings and coordinating various cultural, ritual, and other social gatherings; doing conflict resolution within the yulpa; maintaining the village demographic details; keeping records of government schemes and maintaining liaison with the administration on matters not covered by the panchayat and/or the council; presiding over the harvesting and cultivation timings in the village; keeping a check on the rotation cycle for hosting the ceremonial feast; and ensuring that all the families get water for irrigation and the upkeep and maintenance of irrigation canals. In the pastoral landscape of Changthang, the goba maintains the list of pasture lands, number of livestock with individual families, and boundaries to be adhered to by herders, conducts meetings to decide on migration timings, vests the power to allocate or withdraw access to pasture lands, and resolves conflicts between two herder communities regarding such access.

The local governance system in Spiti (Murali et al., 2021) has similarities and differences. The traditional system continues to be intact with gatpo or nambardar (goba equivalent) and local deities all playing a role in plural systems of governance. The gatpo is a village headman (very rarely, a woman) selected by the entire village periodically. The term itself may be a derivation of gopa (go = head; pa = people of an area). The gatpo belongs to the landed elite that usually comprised only a few families who had the wherewithal and means to represent the village and spend the time necessary to perform all the functions allocated to this position. Like the goba, he is also known as nambardar (an official revenue position appointed by the government). In many (perhaps most) villages in Spiti, even though the panchayat system has been introduced (on which we will say more below), the gatpo continues to act as an interface between the government officials and sarpanches on one hand and the villagers and monasteries on the other. The gatpos traditionally and still play quite a crucial role in pasture management. The gatpo appoints the lukzi who divides which animals go to which pasture. Without the gatpo, no yulva (village assembly) meetings can happen. He decides and consults the dates based on the traditional Tibetan calendar for all important functions and festivals in the village. Gatpo is also the connecting point between the village and the deities of the region (Bajpai & Kothari, 2024).

In common with many traditional systems of governance, these exhibit deep-rooted inequities and discriminations based on gender, caste, and age. Very few gobas, for instance, have been women, and the position has not been accessible to marginalised castes (though this is changing now where the choice of the goba is based on rotation amongst all families).

In the case of Ladakh, an additional governance level is the Ladakh Autonomous Hill Development Council (LAHDC), established in 1995. The LAHDC was born of a demand over several decades by Ladakh's population to have relative autonomy within the state of Jammu and Kashmir. This status has benefited the region in a number of ways. But it has also been severely constrained because the relevant legislation granted only limited administrative, financial, and legal powers to the Council, inadequate use of even these limited powers by LAHDC members, and continued domination by the state government. Additionally, issues of what kind of development would be appropriate have been weakly focused on, with some notable exceptions. In 2019, Ladakh received union territory status without legislative assembly, entailing further weakening of autonomy and a shift of power to the central government in New Delhi. Over the last few years, Ladakh's population has repeatedly mobilised to demand 6th Schedule and full statehood status, to safeguard the local culture, ecology, and traditions from being swamped by influx of people and 'development' projects decided on by outside forces (Kothari, 2024c). Getting such a status would also be crucial to sustain local governance systems.

10.2.4 Van Gujjar Pastoralists in Uttarakhand

The Van Gujjars inhabit the Western Himalayan states of Himachal Pradesh, Uttarakhand, northern Uttar Pradesh, and the union territory of Jammu and Kashmir. They are a nomadic community and rear an indigenous breed of buffalo (called 'gojri'). They are a semi-nomadic pastoralist community who migrate to the upper Himalayan Alpine meadows called 'bugyals' during summers and the lower Gangetic plains during winter. Over time, due to various restrictions imposed on their migration patterns, many have also begun to practice short-term migration to wetlands nearby in Bijnor district, Uttar Pradesh; some have also settled into sedentary agro-pastoralism. After the nationalisation of forests in colonial times, till the designation of these forests as protected areas under the Wild Life (Protection) Act 1972, the forests were divided and demarcated for each family, usually under the leadership of the family head. These heads (men or women) were licensees and were called Lambardars. This system has been officially discontinued, and the presence of Van Gujjars in these forests is now termed 'illegal' by the government.

Traditionally, Van Gujjar society is organised as a loose collective of families and clans, within which are embedded systems of resource governance, distribution, and management. Society is organised on the principles of caring and sharing, and decisions related to collective actions or conflicts are addressed by the Painchi, who are usually people who stand out in the community because of their wisdom, ethical standing, respect based on their personal and social conduct, and ability to resolve conflicts in a fair and unbiased manner. The Painchi system is an important institutional mechanism to maintain peace and harmony. The system functions on the principles of respect for

nature, strong sense of community, and non-violent dialogue and deliberations. The decision of the Painch is binding for all, and non-adherence could
lead to penalty and, sometimes, social boycott. However, the individuals or
groups who are not satisfied with the decision may call for a larger hearing,
where Painch from other hamlets are also invited. Each of the conflicting parties can bring their own Painch. The Painch also ensures that support is provided, in whatever kind, to individuals or families who may be in need, for
example, during health emergencies, marriages, and so on.

Van Gujjars believe that issues arising in the community need to be resolved
by themselves and by leaders who are aware of the local contexts, as well as the
worldviews and values of the community. The Painchi system along with
other mechanisms of self-governance ensures this (Pathak Broome & Chettri,
2023; and personal conversations with community elders by Akshay Chettri
and Neema Pathak Broome between 2021 and 2024).

10.2.5 Kunariya Panchayat in Kachchh

Kunariya panchayat, a cluster of three settlements in Kachchh district,
Gujarat, with a population of about 3500, is largely dependent on agriculture
(farming and animal husbandry), crafts, and labour. In 2016, the then sar-
panch (village head), Suresh Chhanga, initiated a process of implementation
of 73rd Constitutional Amendment (mentioned above). He began a series of
consultations in all the village wards, meetings of small focussed groups,
organisation of public events on nationally important days, and conducting
eight or nine Gram Sabhas (village assemblies) a year. The panchayat focussed
on empowering the smallest unit of decision-making, i.e. the wards. It initiated an ambitious programme to make the village self-sufficient in water,
reviving neglected wetlands and taking up watershed management. It also
focussed on effective implementation of Gram Panchayat Development Plan
(GPDP) by not only making it participatory but also trying to bring in the
concerns and needs of various (especially the marginalised) sections and
bringing in both local and external scientific knowledge as a base (Kothari,
2021a, b).

There has been a conscious attempt to deal with entrenched traditional
inequities of gender and caste. This includes a Balika Panchayat, consisting of
young girls up to the age of 16, as a forum where they can bring their concerns and needs to be conveyed to the main village panchayat (YouTube,
2022). Other initiatives include regeneration of grasslands for better fodder
availability and activation of the local government schools for higher quality

and creative learning (leading to the migration of children back to them from private schools).

As a result of such mobilisation of the village community, Kunariya was able to survive the COVID-19 pandemic period much better than many of its neighbours, both in terms of health, safety and continuing economic activities including those under the Mahatma Gandhi National Rural Employment Guarantee Act.

10.2.6 Communitisation in Nagaland

The state of Nagaland in north-east India consists almost entirely of Adivasis (Tribal communities). It has 16 (although people of Nagaland refer to up to 18) distinct tribes and many subtribes, each with distinct linguistic, cultural, customary, social, and political characteristics.

The main livelihood is traditional shifting cultivation with some amount of settled terrace farming. These are known for sustaining high agro-biodiversity, which along with forests sustain food and nutritional security (Changkija, 2017). The life, livelihoods, cultures, emotions, and identity of the majority of people from all tribes are deeply linked with their lands and forests.

Nagaland has a special constitutional status under Article 371 (A) of the Indian Constitution, meant to safeguard the cultures, traditions, and ways of life of the Naga people. This status enables it to reject or accept national laws relevant to the customary practices, land, and resources of its people. Within this context, Naga society has a layered system of governance. At the village level customary institutions differ from tribe to tribe: from inclusive and democratic decision-making systems of the Angamis, Chakesang, Ao, and Rengma to those where lands, resources, and decision-making around them rest exclusively in the hands of hereditary male village heads. Almost all villages have informal social collectives including mother's groups, self-help groups, student unions, youth groups, age groups, land right-holders unions, and collectives of Church functionaries. With more and more villages (as of 2024, over 400 documented) (Kalpavriksh, Kenono Foundation and LEMSACHENLOK, 2024) formally declaring parts of their traditional forest lands as Community Conserved Areas (CCAs), and the formation of CCA Management Committees (Pathak & Kothari, 2009), a state level forum called Nagaland Community Conserved Areas Forum has also been constituted (Nagaland Community Conserved Areas Forum, 2025).

The state-instituted governance systems are uniform across the state; however, there is a complicated yet functional relationship and overlap between these and the customary systems. For example, every Naga village has a village council (VC) constituted under the Nagaland Village Council Act of 1978 that handles the administration of judicial matters. The VCs, composed of male village elders, have a uniform formal structure across Nagaland, but their functioning differs from tribe to tribe depending on their own customary governance systems.

Under each VC, there exists a Village Development Board (VDB), a statutory body meant to deliver rural development programmes. There are also other statutory institutions in many villages such as Biodiversity Management Committees (BMCs) and institutions set up under the communitisation process of Nagaland (see below). Finally, the state-instituted governance systems are uniform across different districts of Nagaland. There is a complicated relationship and overlap between these and the customary social formulations and collectives.

One of the main problems facing rural India is the neglect of traditional or new local systems of education, health, water harvesting and sharing, and power generation through micro hydel. Very few of these local systems have found their way in the formal service sector or encouraged in any other manner. In Nagaland, this changed to some extent when, in the 1990s and 2000s, an innovative participatory exercise was carried out called 'Imagine Nagaland'. An understanding of the ineffectivity of the government's service delivery system, as also analysis of the problems of privatised services, led to innovative policy changes. The Nagaland Communitisation of Public Institutions and Services Act of 2002 empowered village institutions to manage education, water supply, roads, forests, power, sanitation, health, and other welfare and development schemes. Implementation of this law led to some transformative changes in the school results, health delivery system, and electricity revenue collections. In recent years, however, the process has lost some momentum and appears to be awaiting renewal (Pandey, 2010; Pathak Broome, 2014).

10.2.7 Urban Governance in Bhuj Town, Kachchh

Urban governance has been a major challenge across India's burgeoning towns and cities. The 74th Constitutional Amendment mandated decentralised governance by wards, but this has rarely been implemented. An attempt at trying to bring this to the ground can be seen in the case of Bhuj town in Kachchh (Gujarat). In 2014–2015, as part of a programme 'Homes in the City', five local civil society organisations (Hunnarshala, Kutch Mahila Vikas Sangathan, Arid Communities and Technologies, Sahjeevan, and SETU Abhiyan), have enabled citizens in both poorer and middle-class families to take much greater control over their settlements, including in housing, water self-provisioning, handling of waste, creating greater safety for women and girls, greater livelihood and energy security, and other aspects (Bajpai & Kothari, 2020). Local committees have handled various aspects of this transformation, and the town administration has been made more accountable and responsive. Special efforts have been made to enable and empower marginalised sections including women and girls.

The Bhuj example brings out that transformative processes at the ground are supported by a multidimensional approach. In this, the work of the five CSOs listed earlier, as also others like Sakhi Sangini (a collective of poor women), Shahri Seri Pheriya Sangathan (of street vendors), Bhuj Shahar Pashu Uchherak Maldhari Sangathan (of pastoralists), and Jal Strot Sneh Samvardhan Samiti (of water conservation activists), is of crucial significance. These organisations through their respective focus areas are contributing to the Ward Committee planning process.

10.3 Towards Swaraj

What we see in the above examples are processes of making democracy deeper, more inclusive, and more comprehensive in its application but with continuing challenges. Communities and collectives have asserted their customary forms of governance but have also been aware of the need to deal with traditional inequities and discriminations (with varying degrees of success). In documenting and understanding these, Kalpavriksh's understanding of what democracy consists of, or could consist of, has evolved considerably beyond the liberal electoral forms we are used to. We have observed that the processes have distinct elements of moving towards the vision of what Mahatma Gandhi popularised, swaraj. Loosely defined as 'self-rule', this notion is much deeper than liberal democracy, in that it entails the radical re-claiming of power by the public, from those centralising power either outside or within the community, along with the exercise of that power in responsible ways. In other words, it is a form of freedom and autonomy that does not impinge on the freedom and autonomy of others, and it contains not unbridled rights but those that are tempered by restraints in behaviour and action (Kothari, 2024a, b).[4] Several crucial elements of this can be discerned in the cases described:

---

[4] One of us has taken this further into the notion of 'Eco-Swaraj' or a 'radical ecological democracy', in which the sense of responsibility extends to all of nature.

1. *Capacity and ability to participate:* Radical democracy entails processes of building capacity, especially among those historically and currently marginalised. In the examples given above, the objective is not just to create space for participation but also how can that space be deepened and strengthened and for meaningful participation of the marginalised sections within the communities–women, oppressed castes, religious, and other minorities. Through various methods, communities have tried to create opportunity and open equitable access to forums of decision-making and possibilities of distributed leadership.

2. *Autonomy and self-rule:* These initiatives attempt to enhance the local power to make decisions (individual and collective) and assert the right to free, prior informed consent. In some of the examples given above (e.g. Goba system, Dzumsa, Painchi), despite the presence of constitutionally mandated new institutions such as Panchayats, people continue to bestow tremendous amount of trust in their own traditional systems of decision-making; in other cases (e.g. Kunariya), they are moulding the new institutions towards greater self-rule.

3. *Decision-making at smallest scale:* Radical democracy entails face-to-face interaction and meaningful participation in key larger-scale decisions. The role or power of a Gram Sabha to take all decisions relevant to it is the key idea of swaraj. In Korchi, for example, the Maha Gramsabha (MGS) is important, but it is the Gram Sabha that is the centre of decision-making.

4. *Accountable, transparent, and representative democracy at larger scale.* Institutions for decisions at a larger level, comprising delegates of the individual villages or collectives, are made accountable to the units of direct democracy. For instance, the Korchi Maha Gramsabha executive committee is responsible for reporting back to the individual Gram Sabhas; in Nagaland the state-level forum of CCAs takes its mandate from individual CCAs that are self-governing.

5. *Collective, responsible governance of commons.* The institutions of decision-making have collective, democratic processes of deciding how to manage, use, and conserve the commons. In the case of Korchi, Nagaland, Mendha, Kibber, Ladakh, Spiti, and Van Gujjars, decision-making is place-based, i.e. responding to and based on the needs, rhythms, and movements of the land and of nature as a whole, ensuring that they are sustained and not over-exploited, as also related to in ethical, often spiritual ways.

6. *Responsibility, solidarity, reciprocity, including the non-human:* Autonomy tempered by the need to respect others' political rights and interests is part of community ethics in some of the examples, if not all. "Changla Jeevan Jage Mayan Saathi Sapalorukoon Apu Apuna Jababdarita Jaaniv Ata Pahe

(To achieve well-being everyone needs to know what their responsibility is)" says Izamsai Katengey, a Gond Adivasi (indigenous) activist from Korchi. "Why do we oppose this project, you ask. Let us assume that we Adivasis will have to leave the forest if the mining company displaces us. But our forest deities will have no other place to go. I might shift to a city with my deity, but our collective deity of 33 villages resides in these hills, so many birds, animals and other species live in these forests, where will they go?" In Ladakh and Spiti, the rights of communities are intricately linked with their responsibilities towards the landscapes they live in, responding to the requirements of other beings they co-exist with.

7. *Informed and inclusive decision-making:* Fully informed decision-making including the ability to create and/or have access to existing forums of information and knowledge creation, respectful of marginalised subgroups, consensus as preferred mode, and avoidance of majoritarianism and competitive voting are critical dimensions of radical democracy. When demanding justice, it is not just about one sphere of all life, it also means transforming internal systems of injustice. Similarly, in the city of Bhuj, when designing urban planning processes, there is a conscious attempt to bring in the voice of the most marginalised in cityscapes, like the sex workers, city vendors, city pastoralists, and workers. Given the history of marginalisation, this is never easy, but several of these decision-making processes exhibit the wisdom to be not reduced to majoritarianism or having only the voices of the powerful being heard.

8. *Plurality:* The communities described above display a plurality or diversity of political representation, beliefs, interests, and ways of being (including legal and other forms of institutional pluralism). In Kibber, for instance, there is a co-existence of polycentric governance models with consultations among communities as well as deities and spirits of landforms. From a different perspective, in Korchi, Ladakh, and Uttarakhand, communities work with diverse institutions including their own systems of governance as well as state institutions like Panchayats and relevant state departments, as adaptation to changes taking place around them.

9. *Equity and non-discrimination:* Communities and collectives that are attempting radical democratic transformations, especially in their relations with the state, also realise the need to democratise internally. In the Indian context, patriarchy and sexual discrimination, casteism, ageism, ableism, and ethnic discrimination are deeply entrenched, though typically less so in Adivasi societies than others. Many of these initiatives have tried to proactively deal with these, such as in the case of Kunariya and the Korchi Maha Gramsabha; others such as goba in Ladakh are only now recognising the need to.

In many of the examples above, one also sees the move towards some kinds of redistribution to reduce economic inequities. For instance, a part of the income generated from the use of forest in Korchi's villages is put into a village fund, which is used for the welfare of those who are not well-off; during the COVID-19 pandemic this fund was crucial in sustaining village residents who had gone out for work and had to return as their jobs were shut down and did not have a land base to fall back on. In Kunariya the panchayat manages to use some of its designated funds to help those without land or in other ways economically weak. Such redistribution is even more meaningful if empowered local institutions are also legally supported to gain control over the local means of production, for example, through the Forest Rights Act in case of Korchi, or to access relevant government funds, for example, the Mahatma Gandhi Rural Employment Guarantee Act in the case of Kunariya.

However, given the long-standing and deep-rooted nature of many of these inequities and discriminations, they remain challenges to the goals of justice and equity.

10.4 Conclusion: Local Governance and the Flower of Transformation

For Kalpavriksh, a crucial finding of the case studies presented above is the intersectional nature of transformation. While a key focus of the studies has been on political governance, i.e. the nature of decision-making and the relations of power, this is clearly connected to ecological, economic, social, and cultural aspects of the lives of the relevant communities, as depicted in the Flower of Transformation mentioned above. For instance, traditional or new forms of localised democratic functioning are closely related to managing and sustaining the natural and land commons (pastures, forests, wetlands), including resistance to externally imposed projects that are ecologically destructive (e.g. mining). The continuation or strengthening of local governance systems has a two-way relationship to the continuation or strengthening of the local economic base. For instance, local mobilisation based on cohesive collective power leads to claims for collective legal rights (e.g. regarding forests or other commons), and getting such rights recognised leads to further consolidation of such power. The continuation or evolution of knowledge and cultural systems form another foundation of such decision-making—this includes traditional and new ecological knowledge and the customs, beliefs, and spiritual traditions relating to the commons.

All of the above are related to the social dynamics of communities, including their internal gender, caste, class, and other structures and relations and their relations with other communities. This includes inequities and discriminations of various kinds, which are often the slowest to transform; but it also includes relations of mutual aid, solidarity, and cooperation. While there are clear signs of reduced inequalities and marginalisation, it is unlikely that in any of the examples cited above, holistic or comprehensive transformations in all the five spheres or petals of the Flower have taken place. Rather, they are all at various stages, with substantial distance having been covered in some petals and not so substantial in others. Given that Kalpavriksh has not conducted comprehensive studies in any of these, i.e. covering all spheres in detail, it is not possible to say much more at this stage, other than that such intersectionality, is clearly visible.

More in-depth and wider studies (by or with the relevant communities) are needed to get a better understanding of how the Flower of Transformation is being manifested at each of these sites, where there remain weaknesses or gaps and, therefore, what more is needed to move towards more holistic transformation. It is evident that informal and customary governance systems are far more prevalent than commonly known. Though not all of these are necessarily examples of radical democracy, their continued relevance necessitates adequate recognition by both civil society and government. There are elements in the Constitution of India, and in several laws, that offer some such recognition at a broad level, but they do not fully recognise and provide legal backing to specific systems and are implemented haphazardly. In many areas, there are considerable overlaps in key functions between the traditional heads and the sarpanch/panchayat and other state institutions. In the absence of legal clarity on the divisions of their functions, there is conflict or confusion in case of overlapping jurisdiction such as in agriculture, water management, livestock maintenance, management of rituals, and festivals. Modern institutions have also at times co-opted or displaced traditional ones, at times without replacing many of their functions—most significantly the governance of the commons. It is thus important that there is adequate and appropriate policy recognition of the plurality of institutions that exist locally and clarity to ensure that their inter-relations are complementary.

We believe that a deeper inquiry into these forms of self-governance, along with efforts towards their transformational self-strengthening, could help in finding pathways out of current sociocultural, ecological, economic, and political crises.

References

Alternatives Transformation Format. (2017). *A process for self-assessment and facilitation towards radical change prepared by Kalpavriksh for ACKnowl-EJ.* https://vikalpsangam.org/wp-content/uploads/migrate/Resources/alternatives_transformation_format_revised_20.2.2017.pdf

Arnold, D. (1989). Rebellious Hillmen: The Gudem-Rampa Risings, 1839–1924. In R. Guha (Ed.), *Subaltern studies: Writings on South Asian history and society* (Vol. 1). Oxford University Press.

Bajpai, S., & Kothari, A. (2020). *Towards decentralised urban governance: The case of Bhuj city, Kachchh (India).* Kalpavriksh.

Bajpai, S., & Kothari, A. (2024). *The Gatpo of Spiti: Current relevance of a traditional governance system.* Kalpavriksh and Nature Conservation Foundation (in press).

Bajpai, S., & Kothari, A., with Namgail, T., Sonam, K., & Deachen, K. (2022). *The Goba of Ladakh: Current relevance of a traditional governance system.* Kalpavriksh, Snow Leopard Conservancy – India Trust, Nature Conservation Foundation, Local Futures and Ladakh Arts and Media Organisation.

Bhaskar, P., Ghosh, P., & Chaudhuri, D. (2021, August 3). *Collapse of Adivasi self-governance system in Jharkhand: Need to implement PESA in letter and spirit.* Down to Earth.

Bijoy, C. R. (2024). India and indigenous peoples political autonomy and democratising governance. *Pax Lumina, 5*(3), 35–39.

Changkija, S. (2017). Sustainable hill agricultural diversity in Nagaland. *International Journal of Agriculture, 35*(1), 121–140.

Government of India (GoI). (1996). Fifth Schedule [Article 244(1)] Provisions as to the Administration and Control of Scheduled Areas and Scheduled Tribes. Amendment to the Constitution of India. https://www.mea.gov.in/images/pdf1/s5.pdf

Guha, R. (1989). *The unquiet woods.* Oxford University Press.

Kalpavriksh, Kenono Foundation & LEMSACHENLOK. (2024). *Strengthening Community Conserved Areas in Nagaland via Field intervention, Technology and Policy Support - Final Report for The Nature Conservancy.* https://kalpavriksh.org/strengthening-community-conserved-areas-in-nagaland-via-field-intervention-technology-and-policy-support/

Kothari, A. (2021a). Kunariya: A model panchayat in Gujarat. *Frontline.* 08/10/22. https://ashishkothari.in/wp-content/uploads/2023/04/Kunariya-A-model-panchayat-in-Gujarat-Frontline.pdf

Kothari, A. (2021b). These alternative economies are inspirations for a sustainable world. *Scientific American, 324*(6), 60–69. https://www.scientificamerican.com/article/these-alternative-economies-are-inspirations-for-a-sustainable-world/

Kothari, A. (2024a). In search of alternatives to development: Learning from grounded initiatives. In H. Melber, U. Kothari, L. Camfield, & K. Biekart (Eds.),

Challenging global development: Towards decoloniality and justice. Palgrave Macmillan.

Kothari, A. (2024b). *Ladakh's mass agitation heats up a cold desert.* meer.com, 16 March. https://ashishkothari.in/wp-content/uploads/2024/04/Ladakhs-mass-agitation-heats-a-cold-desert-Meer-13.3.2024.pdf

Kothari, A. (2024c). *Ladakh's mass agitation heats a cold desert.* meer.com, 13 March. https://www.meer.com/en/79171-ladakhs-massagitation-heats-a-cold-desert

Kothari, A., & Das, P. (2016). *Power in India: Radical pathways.* In State of power 2016, Transnational Institute.

Mehta, B. G. (1957). The report of the team for the study of the community projects and National Extension Service, Chaired by Balwantray G. Mehta in 1957. www.panchayatgyan.gov.in. Retrieved from https://web.archive.org/web/20171113110803/http://www.panchayatgyan.gov.in/documents/30336/0/Balvantray_G_Mehta_Committee_report.pdf/563c2a17-a13f-463a-ba4d-339916c32d57

Muhlberger, S. (1998). *Democracy in ancient India.* World History of Democracy site. http://www.nipissingu.ca/department/history/muhlberger/histdem/indiadem.htm

Murali, R., Bijoor, A., & Mishra, C. (2021). *Gender and the commons: Water management.*

Nagaland Community Conserved Areas Forum. (2025). Nagaland CCA Forum. *Community conserved areas.* Accessed May 2, 2025, from https://nccaf.communityconservedareas.org/

Pandey, R. S. (2010). *Communitisation: The third way of governance.* Concept Publishing Company.

Pathak Broome, N. (2014). *Communitisation of public services in Nagaland: A step towards creating an alternative model of delivering public services.* Kalpavriksh and Heinrich Boll Foundation.

Pathak Broome, N. (2018). Mendha-Lekha: Forest rights and self-empowerment. In M. Lang, C. Konig, & A. Regelmann (Eds.), *Alternatives in a world of crisis.* Global Working Group Beyond Development. Rosa Luxemburg Stiftung, Brussels Office and Universidad Andina Simon Bolivar, Ecuador.

Pathak Broome, N., & Chettri, A. (2023). *Traditional governance systems of the van Gujjars in Uttarakhand (India).* https://radicalecologicaldemocracy.org/traditional-governance-systems-of-the-van-gujjars-in-uttarakhand/

Pathak Broome, N. (2009). *Community Conserved Areas in India.* Kalpavriksh. See also, for cases from India, https://communityconservedareas.org; and more globally, www.iccaconsortium.org

Pathak Broome, N., Bajpai, S., Shende, M., & Raut, M. with Jamkatan, G. K., Katenge, I., Salame, Z., Halami, S., Deshmukh, S., & Gogulwar, S. (2022). *Forest resource rights, gram sabha empowerment and alternative transformations in Korchi.* Kalpavriksh, Pune with Maha Gramsabha Korchi and Amhi Amchya Arogyasathi, Kurkheda.

Pathak, N., & Kothari, N. (2009). Nagaland: A quiet revolution. In N. Pathak (Ed.), *The directory of community conserved areas in India.* Kalpavriksh.

Skaria, A. (1999). *Hybrid histories: Forests, frontiers and wildness in Western India*. Oxford University Press.

Thapar, R. (2002). *Early India: From the origins to AD 1300*. University of California Press.

Vikalpsangam. (2024). *The search for radical alternatives: Key aspects and principles, framework of the Vikalp Sangam process 7th avatar*. https://vikalpsangam.org/wp-content/uploads/2024/10/Alternatives-Framework-7th-Avatar-digital-v1.4.pdf

Vikalp Sangam. (2025). *About*. Vikalp Sangam. Accessed May 2, 2025. https://www.vikalpsangam.org/about

YouTube. (2022). *Try Not To Laugh Watching Funny Fails Videos 2022*. YouTube video, 10:05. Posted by "Funny Fails," January 15, 2022. https://www.youtube.com/watch?v=MB1Y_ru1sxM

11

The Role of the Rights of Nature in Establishing Eco-social Contracts

Lauren Tarr, Catherine Haas, and Caitlyn E. Sutherlin

11.1 Introduction

A growing global chorus is asking, *how do we live in harmony with nature?*[1] A key element in addressing this issue is overcoming frameworks that treat Nature as a commodity for human exploitation. How can we live in harmony with nature if we do not first recognise nature as kin, neighbour, and community? The Rights of Nature (RoN) movement provides a transformative framework to address this challenge by recognising Nature as a rights-bearing entity and fostering a shift from anthropocentric to ecocentric perspectives. By rethinking our relationship with Mother Earth, RoN can help in

---

[1] The United Nations General Assembly established a Harmony with Nature programme in 2009 based on this vision. In 2022, the Kunming-Montreal Global Biodiversity Framework (GBF) was adopted and set a goal for the world to be living in harmony with Nature by 2050.

L. Tarr (✉)
Department of Environmental Studies, State University of New York College of Environmental Science and Forestry (SUNY ESF), Syracuse, NY, USA
e-mail: Letarr@esf.edu

C. Haas
Eco Jurisprudence Monitor (EJM), Asheville, NC, USA

C. E. Sutherlin
Global Alliance for the Rights of Nature (GARN) North America Hub, Houghton, MI, USA

P. Huntjens et al. (eds.), *Eco-Social Contracts for Sustainable and Just Futures*,
https://doi.org/10.1007/978-3-031-99109-7_11

developing eco-social contracts[2] where well-being and prosperity of all life are secured.

This chapter explores how the Rights of Nature movement supports the development of new eco-social contracts, establishing more harmonious relationships for humans with the wider Earth community. Through case studies and theoretical analysis, it tracks the growth of the movement and highlights examples of application in different contexts.

We approach this topic as researchers from the United States, of European descent, engaged in the global Rights of Nature (RoN) movement. We note our positionality[3] because part of eco-social contract work, this healing of our relationships to the world and to each other, requires understanding our histories, responsibilities, and roles within communities. This chapter highlights several RoN cases worldwide, shaped by unique socio-environmental histories. As you read, we encourage you to reflect on your own positionality, to consider how you have learned to see your place in the world and relationship with nature, and to hold in mind that we all have the capacity to learn from one another and a responsibility to foster mutual flourishing across the Earth community.

11.2 Ecological Jurisprudence and the Rights of Nature

Jurisprudence is the *philosophy* of law. Examining jurisprudence identifies what values are reflected in laws. These values are both influenced by, and actively shaping, a society's social contracts. A nation's jurisprudence can reflect and perpetuate social hierarchies and power structures and, in terms of eco-social contracts, can be used to shape human–nature relationships.

Western jurisprudence is heavily anthropocentric (Deckha, 2013; Burdon, 2012). This is even seen in environmental laws, which often regulate Nature as property and prioritise economic growth and development (M'Gonigle & Takeda, 2013). Anthropocentric systems of colonialism, capitalism, and

---

[2] Huntjens & Kemp (2025), and also used as IPBES definition in the Transformative Change Assessment (IPBES, 2024), define natural social contracts or eco-social contracts as "the collective power of societies, people and nature for dealing with the polycrisis of the twenty-first century (including inequalities, injustices, distrust, climate and ecological crises) through implicit or explicit collective agreements across multiple governance levels among members of a society to cooperate with one another and abide by certain rules or norms targeted at sustainability, equity and justice."

[3] Positionality statements are a methodological choice to recognise that all knowledge is informed by the location and position of the knowledge producer. For more on the importance of researcher positionality, see Harroway's (1988) discussion on situated knowledges.

industrialisation have driven extreme environmental and human crises (Moore, 2017). These systems reinforce a separation between humans and nature, shaping laws and systems that prioritise short-term human interests. This imbalance has been exacerbated in the last few centuries by colonisation, which disseminated Western legal systems and ideologies that entrenched a human-nature divide (Davis & Todd, 2017; Plumwood, 2012). This human-centred paradigm, codified into legal frameworks, has caused significant harm to both humans and the broader Earth community.

Ecological jurisprudence moves beyond anthropocentric jurisprudence and instead encompasses legal frameworks and philosophies that better align human laws with ecological systems (Pelizzon, 2025). This includes the Rights of Nature movement and other Earth-centred laws that recognise the inherent value of Earth's systems. In addition to Rights of Nature, other growing ecological jurisprudence approaches include establishing ecocide as a crime, pursuing more place-based ecological governance, and supporting rights-based climate litigation like the Rights of Future Generations, all of which support the flourishing of our wider ecological communities, of which humans are one part.

This chapter will focus on one aspect of ecological jurisprudence: the Rights of Nature movement. Rights of Nature laws recognise the inherent Rights of Nature. The Global Alliance for the Rights of Nature (GARN) describes it as:

> Rather than treating Nature as property under the law, Rights of Nature acknowledges that Nature in all its life forms has the right to exist, persist, maintain and regenerate its vital cycles. And we—the people—have the legal authority and responsibility to enforce these rights on behalf of ecosystems. The ecosystem itself can be named as the injured party, with its own legal standing rights, in cases alleging rights violations. ("What are the Rights of Nature - Global Alliance for the Rights of Nature (GARN)," n.d.)

11.3   Eco-social Contracts and the Rights of Nature

Building on the principles of ecological jurisprudence and Rights of Nature, we will explore how these concepts can help answer the call for establishing new eco-social contracts. Knowing that the dominant social contracts of today are not meeting our needs, we must envision new agreements that foster both human and ecological flourishing (Huntjens, 2021). Living in harmony with nature requires recognising humans as one part of a broader ecological community and including all ecological members in our social responsibilities and agreements (Arrows & Narvaez, 2022; Berry, 1999).

There are governance systems (many Indigenous) that maintain relationships of reciprocity and respect with the broader Earth community. These reflect kinship ontologies that embrace the mutual connections of humans with other animals, plants, waters, minerals, spirits, ancestors, future generations, and all elements of our complex web of relations (Van Horn et al., 2021; Salmón, 2000). These types of ecocentric values can also be found in faith systems around the world. For example, Christianity can be interpreted as encouraging environmental stewardship, Buddhism highlights the interconnected nature of all life, Hinduism recognises nonhuman animals as knowledge sharers to achieve wisdom and enlightenment, Baha'i writings equate nature with reflection of the divine, Judaism places importance on the continuation of species, and Islam emphasises balance in ecosystems (UNEP, 2020). Eurocentric colonisation, spreading governance systems based in dualistic and hierarchical worldviews, disrupted these kinship relationships in many parts of the world (Plumwood, 2012). Pursuing eco-social contracts requires addressing the enduring impacts of colonisation that continue to shape our human-environment relationships.

The Rights of Nature movement offers one path to shift from an anthropocentric to an ecocentric paradigm. For societies with anthropocentric jurisprudence, RoN represents a move towards recognising Nature as a community member instead of a commodity. Some Indigenous nations, whose ontologies are already ecocentric, have utilised RoN laws to codify their long-standing "contracts" with Nature. However, not all Indigenous groups support this rights-based framework. Some contend that the RoN movement itself is a Western construct, as many Indigenous cultures have long recognised Nature as possessing inherent rights, though not in the strictly Western judicial sense (Kauffman & Martin, 2017: 131–132). The concept of "rights," particularly the notion of legal personhood for Nature, is rooted in Western legal traditions and may not align with the teachings of many Indigenous peoples.

This tension highlights a broader issue: Indigenous knowledge systems and ontologies are often marginalised or reframed within Western paradigms, despite growing recognition of their validity in conservation and environmental stewardship (Martinez et al., 2023). These ways of knowing are not merely cultural differences but are ontological differences in how nature is regarded (De la Cadena & Blaser, 2018). Knowledge relates to power when it sustains one ontology over others, dismissing other ways of knowing as incomplete or unsubstantial (Foucault, 1975; Gaventa & Cornwall, 2006). When Indigenous ways of knowing are framed as cultural expressions rather than distinct world-making practices, they are subsumed into a Western perspective, contributing to the destruction of Indigenous worlds (De la Cadena & Blaser, 2018). This insight forces us to reconsider environmental debates, including RoN, as ontological conflicts, not just cultural or epistemological ones. Recognising this multiplicity of worlds is essential to building equitable and inclusive socio-environmental governance.

In this context, we will explore how Rights of Nature offers one piece of the pathway for recognising these ontological differences and imagining new eco-social contracts. The goals of eco-social contracts as defined in the United Nations Research Institute for Social Development (UNRISD) issue brief from 2021 (UNRISD, 2021) are summarised below (see Chap. 14):

1. *Human rights for all*—including all members of human societies, particularly those often excluded or marginalised from previous social contracts.
2. *A progressive fiscal contract*—raising funds for climate action, implementation of Sustainable Development Goals (SDGs), and fairly distributing the financial burdens.
3. *Transforming economies and societies*—transforming societies to halt climate change and environmental destruction and promoting equality and social inclusion.
4. *A contract with Nature*—recognising that humans are part of a wider global ecosystem and pursue harmony with nature.
5. *Addressing historical injustices*—remedying historical injustices through a decolonised approach informed by Indigenous knowledges and capacities from the Global South.
6. *A contract for gender justice*—equal respect and rights for all gender expressions and sexual orientations.
7. *New forms of solidarity*—requires bottom-up approaches to transformative change, uniting social movements around shared goals.

Rights of Nature laws can play an important role in actualising several of those goals (Mohamed & Huntjens, 2023). For example, the eco-jurisprudence movement, encompassing concepts like *buen vivir* that prioritise local ecological or traditional knowledge and Rights of Nature, provides a framework for cocreating knowledge and designing adaptation strategies that encourage humans to live in harmony with their environment. The incorporation and prioritisation of local traditional knowledge are key for reducing vulnerabilities, increasing resilience and adaptive capacity by incorporating diversity, ecosystem services, social capital, self-organisation, and learning from experience into decision-making (Eriksen et al., 2021; Hosen et al., 2020). These practices support the eco-social contract goals of transforming societies to halt climate change, creating new forms of solidarity, and creating new contracts with Nature. Later in this chapter, we will explore case studies demonstrating how these goals are being realised through RoN initiatives around the world. But first, this next section will explain the growth of the movement overall.

11.4 The Growth of the Rights of Nature Movement

The rapidly growing Rights of Nature movement shows that the demand for new eco-social contracts is not merely theoretical. The global rise of RoN initiatives illustrates a practical response to the urgent need to uphold ecological health alongside human welfare. This section will examine how the movement has evolved and gained momentum across various legal landscapes.

11.4.1 The Eco Jurisprudence Monitor: A Timeline of the Rights of Nature

The Eco Jurisprudence Monitor is a database tracking the growth of ecological jurisprudence. The Monitor highlights different expressions of ecological jurisprudence, provides details on individual laws, compiles legal documents and resources, and offers tools for analysis (Kauffman et al., 2024).

This chapter focuses on the Rights of Nature movement within ecological jurisprudence, which consists of two broad categories: judicial and nonjudicial actions. Judicial actions establish Nature as a subject of rights through various legal avenues: constitutions, legislation, court rulings, and Indigenous law. Nonjudicial actions, on the other hand—such as policies, conferences, and declarations—propel the movement forward ideologically but do not directly result in legal realisation of rights to Nature.

Using the Eco Jurisprudence Monitor, we can trace the origins and expansion of the contemporary RoN movement across three stages (Fig. 11.1). The first stage (1972–2001) focused on theoretical foundations, primarily rooted in Western ideas. The second stage (2006–2010) marked a shift to implementation, largely driven by Indigenous contributions that transformed RoN into a legal model. The third stage (2014–present) is defined by globalisation and plurality, evolving into a broad movement that reflects the diverse peoples, worldviews, norms, and goals driving RoN forward.

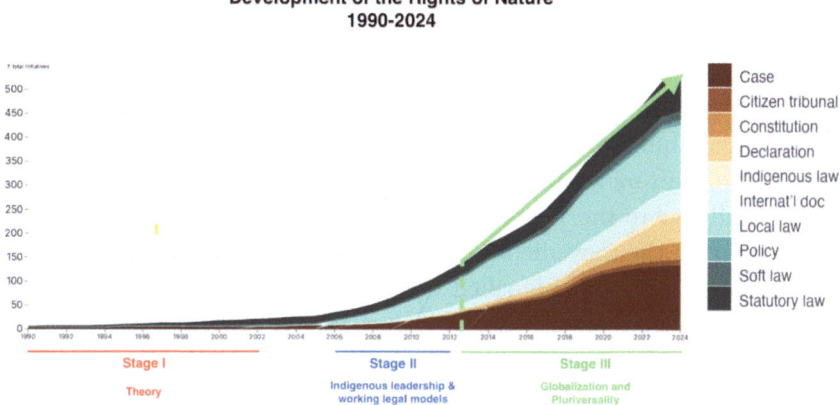

Fig. 11.1   Rights of Nature timeline and growth by legal provision, 1990–2024. Note: This illustrates the timeline of RoN's development, highlighting the transition from theory to implementation and the rapid growth of RoN actions in recent years. The figure also highlights the different legal pathways within RoN and their evolutions in the broader movement. Source: Distributed by the Eco Jurisprudence Monitor. 2024. https://ecojurisprudence.org

Stage I (1972–2001): Debates and Reimaginings

The Rights of Nature (RoN) contemporary movement began in 1972 with a narrow legal question in the United States: Could trees have legal standing? Christopher Stone (1972) argued that granting legal rights to natural entities like trees was not fundamentally different from granting such rights to non-human entities like corporations. He saw this designation as more than a legal shift—it was a step towards reshaping humanity's connection to nature, fostering empathy and a deeper societal transformation. Over the next two decades, the conversation broadened from legal standing to a larger debate on environmental ethics, introducing the concept of Nature's rights.

These early discussions framed Rights of Nature as a debate between ecocentric and anthropocentric ethics, encapsulated by the question: Does Nature have the right to be protected or do humans have a responsibility to protect Nature (Berry, 1999)? Roderick Nash's *The Rights of Nature: A History of Environmental Ethics* (1989) critiqued Western anthropocentrism and property norms, but his analysis largely overlooked global perspectives rooted in nonhierarchical worldviews and alternative conceptualisations of Nature and "rights." Moreover, he neglected the role of spirituality in shaping human–nature relationships, particularly as experienced by other cultures around the world. By centring Western historical perspectives, Nash underscores the persistent limitations of environmental ethics in addressing more diverse and holistic understandings of human-environment relationships.

In 1999, Thomas Berry reinterpreted the RoN debate by integrating morality, cosmology, and Indigenous wisdom. His spiritual critique of anthropocentrism and human jurisprudence brought new depth to the RoN conversation, shifting it from a purely legal concept in the United States into a moral challenge to the Western anthropocentric worldview. This marked a turning point in the movement's evolution. Berry's work and the subsequent global expansion of RoN signalled the end of its theoretical phase and the beginning of a new era focused on legal implementation, often led and inspired by Indigenous worldviews.

Stage II (2006–2010): The Birth of a Movement

Building on Thomas Berry's concepts, the Rights of Nature (RoN) movement emerged as a response to the limitations of Eurocentric legal systems while still largely operating within a rights-based framework. However, it is also deeply inspired by—and often grounded in—Indigenous cosmologies, wisdom, and ways of life.

Between 2006 and 2010, the first RoN laws were enacted in the United States, Ecuador, and Bolivia. During this time, Indigenous leadership was instrumental in transforming RoN from Western abstractions into actionable legal frameworks that now drive the global movement.

In 2008, Ecuador became the first country to enshrine the Rights of Nature in its national constitution. The preamble recognises humans as part of nature and calls upon "the wisdom of all cultures that enrich us as a society" to form a new framework for "coexistence in diversity and harmony with nature, to achieve a good way of living, the *sumak kawsay*" (The Republic of Ecuador National Assembly, 2008a, b). *Sumak Kawsay* (translated as *buen vivir*) is a term in the Kichwa language referring to an ancestral Andean concept highlighting the importance of solidarity, community ties, harmony with nature, and dignity. The express involvement of Indigenous groups in its drafting, alongside the incorporation of Indigenous understandings of human–nature relationships, underscores the critical role of Indigenous worldviews to the development of RoN legal frameworks (Kauffman & Martin, 2017).

In 2010, Bolivia followed suit by passing Law 071, The Rights of Mother Earth, in alignment with its constitution's call for a development model based on *Suma qamaña* (translated as *vivir bien*). *Suma qamaña* is rooted in Indigenous Aymara traditions (the largest Indigenous group in Bolivia) and emphasises the need to live in harmony with Mother Earth. Calzadilla and

Kotzé (2018) describe it as a decolonial challenge to the neoliberal, consumerist paradigm that exploits natural resources and marginalised communities, particularly women, children, and Indigenous peoples while opposing an anthropocentric, market-driven worldview (Calzadilla & Kotzé, 2018, 7–8).

These developments established Ecuador and Bolivia as leaders in the Rights of Nature movement, centring Indigenous worldviews within the dialogue.

Stage III (2014–Present): Globalisation and Plurality of the RoN

Following the adoption of the world's first Rights of Nature laws, the movement experienced rapid globalisation. Ecuador and Bolivia's groundbreaking RoN laws, characterised by their sweeping language and diverse collaboration, opened the floodgates for RoN actions to proliferate worldwide. Figure 9.1 shows this dramatic increase in RoN actions depicted in the years following 2010.

Since 2014, the third stage of the Rights of Nature movement has witnessed a global expansion of judicial and nonjudicial actions, evolving into a diverse international movement (see Fig. 11.2). However, not all communities approach RoN with the same legal frameworks, as these initiatives reflect varying worldviews on the human–nature relationship. RoN jurisprudence includes concepts such as animal rights, environmental personhood, and sometimes Indigenous laws that integrate traditional ecological knowledge

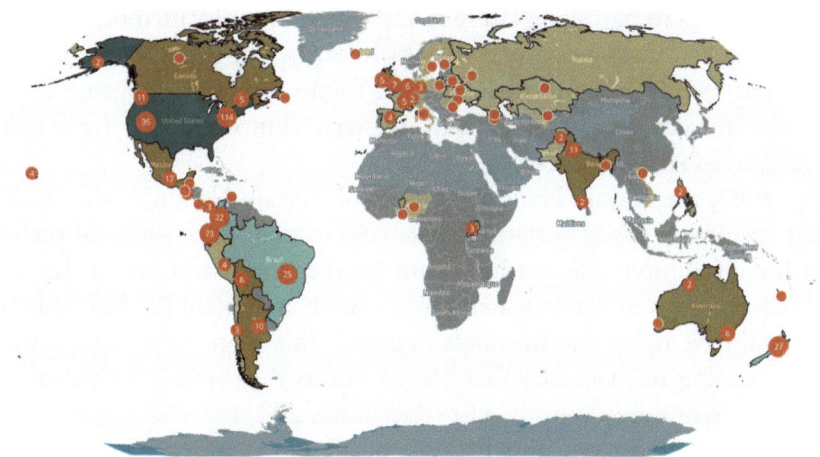

Fig. 11.2   Global distribution of the Rights of Nature. Source: Distributed by the Eco Jurisprudence Monitor. 2024. https://ecojurisprudence.org

and emphasise responsibilities to nature as a living entity. These ideas are expressed in constitutions, court cases, national and local laws, international documents, and declarations. This growth underscores the growing call for a paradigm shift away from anthropocentric relationships with Earth.

11.5 Rights of Nature Frameworks and Eco-social Contracts

The Rights of Nature framework can be used to establish new eco-social contracts, support ecocentric values,[4] and address environmental crises. In the following section, we provide examples from four angles: (1) RoN laws codifying existing ecocentric values; (2) RoN laws promoting ecocentric values within anthropocentric systems; (3) RoN laws in plural-legal contexts, navigating multiple ontologies where ecocentric values are both preexisting and emergent; and (4) the importance of cocreating knowledge across diverse ontologies for climate change adaptation and resilience.

11.5.1 Rights of Nature as Codifying Existing "Contracts" with Nature: Examples from the Navajo and Ecuador

One function of Rights of Nature laws is to codify a society's existing positive "contracts" with Nature. By this we mean there are prominent preexisting cultural values recognising Nature as a fellow community member, as opposed to viewing Nature as property owned by humans. The Rights of Nature movement has seen Indigenous nations using Rights of Nature language as a way to translate ecocentric values into Western (Euro-colonial) legal systems (Rodríguez-Garavito, 2024).

In 2002, the Navajo Nation adopted the Navajo Nation Code, codifying their customary law into statutes. The laws recognise the rights of the wider Earth community: "All creation, from Mother Earth and Father Sky to the animals, those who live in water, those who fly and plant life have their own laws and have rights and freedoms to exist." This is one of the first cases of a nation stating in their laws that Nature (or as they state "all creation") has laws, rights, and freedoms to exist (Racehorse, 2023).

One of the most celebrated Rights of Nature victories is its inclusion in Ecuador's constitution. The involvement of Indigenous groups in its

---

[4]Ecocentric values, in this case, would guide a person to relate to Nature as a respected member of a shared community (as opposed to seeing Nature as an object to exploit).

development, along with referencing Indigenous concepts on human–nature relationships, enumerates the importance of Indigenous worldviews for developing new legal frameworks (Kauffman & Martin, 2017). Incorporating *Sumak Kawsay* (*buen vivir*) as the ideological grounds for establishing rights for both humans and Nature enables Ecuador to use RoN "as a tool for achieving an alternative model of sustainable development that challenges dominant neoliberal approaches" (Kauffman & Martin, 2017: 130). Since 2008, there have been over 60 RoN cases brought to the Ecuadorian court (Eco Jurisprudence Monitor 2024). Most significantly was the 2023 Los Cedros case in which the Rights of Nature were used to halt mining in a forest (Peck et al., 2024).

11.5.2 Rights of Nature as Creating New Eco-social Contracts: US and European Contexts

In societies with anthropocentric jurisprudence (such as the United States and European nations), RoN laws challenge societal norms by redefining eco-social contracts. Natalia Greene, Director of the Global Alliance for the Rights of Nature, highlighted the difficulty of implementing RoN in this context, stating, "Do not get discouraged if it's not happening as fast here…[the United States] is in the belly of the beast of capitalism" (Greene, 2024). European countries face a similar systemic challenge, with Enlightenment thought creating norms that emphasise human rights while indirectly reducing Nature to commodity (Ruales et al., 2024: 7).

Western capitalist societies have laws and cultural norms that treat Nature as property to be owned and used, not as a community member to be consulted. To address environmental degradation, key environmental laws were developed in the 1960s and 1970s in the United States, such as the Clean Air Act, Clean Water Act, and the National Environmental Protection Act. Groundbreaking for their time, they served as models for environmental laws around the world but remained largely focused on protecting Nature for human benefit (M'Gonigle & Takeda, 2013). It was not until 2006 that the United States saw its first attempt at a Rights of Nature ordinance. Since then, over 150 initiatives have been proposed in the United States, typically at the local level, with limited legal victories but strong symbolic value sparking transformative conversations (Huneeus, 2022). For instance, Toledo, Ohio, recognised the Rights of Lake Erie in 2019, but the law was quickly overturned in court (Kilbert, 2020). Despite the significant legal setback, the initiative still sparked other efforts towards the same goal, inspiring a 2024

New York state bill proposal advocating for the Rights of the Great Lakes. Alongside the enthusiasm for the movement, there is also a countercurrent of resistance—five US states have introduced legislation since 2019 to ban future Rights of Nature initiatives (Dwyer et al., 2024).

In recent decades, Europe has implemented extensive environmental legislation, including the EU Green Deal (2019), which aims for climate neutrality by 2050. However, like the United States, these laws often reflect anthropocentrism, prioritising economic interests over Nature's intrinsic value (Hoek et al., 2023). Inspired by the global Rights of Nature movement, Europe is seeing growing interest in RoN approaches. In 2022, Spain's Mar Menor lagoon became Europe's first ecosystem granted legal personhood, backed by over 600,000 signatures advocating for its protection and restoration (Stokstad, 2022). This law has sparked similar proposals across Europe but faces legal challenges akin to those in the United States.

11.5.3 Rights of Nature in Plural-Legal Systems

Rights of Nature laws often operate within settler-colonial contexts where Indigenous peoples' legal systems exist alongside imposed settler legal frameworks. Legal pluralism—the coexistence of multiple legal systems—provides a framework to assess RoN's adaptability and impacts, particularly in settings where Indigenous and Western legal traditions intersect (Dancer, 2021; O'Donnell et al., 2020). Dancer (2021) highlights the importance of integrating legal pluralism into sustainable development frameworks to avoid "double colonisation" of people and Nature and emphasises that colonial law is not the sole legal system.

Legal pluralism is unfolding around Rights of Nature initiatives. For example, RoN initiatives in countries such as New Zealand, Australia, the United States, Canada, Uganda, Nigeria, and South Africa have linked Indigenous and customary laws with statutory frameworks. This dialogue between legal systems offers both challenges and opportunities for recognising Nature's rights. Below we will briefly describe a case in Turtle Island (North America) that crosses several different nations.

Case Study: St. Lawrence River

The St. Lawrence River, situated in North America between the Great Lakes and the North Atlantic Ocean, exemplifies RoN emerging in a plural-legal space. The River crosses several nations in the Haudenosaunee Confederacy, the United States, Canada, and First Nations. A few of these communities have initiatives recognising the River's rights. These are briefly described in Table 11.1.

Table 11.1 Rights of Nature initiatives around the St Lawrence River

Initiative title and year	Location	Description	Sources
St. Lawrence River/ Kaniatarowanénhne Bill of Rights and Responsibilities (2022)	New York (US); Mohawk Nation (Haudenosaunee)	Combines a declaration and local ordinance to recognise the inherent rights of the River and human responsibilities towards Nature. Highlights that Indigenous communities have worked in harmony with the St. Lawrence River Watershed since time immemorial. Guided by the Kanianerenkó:wa (Haudenosaunee Great Law of Peace) and the Ohén:ton Karihwaténhkwen (Thanksgiving Address), it instructs communities to honour, respect, and care for the River's delicate balance	Eco Jurisprudence Monitor (2024d)
An Act to Confer Rights on the St. Lawrence River (2022)	Quebec, Canada	Provincial-level law acknowledging Indigenous nations' relationships with the River. Proposes a guardianship committee with Indigenous representatives. Notes existing Canadian laws that could be amended to grant legal status to the St. Lawrence River as a rights-bearing entity	Eco Jurisprudence Monitor (2024c), Lessard-Therrien, (2022)
St. Lawrence River Capacity and Protection Act (2022)	Canada (National Level)	National-level law proposed to recognise the legal personality and rights of the St. Lawrence River. Highlights the role of First Nations communities as traditional custodians and legal guardians, ensuring protection and promotion of the River's rights and interests	Eco Jurisprudence Monitor (2024b)
Resolution on the Rights of the St. Lawrence River (2023)	Assembly of First Nations Quebec-Labrador (AFNQL)	Establishes a First Nation alliance to protect the St. Lawrence ecosystem, aiming to grant legal personhood to the River. Proposes governance determined by First Nations and confirms the Chiefs of the AFNQL's role in all decisions affecting the River's present and future	Eco Jurisprudence Monitor (2024a)

Source: Authors

The Rights of the St. Lawrence River cases illustrate both legal and ontological pluralism. Dancer (2021) emphasises that legal pluralism is crucial to the ecological jurisprudence movement, as no single legal system can fully address the complexity of human–nature relationships. A pluralistic approach fosters collaboration across political and cultural boundaries, enabling diverse nations to establish Rights of Nature frameworks. It aligns with eco-social contract goals of addressing historical injustices and establishing a contract with Nature.

Lessons from this plurality approach, and empowerment of local communities, can be extended to climate adaptation strategies.

11.5.4 Knowledge Plurality in Climate Adaptation

Operationalising legal pluralism is essential for promoting Indigenous environmental justice and advancing Rights of Nature initiatives, particularly in the face of climate change (Parsons et al., 2021). This plurality extends beyond legal processes to the co-creation of knowledge, embracing diverse perspectives to confront the challenges of a changing climate. The co-creation of knowledge, which combines local traditional knowledge (LTK)[5] with Western science, is a bottom-up approach that integrates local values and interests with scientific expertise (Romero et al., 2018). This process can lead to more effective and sustainable adaptation strategies while avoiding the creation of new vulnerabilities (Eriksen et al., 2021; Hosen et al., 2020).

By reintegrating traditional practices once dismissed, coproduction of knowledge decolonises science and leverages modern technology to address climate change adaptations, including disruptions to long-standing patterns (Mardero et al., 2023; Mbah et al., 2021). By merging traditional ecological knowledge (TEK) with Western scientific approaches, communities can craft adaptation strategies that address human needs while respecting the inherent rights of ecosystems—a central tenet of RoN.

However, systemic challenges remain. For example, in Canada, Indigenous ways of knowing were stripped of their cultural significance when they were attempted to be included in national adaptation and mitigation policies, which resulted in the distortion or miscommunication of their knowledge in order to adhere to Euro-Canadian standards of legitimacy (Ellis, 2005). Ellis

---

[5] Where LTK can encompass traditional environmental knowledges, Indigenous knowledge, and local and ancestral knowledge.

Table 11.2 Eco-social contract goals and the Rights of Nature

Eco-social Contract Goal	Connection to Rights of Nature Movement
Human Rights	• Environmental and human flourishing are intertwined, and communities have been using Rights of Nature laws to protect some of their rights and well-being
Creating a Contract with Nature	• Shifts perspectives from Nature as an object to Nature as a subject • Human guardians or individuals can advocate for Nature • Fosters reciprocity, respect, and kinship
Addressing Historical Injustices	• Aims to repair broken relationships between humans and Nature by recognising Nature's inherent value • Addresses some harms of colonialism by supporting Indigenous rights, relationships, knowledge, sovereignty, and other traditional ways of knowing
Supporting New Forms of Solidarity	• Supports environmental justice and community rights • Embraces legal and ontological plurality

Source: Authors

(2005) argued that Indigenous knowledge has the potential to challenge the very institutions of Western society, and for that reason, it is often reduced to being a supplementary source of information, stripped of its value and thus its power and significance. These systemic challenges underscore the need for eco-social contracts that address epistemic injustice and honour diverse ontologies.

11.6   Using Rights of Nature to Meet Eco-social Contract Goals

The previous sections described the growth and application of the Rights of Nature. A summary of how the RoN movement touches on four key eco-social contract goals is seen in Table 11.2.

11.7   Future of the Rights of Nature Movement

The Rights of Nature movement is expanding. By 2024, over 400 Rights of Nature initiatives have been introduced in over 35 countries, with at least 5 at the national level (Eco Jurisprudence Monitor, 2024e). As the movement grows, questions remain about legal recognition, enforcement, and its broader impact. In some contexts, RoN is a stepping stone towards establishing cultural paradigm shifts towards ecocentrism. Using the language of "rights" gives the RoN debate a legal tone, but the Rights of Mother Earth is more than a debate of legal frameworks. In its deepest conception, RoN theory is not just asking whether or not we should give new environmental protections to Nature but posing a fundamental challenge to Western ontology by questioning whether Nature holds inalienable and inherent rights of the same value and of an equal hierarchical consideration as human rights (Sólon, 2018).

This call for systemic change is necessary for the creation of new eco-social contracts, and as Huntjens and René Kemp (2022) describe, these major transitions will take time and happen through connected actor-coalitions and interdependent systemic leverage points. The legal recognition of RoN still frequently clashes with laws that prioritise resource exploitation, economic growth, and property rights, but there are opportunities for social change outside the legal system. Through social learning[6] and knowledge co-creation across multiple ontologies, shared values can be realised both for humanity and for Mother Earth. Even in the case of legal setbacks, there exists an advancement in the social learning process to progress the movement. In the United States, for example, the Lake Erie Bill of Rights was overturned after a lawsuit from corporate interests (Kilbert, 2020). Despite this outcome, it still inspired the Great Lakes Bill of Rights initiative. The torch is carried forward to bring the movement to new constituents and to examine Nature within new positions of local or regional governance.

The Rights of Nature movement is also gaining traction in global governance, such as through the United Nations Harmony with Nature programme and the International Union for Conservation of Nature. A notable global success for the Rights of Nature is seen through the Kunming-Montreal Global Biodiversity Framework (GBF) adopted at COP15 in 2022 after a 4-year consultation process. This landmark framework supports the Sustainable Development Goals and acknowledges diverse value systems, including Nature's intrinsic value and the RoN. Section C, Point 7(b), of the Kunming-Montreal GBF states: "Nature embodies different concepts for different people, including biodiversity, ecosystems, Mother Earth, and systems of life. These diverse value systems and concepts, including the rights of

---

[6] Social learning refers to a change that occurs within wider social units, such as the values or norms of an organisation, as a result of social networks and interactions (Reed et al., 2010). Social learning can be a way of integrating LTK into adaptation strategies as a method of increasing climate change adaptation resilience and reducing vulnerability (Eriksen et al., 2021; Hosen et al., 2020).

Nature and rights of Mother Earth, are integral to its successful implementation" (Secretariat of the Convention on Biological Diversity, 2024).

All countries committed to developing National Biodiversity Strategies and Action Plans (NBSAPs) are set to implement the GBF. These plans outline national targets to integrate the Framework's principles, making them critical tools for translating global commitments into local action. Spain, Australia, Fiji, Aotearoa New Zealand, and Slovenia have incorporated intrinsic and ecocentric values into their NBSAPs. Some go further: Aotearoa, New Zealand, explicitly supports Rights of Nature, while Slovenia emphasises eco-centrism and other ecocentric biodiversity laws. These approaches reflect the growing recognition that valuing Nature's intrinsic worth supports advancing Rights of Nature initiatives (Earth Law Center, 2024).

The GBF integrates these principles into actionable targets. Target 19 commits funding for "Mother Earth-centric actions," encompassing ecocentric and rights-based measures, positioning the global Rights of Nature movement squarely within its scope. However, as noted by the Intergovernmental Science-Policy Platform on Biodiversity and Ecosystem Services (IPBES), biodiversity policies remain largely influenced by economic considerations. IPBES identifies over 50 ways of valuing Nature, emphasising the need to prioritise alternative values such as responsibility, reciprocity, and respect to effectively address biodiversity loss (IPBES, 2022). By incorporating diverse worldviews, the GBF provides a foundation for reshaping environmental laws to reflect humanity's broader relationship with Nature.

While the Rights of Nature movement is spreading, it also faces resistance due to its challenge to resource-extractive industries. RoN laws, like other attempts at environmental protection, are seen as a threat to profit-driven economic models that prioritise short-term monetary gain over long-term ecological well-being. For example, *buen vivir* emphasises establishing development priorities within the context of local norms and cultural values (Kauffman & Martin, 2017). It can be an alternative pathway to sustainable development by which development is determined through a collaborative creation of ideas and processes among communities and will differ based on social and environmental contexts of those communities (Chassagne, 2019; Kauffman & Martin, 2017). It is a framework that encourages and protects diversity—particularly in the suggestion of a "plurinational State, meta-ecological citizenship and people-centered development" (Fatigato, 2023: 12–13)—and stands in stark contrast to the ideal of economic growth and neoliberal development, which further divided the world according to developed vs. developing (Fatigato, 2023). Implementing the *buen vivir* concept of development and adaptation differs from models that are constructed from the top-down that prioritise economic development. Recognising Nature as a rights-bearing entity, as opposed to commodifying it as property to own or as a resource to consume, challenges dominant consumptive capitalist structures. These tensions are complicated to overcome and are an ongoing challenge for the Rights of Nature movement.

11.8 Conclusion

The Rights of Nature movement answers the call "How do we live in harmony with Nature?" by creating a foundation where Nature's inherent rights are recognised. It is part of a growing ecological jurisprudence that realigns human laws with ecological systems. Through this, it supports the creation of new eco-social contracts that support both ecological integrity and social justice.

The rapid growth of the Rights of Nature movement shows the widespread demand for new eco-social contracts. The polycrisis of today illustrates the need to heal our relationships with nature and with one another. As one tool for eco-social contracts, RoN offers a transformative framework for societies to codify ecologically responsible relationships with nature. And as a global movement, RoN honours the importance of legal and ontological pluralities in this work.

The value in the Rights of Nature movement does not strictly lie within its legal abilities and advances, but the mere pursuit of Rights of Nature in some regions encourages citizens to critically examine their own relationship with their environment. Therefore, perhaps success in the Rights of Nature movement is not simply measured by embedding these rights into law but also by encouraging the recognition of Nature's rights in people's minds.

By embedding Nature's rights into law and thought, RoN provides a practical and ethical foundation for the eco-social contracts needed to secure a sustainable future for all beings on the planet.

References

Arrows, W. T. F., & Narvaez, D. (2022). *Restoring the kinship worldview: Indigenous voices introduce 28 precepts for rebalancing life on planet Earth*. North Atlantic Books.

Berry, T. (1999). *The great work: Our way into the future*. Bell Tower.

Burdon, P. D. (2012). A theory of Earth jurisprudence. *Australian Journal of Legal Philosophy, 37*, 28.

Calzadilla, P. V., & Kotzé, L. J. (2018). Living in harmony with nature? A critical appraisal of the rights of Mother Earth in Bolivia. *Transnational Environmental Law, 7*(3), 397–424. https://doi.org/10.1017/S2047102518000201

Chassagne, N. (2019). Sustaining the 'good life': Buen Vivir as an alternative to sustainable development. *Community Development Journal, 54*(3), 482–500.

Dancer, H. (2021). Harmony with nature: Towards a new deep legal pluralism. *The Journal of Legal Pluralism and Unofficial Law, 53*(1), 21–41. https://doi.org/1 0.1080/07329113.2020.1845503

Davis, H., & Todd, Z. (2017). On the importance of a date, or decolonizing the anthropocene. *ACME: An International Journal for Critical Geographies, 16*(4), 761–780. https://doi.org/10.14288/acme.v16i4.1539

Deckha, M. (2013). Initiating a non-anthropocentric jurisprudence: The rule of law and animal vulnerability under a property paradigm. *Alberta Law Review, 50*, 783–783. https://doi.org/10.29173/alr76

De la Cadena, M., & Blaser, M. (2018). *A world of many worlds*. Duke University Press.

Dwyer, E., Bleksley, A., & Kol, D. (2024, June 19). *Memorandum: State legislature bills banning rights of nature laws*. Earth Law Center.

Earth Law Center. (2024). *Ecocentrism in the global biodiversity framework*. Earth Law Center.

Eco Jurisprudence Monitor. (2024a). *Assembly of first nations Quebec-Labrador resolution recognizing the rights and legal personhood of the St. Lawrence River*. Eco Jurisprudence Monitor. Accessed January 10, 2025, from https://ecojurisprudence.org/initiatives/first-nations-resolution-recognizing-the-rights-of-the-st-lawrence-river/

Eco Jurisprudence Monitor. (2024b). *Canada bill to recognize the rights and legal personality of the St. Lawrence River*. Eco Jurisprudence Monitor. Accessed January 10, 2025, from https://ecojurisprudence.org/initiatives/canada-act-giving-legal-capacity-to-the-st-lawrence-river/

Eco Jurisprudence Monitor. (2024c). *Quebec, Canada bill to recognize the rights of the Saint Lawrence River*. Eco Jurisprudence Monitor. Accessed January 10, 2025, from https://ecojurisprudence.org/initiatives/quebec-canada-bill-to-recognize-the-rights-of-the-saint-lawrence-river/

Eco Jurisprudence Monitor. (2024d). *North Country (U.S.) citizen ordinance and declaration of the rights and responsibilities of the St. Lawrence River / kaniatarowanénhne*. Eco Jurisprudence Monitor. Accessed January 10, 2025, from https://ecojurisprudence.org/initiatives/north-country-u-s-bill-of-rights-and-responsibilities-of-the-st-lawrence-river-kaniatarowanenhne/

Eco Jurisprudence Monitor. (2024e). *Toolkit*. Accessed January 10, 2025, from https://ecojurisprudence.org/initiatives/

Ellis, S. C. (2005). Meaningful consideration? A review of traditional knowledge in environmental decision making. *Arctic, 58*(1), 66–77. https://www.jstor.org/stable/40512668.

Eriksen, S., Schipper, E. L. F., Scoville-Simonds, M., Vincent, K., Adam, H. N., Brooks, N., Harding, B., Khatri, D., Lenaerts, L., Liverman, D., Mills-Novoa, M., Mosberg, M., Movik, S., Muok, B., Nightingale, A., Ojha, H., Sygna, L., Taylor, M., Vogel, C., & West, J. J. (2021). Adaptation interventions and their effect on vulnerability in developing countries: Help, hindrance or irrelevance? *World Development, 141*(2021), 1–16. https://doi.org/10.1016/j.worlddev.2020.105383

Fatigato, M. C. (2023). *Buen Vivir: An opportunity to re-think the development and sustainability model* (No. 10/23). Sapienza University of Rome, DISS.

Foucault, M. (1975). Discipline and punish. In *Social theory re-wired* (pp. 291–299). Routledge.

Gaventa, J., & Cornwall, A. (2006). Challenging the boundaries of the possible: Participation, knowledge and power. *IDS Bulletin, 37*(6), 122–128.

Greene, N. (2024). *Reckoning with the truth to get in right relationship*. Accessed January 10, 2025, from https://www.youtube.com/watch?v=onp0TSv3jxo

Haraway, D. (1988). Situated knowledges: The science question in feminism and the privilege of partial perspective. *Feminist Studies, 14*(3), 575–599. https://doi.org/10.2307/3178066

Hoek, N., Kaststeen, I., van Gils, S., Janssen, E., & van Gils, M. (2023). Implementing rights of nature: An EU natureship to address anthropocentrism in environmental law. *Utrecht Law Review, 19*(1). https://doi.org/10.36633/ulr.880

Hosen, N., Nakamura, H., & Hamzah, A. (2020). Adaptation to climate change: Does traditional ecological knowledge hold the key? *Sustainability, 12*(2), 1–18. https://doi.org/10.3390/su12020676

Huneeus, A. (2022). The legal struggle for rights of nature in the United States. *Wisconsin Law Review*, 133.

Huntjens, P. (2021). *Towards a natural social contract: Transformative social-ecological innovation for a sustainable, healthy and just society*. Springer International Publishing. https://doi.org/10.1007/978-3-030-67130-3

Huntjens, P., & Kemp, R. (2022). The importance of a natural social contract and co-evolutionary governance for sustainability transitions. *Sustainability, 14*(5), 2976. https://www.mdpi.com/2071-1050/14/5/2976

Huntjens, P., & Kemp, R. (2025). The transformation flower approach for eco-social contracting: Comparative insights from eight case studies in the Global South and North. In P. Huntjens, N. Mohamed, K. Hujo, & M. Desai (Eds.), *Eco-social contracts for sustainable and just futures* (Chapter 16, pp. 283–312). Springer Nature.

IPBES. (2022). Methodological assessment report on the diverse values and valuation of nature of the intergovernmental science-policy platform on biodiversity and ecosystem services. In P. Balvanera, U. Pascual, M. Christie, B. Baptiste, & D. González-Jiménez (Eds.). IPBES Secretariat. https://doi.org/10.5281/zenodo.6522522

IPBES. (2024). Thematic assessment report on the underlying causes of biodiversity loss and the determinants of transformative change and options for achieving the 2050 vision for biodiversity of the intergovernmental science-policy platform on biodiversity and ecosystem services. In K. O'Brien, L. Garibaldi, & A. Agrawal (Eds.). IPBES Secretariat. https://doi.org/10.5281/zenodo.11382215

Kauffman, C., Haas, C., Bajpai, S., Leonard, K., Macpherson, E., Martin, P., Pelizzon, A., Putzer, A., & Sheehan, L. (2024). *Eco Jurisprudence Monitor*. https://ecojurisprudence.org

Kauffman, C. M., & Martin, P. L. (2017). Can rights of nature make development more sustainable? Why some Ecuadorian lawsuits succeed and others fail. *World Development, 92*(April), 130–142. https://doi.org/10.1016/j.worlddev.2016.11.017

Kilbert, K. (2020). Lake Erie Bill of Rights: Stifled by all three branches yet still significant. *Ohio St. LJ Online 81*, 227.

Lessard-Therrien, E. (2022). *Bill 990, An Act to Confer Rights on the St. Lawrence River - National Assembly of Québec.* https://www.assnat.qc.ca/en/travaux-parlementaires/projets-loi/projet-loi-990-42-2.html?appelant=MC

Martinez, D. J., Cannon, C. E. B., McInturff, A., Alagona, P. S., & Pellow, D. N. (2023). Back to the future: Indigenous relationality, kincentricity and the North American model of wildlife management. *Environmental Science and Policy, 140*, 202–207. https://doi.org/10.1016/j.envsci.2022.12.010

Mardero, S., Schmook, B., Calmé, S., White, R. M., Rehema, M., Chang, J. C. J., Casanova, G., & Castelar, J. (2023). Traditional knowledge for climate change adaptation in Mesoamerica: A systematic review. *Social Sciences and Humanities Open, 7*(1), 1–15. https://doi.org/10.1016/j.ssaho.2023.100473

Mbah, M., Ajaps, S., & Molthan-Hill, P. (2021). A systematic review of the deployment of indigenous knowledge systems towards climate change adaptation in developing world contexts: Implications for climate change education. *Sustainability, 13*(14811), 1–24. https://doi.org/10.3390/su13094811

M'Gonigle, M., & Takeda, L. (2013). The liberal limits of environmental law: A green legal critique. *Pace Environmental Law Review, 30*(3), 1005. https://doi.org/10.58948/0738-6206.1730

Mohamed, N., & Huntjens, P. (2023). *Dismantling the ecological divide: Toward a new eco-social contract.* UNRISD Issue Brief 15. https://cdn.unrisd.org/assets/library/briefs/pdf-files/2023/ib15-a-new-contract-with-nature.pdf

Moore, J. W. (2017). The capitalocene, part I: On the nature and origins of our ecological crisis. *The Journal of Peasant Studies, 44*(3), 594–630. https://doi.org/10.1080/03066150.2016.1235036

Nash, R. F. (1989). *The rights of nature: A history of environmental ethics.* University of Wisconsin Press.

O'Donnell, E., Poelina, A., Pelizzon, A., & Clark, C. (2020). Stop burying the Lede: The essential role of indigenous law(s) in creating rights of nature. *Transnational Environmental Law, 9*, 403–427. https://doi.org/10.1017/S2047102520000242

Parsons, M., Fisher, K., & Crease, R. P. (2021). Legal and ontological pluralism: Recognising rivers as more-than-human entities. In M. Parsons, K. Fisher, & R. P. Crease (Eds.), *Decolonising blue spaces in the anthropocene* (pp. 235–282). Springer International Publishing. https://doi.org/10.1007/978-3-030-61071-5_6

Peck, M. R., Desselas, M., Bonilla-Bedoya, S., Redín, G., & Durango-Cordero, J. (2024). The conflict between rights of nature and mining in Ecuador: Implications of the Los Cedros cloud forest case for biodiversity conservation. *People and Nature, 6*(3), 1096–1115. https://doi.org/10.1002/pan3.10615

Pelizzon, A. (2025). *Ecological jurisprudence: The law of nature and the nature of law.* Springer Nature.

Plumwood, V. (2012). Decolonizing relationships with nature. In *Decolonizing nature* (pp. 51–78). Routledge.

Racehorse, V. (2023). Indigenous influence on the rights of nature movement. *Natural Resources and Environment, 38*(2), 4.

Reed, M. S., Evely, A. C., Cundill, G., Fazey, I., Glass, J., Laing, A., Newig, J., Parrish, B., Prell, C., Raymond, C., & Stringer, L. C. (2010). What is social learning? *Ecology and Society, 15*(4).

The Republic of Ecuador National Assembly. (2008a). "Preamble". In *Constitution of the Republic of Ecuador*. English translation of the original text of the constitution of 2008: [8]-[9]. https://heinonline.org/HOL/P?h=hein.cow/zzec0028&i=8

The Republic of Ecuador National Assembly. (2008b). Chapter seven: Rights of nature articles 71-74. In *Constitution of the Republic of Ecuador* English translation of the original text of the Constitution of 2008: [44]-[45]. https://heinonline.org/HOL/P?h=hein.cow/zzec0028&i=44

Ruales, J. G., Hovden, K., Kopnina, H., Robertson, C. D., & Schoukens, H. (2024). *Rights of nature in Europe: Encounters and visions.* Taylor and Francis.

Rodríguez-Garavito, C. (2024). *More than human rights: Law, thought and narrative for earthly flourishing.* NYU MOTH Project.

Romero, M., David, S. C., & Pereira, Â. G. (2018). Climate-related displacements of coastal communities in the Arctic: Engaging traditional knowledge in adaptation strategies and policies. *Environmental Science and Policy, 85*(2018), 90–100. https://doi.org/10.1016/j.envsci.2018.04.007

Salmón, E. (2000). Kincentric ecology: Indigenous perceptions of the human–nature relationship. *Ecological Applications, 10*(5), 1327–1332. https://doi.org/10.1890/1051-0761(2000)010[1327:KEIPOT]2.0.CO;2

Secretariat of the Convention on Biological Diversity. (2024). *Kunming-Montreal global biodiversity framework.* Convention on Biological Diversity. https://www.cbd.int/gbf

Sólon, P. (2018). The rights of Mother Earth. In V. Satgar (Ed.), *The climate crisis (South African and global democratic eco-socialist alternatives)* (pp. 107–130). Wits University Press. https://doi.org/10.18772/22018020541.10

Stokstad, E. (2022). This lagoon is effectively a person, new Spanish law says. *Science (New York, N.Y.), 378*(6615), 15–16. https://doi.org/10.1126/science.adf1848

Stone, C. D. (1972). Should trees have standing—Toward legal rights for natural objects. *Southern California Law Review, 45*(2), 450–501.

UNRISD. (2021). *A new eco-social contract. Vital to deliver the 2030 agenda for sustainable development.* UNRISD issue brief 11. UNRISD.

United Nations Environment Programme (UNEP). (2020). *From Islam to Buddhism, faiths have long encouraged stewardship of nature.* UNEP. https://www.unep.org/news-and-stories/story/islam-buddhism-faiths-have-long-encouraged-stewardship-nature

Van Horn, G., Kimmerer, R. W., & Hausdoerffer, J. (2021). *Kinship: Belonging in a world of relations.* Center for Humans and Nature.

What are the Rights of Nature - Global Alliance for the Rights of Nature (GARN). (n.d.). Accessed January 10, 2025, from https://www.garn.org/rights-of-nature/

12

Putting Food Justice at the Centre of an Eco-social Contract

Kiah Smith

12.1 Introduction

Rising global hunger makes it overwhelmingly clear that current food system dynamics represent a breakdown in the social contract. While we produce enough food to feed the global population 1.5 times over (FAO et al., 2022; Holt-Giménez et al., 2012), scholarship widely demonstrates that hunger results from inequitable access to healthy, affordable and sustainable diets, alongside damaging production–consumption practices that generate food waste (Willett et al., 2019). High food prices are driven by supermarket concentration, financialised supply chains and the environmental 'externalities' of industrial food systems, at the expense of local and regional agri-food cultures and economies (Lang & Heasman, 2015; Spencer, 2024). Decisions over food governance are often beyond the control of ordinary people, while industry and finance hold disproportionate power (Anderson, 2024; Clapp et al., 2022). Hunger is further reproduced by rising income poverty and social exclusion, both products of power imbalances and strongly associated with class, race and gender inequalities (Alkon & Agyeman, 2011; Schneider & Chappell, 2017).

K. Smith (✉)
Centre for Policy Futures, University of Queensland, Brisbane, Australia
e-mail: k.smith2@uq.edu.au

P. Huntjens et al. (eds.), *Eco-Social Contracts for Sustainable and Just Futures*,
https://doi.org/10.1007/978-3-031-99109-7_12

This chapter describes how a new eco-social contract can be extended through the principles and practices of *food justice*, drawing on findings from the 'Fair Food Futures' project—an Australian Research Council study that explores how Australian civic food networks (CFNs) envision and work towards food justice across movements and policy spaces, as they seek to scale up transformative food systems change. Food justice aims to eliminate inequalities that cause hunger, ensure that benefits in food systems are shared more fairly and improve people's control over the food system (Gottlieb & Joshi, 2010). Food justice therefore goes beyond availability, access, utilisation and stability of the food system (the definition of food security) to redress *how*, *by whom* and *for whom* food is produced, distributed, consumed and governed. Drawing on empirical insights from future scenario building with civil society food movements, this chapter identifies four principles for eco-social food justice. These principles are informed by the theory and practice of justice in Australia from the perspective of CFNs, a group that has largely been 'left out' of food policymaking despite their key role in providing alternative pathways for food security on the ground. The four principles aim to shift unequal power relations by embedding rights for all, intersectional solidarity, resilience, and 'food as commons' concepts. The chapter concludes with recommendations for putting these principles into practice, paying particular attention to increasing civic participation of marginalised groups in debating and mobilising food justice solutions.

12.2 From Food Security to Food Justice

In 2022, 2.3 billion people (29.3%) globally were moderately or severely food insecure, an increase of 350 million since the outbreak of the COVID-19 pandemic (FAO et al., 2022), which has highlighted the undeniable connection between food security, food system vulnerability and food (in)justice. Food security is defined by the FAO (2001) as 'when all people, at all times, have physical, social and economic access to sufficient, safe and nutritious food that meets their dietary needs and food preferences for an active and healthy life'. This has four pillars: availability (supply, quantity, quality), access (capacity to afford a nutritious diet), utilisation (knowledge, skills, infrastructure) and stability (over time). However, the overemphasis on the pillar of food availability (supply, quantity and quality) has been widely criticised for privileging efforts to increase food production and its inputs (finance, technology, data) and outputs (exports, trade, markets) (Carolan, 2013; Holt-Gimenez, 2017) rather than addressing the root causes of hunger—the poverty and economic hardship that make good food unaffordable for many people. Growing *more food* dominates agricultural policymaking but has not led to reduced hunger. While the other pillars of food security (access, utilisation and stability) suggest that distributional inequities have a role to play, it also remains that people's agency (Clapp et al., 2022) that is, their *right to food, right to farm* and *right to shape their own relationships with the food system* are dimensions of food security that are consistently left out of mainstream food governance. These critiques highlight the need for decision-making to become more participatory and democratic, especially in light of high corporate concentration in industrial food systems (IPES-Food, 2017). In sum, improving food security has less to do with increasing production (or imports/exports), and much more to do with orienting food governance towards policies that improve responsible production, distribution and consumption of healthy and nutritious food from more sustainable and equitable food systems (Andrée et al., 2019). This in turn requires democracy and food justice (Levkoe, 2006) to be placed at the heart of a new eco-social contract.

Although Australia is a wealthy food-exporting nation, domestic hunger is significant. Food insecurity affects one in five Australians, or 3.7 million households (Foodbank, 2023). Indigenous Australians are five to six times more likely to be food insecure than other Australians (ABS, 2015), largely as a direct consequence of settler-colonial relations, Indigenous exclusion from land and government under-investment in remote communities where many Indigenous people live (Staines & Smith, 2021). Women are more likely to experience hunger than men, as do migrants and refugees, people living rurally, older people, young people, renters, people on welfare support payments, families with children and people experiencing homelessness and poverty (McKay et al., 2019; Yii et al., 2019). Global food prices are higher now than they were during the 2008 food price crisis (FAO, 2022); in Australia, the cost of fresh food and overall grocery bills have almost doubled in the past two years (Lewis et al., 2024). Supply chain failures during the COVID-19 pandemic impacted food access, affordability and availability (Nemes Gusztáv et al., 2021; Blay-Palmer et al., 2020) further exposing the limits of global food trade and just-in-time supermarket supply chains. While this was a moment of disruption, it was not an isolated one (Stead & Hinkson, 2022). And while food charities provide short-term hunger relief, they address the symptoms but not the causes of hunger and indicate that something is wrong with the food system (Lindberg et al., 2015; Pollard et al., 2018; Riches, 2018).

Where the current social contract is based on ecological extractivism and 'consumptogenic markets' (Frank et al., 2024) that perpetuate profits in the corporate-industrial food complex, an eco-social contract that puts food justice at the centre of policy reform would be vastly different. Food justice is defined as "ensuring more equitable access to food that is ecologically sustainable, healthy and fairly produced, exchanged and consumed" (Gottlieb & Joshi, 2010:6). With attention to social-ecological resilience, sustainability and transformation, food justice borrows from the concepts of environmental justice and food sovereignty to highlight the range of inequalities created by and affecting the food system, seeing inequalities as unevenly distributed and having roots in wider patterns of injustice associated with capitalist food systems, histories of colonisation and patriarchy (Cadieux & Slocum, 2015; Schlosberg, 2007). Food justice emerged from US environmental and social justice movements, stemming from the recognition of the food system itself as a racialised, gendered and classist 'project' in which "communities of colour and poor communities have time and again been denied access to the means of food production, and due to both price and store location, often cannot access the diet advocated by the food movement" (Alkon & Agyeman, 2011:5). This parallels the history of settler colonialism in Australia through which agriculture and farming have been locked into complex relationships with Indigenous dispossession and sovereignty, and with ongoing racial inequalities in food access for Indigenous people especially in remote communities (Mayes, 2018).

Today, food justice represents a diverse transnational movement, encompassing many values and practices. This is referred to as the 'plurality of justice' (Schlosberg, 2007), whereby food justice is increasingly understood as being "enacted in situated contexts in response to multidimensional, embodied justice… practiced (or performed) often in messy, ongoing, quotidian ways that seek to heal and repair in our imperfect world" (Coulson & Milbourne, 2021: 46). They argue for a 'justice multiple' approach that can "incorporate a diversity of justice framings which are shaped by various spatial, temporal and scalar relations" (Coulson & Milbourne, 2021:45). Conceptually, food justice follows four main theoretical threads. First, distributional justice addresses how benefits and burdens are allocated between different individuals and groups, with a focus on fairness. Solutions aim to eliminate disparities and inequities by paying greater attention to how "the benefits and risks of where, what and how food is grown, produced, transported and distributed, and accessed and eaten are shared fairly" (Gottlieb & Joshi, 2010: 6). Second, relational justice emphasises actions to affect systemic change by addressing oppressive power relations, and advocate "greater control over food production and consumption by people who have been marginalised by mainstream agri-food regimes" (Cadieux & Slocum, 2015:3). Recognising that everyone should be able to express their needs, emancipatory strategies are therefore needed to empower and enable the most marginalised people to challenge control of food system decisions, pluralise the voices that shape governance and improve participatory policymaking as well as participatory outcomes. Both of these forms of justice require *recognition* of the intersectional nature of systemic injustices that stem from historical and ongoing marginalisation of some groups more than others. Through colonisation, Indigenous communities have been subject to laws and policies that have taken away their ability to own land for food production; under patriarchy, women's labour and ecological knowledge has been undervalued; while capitalism and the corporate food regime have consistently and violently displaced smallholder farmers, exploited workers and reproduced class-based food poverty and environmental externalities (Holt-Gimenez, 2011: 313; Cadieux & Slocum, 2015; Moragues-Faus, 2017). In addressing race, class and gender inequalities, food justice thus recognises the intersectional sources of marginalisation and the disproportionate negative consequences of that marginalisation for diverse subjects (de Bruin et al., 2023).

Coulson and Milbourne (2021) further argue that food justice be extended to encompass justice across time and space, and the more-than-human. Third, in the just sustainability approach, considerations of who benefits and who doesn't across time and space is based on the normative assertion that both social and environmental justice need to be sustained for future generations (Agyeman et al., 2016). This view widens obligations across temporalities and acknowledges that translocal power relations are experienced in situated contexts across multiple scales, and thus shape how different actors contribute to causing and solving inequalities: a concept that closely reflects the concept of 'common but differentiated responsibilities' enshrined in Article 7 of the Rio Declaration (1992). Fourth, eco-social justice goes beyond anthropocentric framings of justice to incorporate non-human—plants and animals, biodiversity, technologies and food itself. This type of justice recognises human and non-human collaboration, and that interconnected injustices are embedded and experienced in ecologies and can impede the basic capabilities and functioning of human-ecological systems. Human beings are seen as interdependent with nature, and thus responsible for developing economies and societies oriented towards well-being for both humans and nature (Waddock, 2024). Building solidarity across social and environmental justice movements is therefore increasingly relevant to food justice, as the permaculture principles of *earth care, people care* and *fair share* articulate (Coulson & Milbourne, 2021). This alignment between food justice, ecologies and sustainability is such that:

The vision espoused by many food justice activists goes beyond one in which wealthy consumers vote with their forks in favour of a more environmentally sustainable food system to imagine that all communities, regardless of race or income, can have both increased access to healthy food and the power to influence a food system that prioritises environmental and human needs over agribusiness profit. (Alkon & Agyeman, 2011: 6)

To expand food security, the concept and practice of food justice provides a necessary foundation for enabling a new approach to eco-social justice. With plural theoretical (and normative) drivers, food justice recognises the collective trauma experienced as a result of historical inequality, with traditionally marginalised groups (people of colour, women, migrants, workers) as key actors alongside local and national networks and coalitions (Anderson, 2019; Andrée et al., 2019; Cadieux & Slocum, 2015). For Australian CFNs, the most important driver of food justice is the growth of localised, decentralised and 'regenerative' food networks, although these are somewhat constrained by the dominant model of long supply chains, export-orientation and processed food. The economic model that generates food waste is identified as a key barrier to more 'just' food systems, prompting solutions that re-imagine economies based on degrowth, circular economy and social economic practices. Participatory governance (or lack of) is seen by CNFs as both a lever for, and barrier to, progressing social justice in the food system. In exploring these and other factors impacting food justice—such as new thinking about food charities, food and climate activism, technological innovation and human rights—Australian CFNs widely argue for the need to achieve food justice in a more integrated way (Freeman & Smith, 2024; Smith, 2019).

While global organisations have increasingly acknowledged the importance of social exclusion and inclusion in understanding food system transformation—and recommend rights, participation, empowerment and capacity building to address food-related social inequalities affecting vulnerable groups (women, the poor, Indigenous, smallholders, etc.)—food justice discourse sits outside of these debates, as a more radical framing than food security. This is because, in addition to addressing the immediate problems of hunger, malnutrition, food insecurity and environmental degradation, a much more difficult task is to address the root causes of hunger and change the structure of the food system itself (Holt-Gimenez, 2011). Food justice scholars argue that this is an *ideological* challenge about shifting understandings of causes and expanding visions of possible futures, a *political* challenge about how to govern shared resources to sustain people and planet, and an *economic* challenge to foster innovations for producing, exchanging and consuming food in ways that are more equitable and sustainable (Loh & Agyemon, 2019: 216).

12.3 Case Study: Investigating Fair Food Futures in Australia

Australia has a vibrant and growing 'fair food' movement that contributes diverse practices of eco-social food justice (Smith, 2019; Canal Vieiria et al., 2020). Civic 'fair food' networks encompass a wide range of alternative food provisioning actors (i.e. producers, distributors and consumers) and advocacy coalitions (i.e. grassroots alliances, policy networks), who have long recognised the connection between reducing unequal power relations (including those based on gender, race and class) and improving democratic (i.e. just) access to healthy and affordable food. Over the past decade of researching food system sustainability, resilience and governance, I have looked to 'civic food networks' for evidence of the types of deeper transformations that are needed in Australia (where I live) (Freeman & Smith, 2024; Smith, 2022a, 2022b; Smith et al., 2015). Civic food networks include three main types of practices or actors:

- Alternative/local food provisioning, which includes producing food in community and backyard gardens, and (re)distributing food through community supported agricultures, food coops, food hubs, food swaps and food relief charity networks.
- Eco-social practices that seek to centre sustainability in food systems, by connecting economy, environment and society in approaches based in agroecology, solidarity economy, degrowth and decolonisation.
- Civic food governance, as indicated by the growth of 'fair food' coalitions, networks and research and policy advocacy initiatives, and mechanisms such as food policy councils and local food plans.

Few scholarly studies have sought to connect the situated visions and practices associated with fair food CFNs in Australia with a more "pluralistic, embodied and less universalistic notion of justice" being advanced by Coulson and Milbourne (2021:45), Schlosberg (2007) and others (de Bruin et al., 2023). In response, my *Fair Food Futures* project has engaged over 150 participants between 2019 and 2023—from small-scale producers, alternative distributors, community gardeners, food charity representatives, policymakers, academics and 'food citizens'—in case studies, interviews and participatory workshops, with the aim of co-creating community-led visions of food justice in Australia. The study's methodology combined participatory futures thinking and visual methods with the theory and practice of food utopias (Stock et al., 2015), to explore how people across the country are re-imagining what a better food system might look like, and to identify policy recommendations for food system reform. We asked: *What does your fair food future look like, and how do we get there?* By placing food justice very centrally in our discussions about hunger, food poverty, and systemic issues such as race, class and gender when talking about the problems and solutions to food injustices in Australia, the project has resulted in a series of utopian 'storylines' for reclaiming food economies and addressing systemic injustice. These provide lessons for constructing a new eco-social contract, precisely because their solutions radically shift focus from food security to the interconnected social and environmental inequalities associated with power, privilege, and oppression (of people and planet) within the food system.

12.4 Four Principles for Eco-social Food Justice

In *Fair Food Futures*, our co-designed future scenarios identified four principles that are needed to progress food justice in a more integrated and transformative way. Each principle presents a shared normative narrative for re-organising social-environmental-economic relations, quite different to the neoliberal status quo. They allow us to imagine the urgent transformation of gender-patriarchy, indigeneity-colonialism and ecology-capitalism, and in doing so, re-position CFNs (encompassing individuals and movements) as key implementation agents for a new eco-social contract.

12.4.1 Food and Rights for All

The first principle builds on long-standing food movement advocacy around the Right to Food, reinforcing human rights as the foundational element in any just transition and new eco-social contract (Krause et al., 2022; UNRISD, 2021). Article 25.1 of the Universal Declaration of Human Rights sets out the right to food, stating: *Everyone has the right to a standard of living adequate for the health and well-being of himself and of his family, including food, clothing, housing and medical care and necessary social services, and the right to security in the event of unemployment, sickness, disability, widowhood, old age or other lack of livelihood in circumstances beyond his control.* By "ensuring that all people have the capacity to feed themselves in dignity" (Zeigler, 2011), the right to food requires safeguarding people's entitlements, especially for vulnerable groups, to grow healthy food and access affordable or redistributed food. It also asks people to determine their own food system through democratic, participatory governance and requires that the people with the most at stake are included and empowered (Special Rapporteur on the Right to Food, 2021), thus placing responsibility (and accountability) on nation states to proactively strengthen policies that ensure equitable access to healthy, affordable and culturally appropriate food (Sampson et al., 2021).

Legislating the *Right to Food* into national and state food policies and human rights instruments requires a major redistribution of power within the food system, as well as a redesign of governance processes to better facilitate concrete human rights outcomes. This principle reflects a dominant theme in CFN actors' future aspirations for transforming food systems, whereby they described the unequal distribution of hunger as a significant human rights issue and argued that there has not been enough progress towards pursuing food justice within a human rights framework in Australia. It is clear that the two dominant modes of food provisioning in Australia—supermarket retail and charity-operated food relief—are failing to adequately fulfil people's rights and entitlements to safe, nutritious and sufficient food (Godrich et al., 2021). This was summed up by one participant in our study, who said:

> If we really want to make progress on this in Australia and elsewhere, we need to reframe around the right to food. I think that particularly governments at all levels need to recognise their responsibility, their role in helping people to realise that right within a sustainable, resilient, fair food system. (Food policy expert, January 2020)

A rights-based pathway diverges from market and charity solutions; it enables a root-cause approach that seeks to identify and overcome the inequalities causing food insecurity, support population food rights and food system ecological sustainability and provide a range of co-benefits for health equity, trade, aid and welfare policy (Lindberg et al., 2019). Participants further emphasised the link between everyone's right to food and supporting farmer well-being, decoupling food production from fossil fuels, moving away from charity dependence and supporting the related concept of food sovereignty. While food security focuses on maximising healthy food production and access, food sovereignty extends a deeper vision that "sees food as being integral to local cultures, closes the distance between production and consumption, is based on local knowledge and seeks to democratize the food system" (Wittman et al., 2010:7). This principle provides a pathway to address the structural causes, not just the symptoms, of social inequities, and calls for a redesign of governance to better facilitate concrete human rights outcomes as an urgent priority; this is explained by an urban farmer in our study, who said:

Everyone should have the right to go to bed without hungry stomach and should have access to quality food. At the same time, farmers also get their fair share. Right to Food is core to food sovereignty but we're achieving it by changing food systems, not by foodbanks. (Urban farmer, August 2020)

12.4.2 Intersectional Solidarity and Care

Building on the right to food, CFNs also demand greater recognition of a broad range of intersectional rights—such as the rights of women and children, migrant and refugee rights, labour rights, the right to health and the rights of Indigenous peoples. The second principle adopts an intersectional approach that centres care and reparation to people and planet through improving solidarity and reciprocity. While the provision of food charity is a deeply entrenched strategy for redistribution within neoliberal welfare states (such as in Australia) (Richards et al., 2016), by providing short-term hunger relief and 'filling the gap' in people's ability to purchase food, they do not address the causes of income poverty or symptoms of high food prices within supply chains (Dowler & O'Connor, 2012).

The principle of intersectional solidarity and care diverges from charity within neoliberal market economies. Solidarity pushes beyond benevolence to explicitly challenge capitalist power relations by emphasising (a) reciprocity and accountability over competition, (b) local contexts and livelihoods over macroeconomics; (c) social networks based on democratic structures, and (d) justice, empowerment and active citizenship (Dacheux & Goujon, 2011; Loh & Agyemon, 2019). The term 'intersectionality' highlights the interconnectedness between people's experiences of race, gender, class, income levels, indigeneity, ability or disability, education or location and how they experience food system inequality—particularly access to healthy, affordable food. Reflecting critiques from global decolonisation and feminist movements, this principle commits food system actors to redressing enduring past injustices (e.g. colonialism, racism, sexism) by connecting food access with wider issues such as housing, income and healthcare systems in which women and Indigenous people and people experiencing poverty are also widely disadvantaged. This is reflected in the below quote from one of our participatory workshops in Brisbane:

What we need is some kind of solidarity economy, which then means, okay, what are our other costs of living. So we kept coming back to this idea of intersectional movements, where it's not just about the food movement, it really is about housing security and housing as a commodity, and workers' rights, and women's rights, and all those other social justice intersections. (Community supported agriculture organiser, September 2021)

This form of food justice recognises the differential experiences of inequality faced by people who experience hunger the most, and that these are the groups to whom more care needs to be extended as food systems are transformed. Solidarity between different social groups and across agendas is therefore key to addressing complex food system inequalities at the heart of a new eco-social contract. For CFNs, food, healthcare and reducing poverty go hand in hand— for example, a universal basic income can help ensure that all people can afford a healthy diet, while a stable (and sufficient) income improves people's resilience to crises in general by reducing 'food poverty' during times of both turbulence and stability. With this goes better measures of well-being that connect food, health and other cultural factors, including Indigenous self-determination. This principle requires states to improve policy coherence by bringing together relevant economic, agricultural, consumer, health, gender, environmental and social policies, and requires major reforms to the role of charities and other social support mechanisms such as welfare provisioning and basic income entitlements in an eco-social contract for the future.

12.4.3 Food Systems as a Common Good

The third principle combines distributive and procedural aspects of justice, recognising the continued capacity for the land, air, water, soil, biodiversity and seeds to sustain people's health and well-being while respecting planetary boundaries. Food systems failure (Smith, 2016) is entangled with other systems' failures (energy, water, land, biodiversity, climate). This can be attributed to commodification and extractivism that sets the value of the natural world through the lens of the market, causing the separation of people and cultures from the physical, social and spiritual resources that underpin our foodways. Contesting food and food systems as a private good opens space to re-imagine these as a common good (Vivero-Pol et al., 2019). This also extends an eco-social justice position, whereby a wider array of human social groups and non-human food system actors (such as animals and biodiversity) are intrinsically connected, valued and cared for collaboratively (Waddock, 2024).

There are multiple ways to 're-common' the food system. First, expanding food production in cities, backyards, public spaces such as parks and verge gardens, and making sure that people have the autonomy to gather food for free from these places enables people to reclaim the commons and fight back against the privatisation of public space. For food systems to become hyper-localised in this way, town planning would need to commit to repurposing urban spaces to grow more food, which in turn requires rethinking land access and affordability. As one Brisbane participant explained:

> The question of private property, well I think we should have [better] access to land. At the moment we don't have this collectively—we have little pockets of community gardens and so forth, but I think people should have access to land to be able to have more communal land for food. (Food activist, September 2021)

A more explicit role for First Nations' worldviews and peoples to direct change within democratic food governance is another key step. Recognising Indigenous peoples' claims to land and their rights to forage, hunt and cultivate land for food, which have historically been denied through settler colonialism (Staines & Smith, 2021), was especially important to young people engaged in food justice in our study, who said:

> We are a grassroots movement where we set up guerrilla gardens in public spaces as a way to protest the idea of state-owned land. We're getting the acknowledgement and permission to set up the land by Indigenous Elders. Just as a symbol of resistance to be like 'this is not public land but maybe this is also for everyone and is Indigenous land because sovereignty was never ceded'. (Youth workshop participant, online, November 2021)

Third is public ownership and regulation of food-related technologies, at both production (e.g. seeds, animal welfare, farm data) and consumption (e.g. networks and finance) levels. In this approach to eco-social justice, the de-privatisation of ecological resources, digital technologies and online infrastructure is essential to enabling grassroots democracy where ordinary people can control the distribution of benefits from new technologies that will inevitably change the way food is produced, distributed and consumed. Participants in one workshop shared their vision for 'Technology for the People' thus:

> For more democratic control of technology there would need to be an increasing recognition of the fact that digital infrastructure is common infrastructure. There needs to be some changing understanding of how we build digital technology and who pays for it, and who benefits from it and so on … It shouldn't be possible for governments to put money into digital infrastructure that is proprietary, instead of digital infrastructure that is for the common good and owned by people and by citizens. (Technology workshop participant, online, November 2021)

12.4.4 Resilience Beyond Crisis

This final principle acknowledges that future crises under climate change and growing social inequality are inevitable—in fact, they are part of the global industrial food system itself, which reproduces inequalities and injustices through its very structure (Stead & Hinkson, 2022). Social-ecological resilience has emerged as an important lens to assess sustainable and equitable food systems (Bohle et al., 2010), and is broadly defined as a system's ability to absorb and adapt to shocks, stressors, disturbances and change (Berkes et al., 2002), positioned largely against the context of the climate change crisis (Folke, 2016). Within this literature, it is argued that food systems must be resilient in order to cope during a crisis, and that local food networks are integral to improving resilience (FAO, 2020). However, the principle of resilience beyond crisis means that food systems must also provide a source of transformation. This is a reversal of the dominant focus on 'how to make food systems more resilient', and instead refers to the capacity to create a fundamentally new system that "allows undesirable socioeconomic states (for example a system characterized by deep deficits in income, power, education and social capital) to be transformed into more desirable ones without threatening the integrity of the atmosphere or the ecological systems on which humans depend" (Boyd et al., 2008: 392).

For food to be equally accessible during a disaster as it is in 'normal' times, food justice requires a rapid move to localise food production, distribution and consumption in line with practices of agroecology, regenerative agriculture and circular economy; redirect financial profits to benefit local communities and ecologies; and shift multi-level governance processes to reflect long-term goals. This combines elements of just sustainability (temporal justice) and eco-social justice (i.e. ongoing capacity for the land, air, water, soil, biodiversity and seeds to sustain people's health and well-being while respecting planetary boundaries). In important ways, the food justice visions of many CFNs are highly compatible with the resilience focus of disaster management and the concept of 'resilience beyond crisis', but with additional attention to the redistribution and sharing of power. According to Olsson et al. (2015: 4), power sets the conditions for more flexible, collaborative forms of governance that contribute to long-term resilience of social-ecological systems. In our study, enhancing system resilience was associated with 'degrowth' (Nelson & Edwards, 2021), using fewer resources in line with planetary boundaries and shifting towards short supply chains:

It all revolves around local food economy. So, it is typically small-scale farms [with] multiple sales and marketing streams to provide economic diversity. In terms of climate and other kinds of impacts, they're already embedding resilience. (Peri-urban farmer, August 2022)

12.5 Conclusion: Civil Society—The Catalyst for Change

Australia's food system was already extremely vulnerable before COVID-19. Ecological degradation from high-input, intensive mode of agriculture is ongoing, land use pressures due to climate change and resource constraints are intensifying, as is corporate concentration, and food is wasted at an alarming rate. We know that those already most vulnerable to the impacts of climate change—due to poverty, gender, class, race and other factors—will be most affected by future crises. The pandemic further revealed that:

> The global food system is one constantly scrambling to patch the very cracks and weaknesses it reproduces … The vulnerabilities and inequalities produced as part of business-as-usual in the global food system have been intensified and rendered newly visible by COVID-19, but this intensification has also shone new light on transformational possibilities. (Stead & Hinkson, 2022: 5)

In listening to the values and practices of civil society in Australia, this research reaffirms the urgent need for a new eco-social contract with civic food justice at the centre of food system transformation. There are many ways to practice food justice, but a common thread is the need to make visible the systemic causes of hunger—such as food poverty, ecological collapse, and race, class and gender-based inequalities. These in turn stem from historical and ongoing injustices associated with colonisation, patriarchy and capitalism; this is the main difference between food security discourse (which emphasises production) and food justice (which challenges systemic inequality).

The four principles for eco-social food justice outlined in this chapter were informed by justice theories in the literature and qualitative analysis of the perspectives of 'fair food' CFNs in Australia. The four principles are: 'human rights for all', 'intersectional solidarity and care', 'food as commons' and 'resilience beyond crisis'. From an international perspective, taking a more inclusive and justice-oriented approach to food security directly responds to the United Nations 2030 Agenda and Sustainable Development Goals, which themselves connect 'Zero Hunger' with interconnected environmental (e.g.

climate action), social (e.g. no poverty, education for all) and economic (e.g. reduced inequalities, decent work) shifts. Indeed, considering the interconnectedness between food systems and many other systems where the social contract has also broken down (e.g. social welfare systems that do not ensure quality of life for all, failure to 'decarbonise' economies to mitigate climate change), these four principles can also provide guidance to progress justice, rights and empowerment across multiple other governance spheres. Focus points for implementing these principles into policymaking include:

- Legislating the *Right to Food* into national and state food policymaking, and redesigning governance processes to facilitate human rights and the Rights of Nature.
- Local procurement policies and growing food directly in institutional settings—such as schools, prisons and hospitals.
- Universal basic income schemes for groups known to be particularly food insecure, given the strong link between poverty and food insecurity.
- Connect food with wider policy issues (and social movements) around housing, income, healthcare, gender equality, Indigenous sovereignty, labour rights, environmental/climate policy.
- Reorient land, technology and other food system 'goods' to provide collective benefits for local and regional food system actors who have been marginalised and exploited, such as Indigenous people, women, migrants and workers.
- Support cooperative models such as community gardens and community supported agriculture, as these can improve the affordability of healthy local food, ensure a fair pay for farmers and food workers, reduce vulnerability and increase self-sufficiency.
- Shorten food chains to provide co-benefits for social justice and social-ecological resilience (e.g. reduce food miles).
- Finally, ensure meaningful participation of civil society in food system decision-making at all levels via stronger government support (financing, education and accountability) for food policy councils and participatory budgeting.

References

ABS (Australian Bureau of Statistics). (2015). 4727.0.55.005 – Australian Aboriginal and Torres Strait Islander Health Survey: Nutrition Results – Food and Nutrients, 2012–13.

Agyeman, J., Schlosberg, D., Craven, L., & Matthews, C. (2016). Trends and directions in environmental justice: From inequity to everyday life, community and just sustainabilities. *Annual Review of Environment and Resources, 41*, 321–340.

Alkon, A. H., & Agyeman, J. (Eds.). (2011). *Cultivating food justice: Race, class, and sustainability*. MIT Press.

Anderson, M. (2019). Comparing the effectiveness of structures for addressing hunger and food insecurity. In P. Andrée, J. Clark, C. Levkoe, & K. Lowitt (Eds.), *Civil society and social movements in food system governance* (pp. 124–144). Routledge.

Anderson, M. (2024). *Transforming food systems: Narratives of power*. Routledge.

Andrée, P., Clark, J. K., & Levkoe, C. Z. (2019). *Civil society and social movements in food system governance*. Routledge.

Berkes, F., Colding, J., & Folke, C. (2002). *Navigating social-ecological systems: Building resilience for complexity and change*. Cambridge University Press.

Blay-Palmer, A., Halliday, J., Santini, G., Taguchi, M., & van Veenhuizen, R. (2020). *City region food systems to cope with Covid-19 and other pandemic emergencies*. RUAF. https://ruaf.org/news/city-region-food-systems-to-cope-with-covid-19-and-other-pandemic-emergencies/

Bohle, H. G., Ericksen, P., & Stewart, B. (2010). Vulnerability and resilience of food systems. In J. Ingram, P. Ericksen, & D. Liverman (Eds.), *Food security and global environmental change* (pp. 67–77). Taylor and Francis.

Boyd, E., et al. (2008). Resilience and 'climatizing' development: Examples and policy implications. *Development, 51*, 390–396.

de Bruin, A., de Boer, I., Faber, N., de Jong, G., Termeer, K., & de Olde, E. (2023). Easier said than defined? Conceptualising justice in food system transitions. *Agriculture and Human Values, 41*, 345–362.

Cadieux, K., & Slocum, R. (2015). What does it mean to do food justice? *Journal of Political Ecology, 22*(3), 1–26.

Canal Vieiria, L., Serrao-Neumann, S., & Howes, M. (2020). Daring to build fair and sustainable urban food systems: A case study of alternative food networks in Australia. *Agroecology and Sustainable Food Systems, 45*, 1–22. https://doi.org/10.1080/21683565.2020.1812788

Carolan, M. (2013). *Reclaiming food security*. Earthscan/Routledge.

Clapp, J., Moseley, W. G., Burlingame, B., & Termine, P. (2022). Viewpoint: The case for a six-dimensional food security framework. *Food Policy, 106*, 102164.

Coulson, H., & Milbourne, P. (2021). Food justice for all? Searching for the justice multiple in UK food movements. *Agriculture and Human Values, 38*, 43–58.

Dacheux, E., & Goujon, D. (2011). The solidarity economy: An alternative development strategy? *International Social Science Journal, 62*(203–204), 205–215.

Dowler, E., & O'Connor, D. (2012). Rights-based approaches to addressing food poverty and food insecurity in Ireland and UK. *Social Science and Medicine, 74*, 44–51.

FAO. (2001). *The State of Food and Agriculture 2001. No. 33.* Food and Agriculture Organisation.

FAO. (2020). *Urban food systems and COVID-19: The role of cities and local governments in responding to the emergency.* Food and Agriculture Organization. http://www.fao.org/3/cb0407en/CB0407EN.pdf

FAO. (2022). *FAO food price index.* Food and Agriculture Organization. https://www.fao.org/worldfoodsituation/foodpricesindex/en/

FAO, IFAD, UNICEF, WFP and WHO. (2022). *In Brief to The State of Food Security and Nutrition in the World 2022. Repurposing food and agricultural policies to make healthy diets more affordable.* Food and Agriculture Organization.

Folke, C. (2016). Resilience (republished). *Ecology and Society, 21*(4).

Foodbank. (2023). *Foodbank Hunger Report 2023.* https://reports.foodbank.org.au/foodbank-hunger-report-2023/

Frank, N., Arthur, M., & Friel, S. (2024). Shaping planetary health inequities: The political economy of the Australian growth model. *New Political Economy, 29*(2), 273–287.

Freeman, C., & Smith, K. (2024). From coalitions to social movements: Lessons from civic food coalition formation in Australia. *International Journal of the Sociology of Agriculture and Food, 30*(1), 1–17.

Godrich, S., Barbour, L., & Lindberg, R. (2021). Problems, policy and politics-perspectives of public health leaders on food insecurity and human rights in Australia. *BMC Public Health, 21*, 1132.

Gottlieb, R., & Joshi, A. (2010). *Food justice.* MIT Press.

Holt-Gimenez, E. (2011). Food security, food justice or food sovereignty? Crises, food movements and regime change. In A. Alkon & J. Agyeman (Eds.), *Cultivating food justice: Race, class, and sustainability* (pp. 309–330). MIT Press.

Holt-Gimenez, E. (2017). *A foodie's guide to capitalism: Understanding the political economy of what we eat.* Monthly Review Press.

Holt-Giménez, E., Shattuck, A., Altieri, M., Herren, H., & Gliessman, S. (2012). We already grow enough food for 10 billion people … and still can't end hunger. *Journal of Sustainable Agriculture, 36*(6), 595–598.

IPES-Food. (2017). *Too big to feed: Exploring the impacts of mega-mergers, concentration, concentration of power in the agri-food sector.* IPES-Food.

Krause, D., Stevis, D., Hujo, K., & Morena, E. (2022). Just transitions for a new eco-social contract: Analysing the relations between welfare regimes and transition pathways. *Transfer, 28*(3), 367–382.

Lang, T., & Heasman, M. (2015). *Food wars.* Routledge.

Levkoe, C. L. (2006). Learning democracy through food justice movements. *Agriculture and Human Values, 23*(1), 89–98.

Lewis, M., Nash, S., & Lee, A. (2024). Cost and affordability of habitual and recommended diets in welfare-dependent households in Australia. *Nutrients, 16*(5), 659.

Lindberg, R., Barbour, L., & Godrich, S. (2019). A rights-based approach to food security in Australia. *Health Promotion Journal of Australia, 32*, 6–12.

Lindberg, R., Whelan, J., Lawrence, M., Gold, L., & Friel, S. (2015). Still serving hot soup? Two hundred years of a charitable food sector in Australia: A narrative review. *Australian and New Zealand Journal of Public Health, 39*, 358–365.

Loh, P., & Agyemon, J. (2019). Urban food sharing and the emerging Boston food solidarity economy. *Geoforum, 99*, 213–222.

Mayes, C. (2018). *Unsettling food politics: Agriculture, dispossession and sovereignty in Australia*. Rowman and Littlefield.

McKay, F. H., Haines, B. C., & Dunn, M. (2019). Measuring and understanding food insecurity in Australia: A systematic review. *International Journal of Environmental Research and Public Health, 16*(3), 476.

Moragues-Faus, A. (2017). Problematising justice definitions in public food security debates: Towards global and participative food justices. *Geoforum, 84*, 95–106.

Nelson, A., & Edwards, F. (Eds.). (2021). *Food for degrowth: Perspectives and practices*.

Nemes Gusztáv, N., Chiffoleau, Y., Zollet, S., Collison, M., Benedek, Z., Fedele, C., Dulsrud, A., Fiore, M., Holtkamp, C., Kim, T.-Y., Korzun, M., Mesa-Manzano, R., Reckinger, R., Ruiz-Martínez, I., Smith, K., Viteri, M. L., Tamura, N., & Orbán, E. (2021). The impact of Covid-19 on alternative and local food systems and the potential for the sustainability transition: Insights from 13 countries. *Sustainable Production and Consumption, 28*, 591–599.

Olsson, P., Galaz, V., & Boonstra, W. (2015). Sustainability transformations: A resilience perspective. *Ecology and Society, 19*(4), 1.

Pollard, C. M., Mackintosh, B., Campbell, C., Kerr, D., Begley, A., Jancey, J., Caraher, M., Berg, J., & Booth, S. (2018). Charitable food systems' capacity to address food insecurity: An Australian capital city audit. *International Journal of Environmental Research and Public Health, 15*(6), 1249. https://doi.org/10.3390/ijerph15061249

Richards, C., Kjaernes, U., & Vik, J. (2016). Food security in welfare capitalism: Comparing social entitlements to food in Australia and Norway. *Journal of Rural Studies, 43*, 61–70.

Riches, G. (2018). *Food bank nations: Poverty, corporate charity and the right to food*. Routledge.

Sampson, D., Cely-Santos, M., Gemmill-Herren, B., Bain, N., Bernhart, A., Besner Kerr, R., Blesh, J., Bowness, E., Feldman, M., Goncalves, A., James, D., Kerssen, T., Klassen, S., Wezel, A., & Wittman, H. (2021). Food Sovereignty and Rights-based approaches strengthen food security and nutrition across the globe. *Frontiers in Sustainable Food Systems, 5*.

Schlosberg, D. (2007). *Defining environmental justice: Theories, movements and nature.* Oxford University Press.

Schneider, M., & Chappell, J. (2017). The new three-legged stool: Agroecology, food sovereignty, and food justice. In M. Rawlinson & C. Caleb Ward (Eds.), *The Routledge handbook of food ethics* (pp. 419–429). Routledge.

Smith, K. (2016). Food systems failure and prospects for the future. In M. Shucksmith & D. Brown (Eds.), *Routledge handbook of rural studies*. Routledge.

Smith, K. (2019). Localising SDG2 Zero Hunger through "fair food" in Australia. *Asian Development Perspectives., 10*(2), 135–148.

Smith, K. (2022a). Civic food utopias in Australia: The challenge of justice and representation. *Sociologia Ruralis*, 1–20.

Smith, K. (2022b). *Future scenarios for food justice and Zero Hunger in Australia: A synthesis for policy action.* Policy brief #1, https://fairfoodfutures.com/explore-the-scenarios/ (1 July 2022).

Smith, K., Lawrence, G., MacMahon, A., Muller, J., & Brady, M. (2015). The resilience of long and short food chains during the Queensland floods of 2011. *Agriculture and Human Values, 33*, 45–60.

Special Rapporteur on Right to Food. (2021). Promotion and protection of human rights: human rights questions, including alternative approaches for improving the effective enjoyment of human rights and fundamental freedoms, 76th Session of … item 75(b), 27 July 2021. https://documents.un.org/doc/undoc/gen/n21/208/03/pdf/n2120803.pdf

Spencer, L. (2024). Late to the table: Australian law and policy on food security. *Alternative Law Journal, 49*(1), 26–32.

Staines, Z., & Smith, K. (2021). Workfare and food in remote Australia: 'I haven't eaten… I'm really at the end… *Critical Policy Studies*.

Stead, V., & Hinkson, M. (Eds.). (2022). *Beyond global food supply chains: Crisis, disruption, regeneration.* Palgrave Macmillan.

Stock, P., Carolan, M., & Rosin, C. (2015). *Food Utopias: Reimagining citizenship, ethics and community.*

UNRISD. (2021). *A new eco-social Contract: Vital to deliver the 2030 Agenda for Sustainable Development.* Issue Brief 11. UNRISD.

Vivero-Pol, J. L., Ferrando, T., Olivier De Schutter, O., & Mattei, U. (2019). *Routledge handbook of food as a commons.* Routledge.

Waddock, S. (2024). Holistic eco-social imaginaries for a life-centered future. *Sustainability Science, 19*, 2119–2134.

Willett, W., Rockström, J., Loken, B., Springmann, M., Lang, T., Vermeulen, S., Garnett, T., Tilman, D., DeClerck, F., Wood, A., Jonell, M., Clark, M., Gordon, L. J., Fanzo, J., Hawkes, C., Zurayk, R., Rivera, J. A., De Vries, W., Majele Sibanda, L., & Murray, C. J. L. (2019). Food in the anthropocene: The EAT–Lancet commission on healthy diets from sustainable food systems. *The Lancet (British Edition), 393*(10170), 447–492.

Wittman, H., Desmarais, A., & Wiebe, N. (2010). *Food sovereignty: Reconnecting food, nature and community.* Fernwood Publishing, Food First, Pambuzuka Press.

Yii, V., et al. (2019). Population-based interventions addressing food insecurity in Australia: A systematic scoping review. *Nutrition and Dietetics, 77,* 6–18.

Zeigler, J. (2011). *The fight for the right to food: Lessons learned.* Palgrave.

Part III

Processes for Renegotiating and Designing New Eco-Social Contracts in Practice

13

A Planet in Peril, a People in Power: The Eco-Social Revolution

Kumi Naidoo

In the extraordinary tapestry of our shared human journey, a daunting portrait emerges—a symphony of crises that reverberate through the fibres of our global society. This dissonance, a convergence of a climate emergency, staggering inequality, the throes of rising authoritarianism, and a flawed global economic framework, mirrors a world at a pivotal crossroads. In the face of such overwhelming realities, we remain tasked with striking the delicate balance between presenting a truthful narrative of these tumultuous times and sparking a fervour for change that does not paralyse but instead galvanises action.

This task is not just pertinent; it is an urgent moral imperative. The COVID-19 pandemic unveiled systemic frailties and accentuated the urgency to redefine our interconnectedness with each other and with nature. We stand at the precipice of what could be the most consequential decade, one demanding transformative choices regarding climate and equality. Despite significant albeit insufficient (and declining) resources, allocated to address these challenges, efforts remain fragmented and inadequate because we often misalign our focus—prioritising centralised power and profit over grassroots empowerment and sustainable well-being.

The very essence of our future lies in recalibrating this focus. We must transition from systems that devalue life to economies that cherish sustainable well-being and comprehensive prosperity. It's not merely an ideal; real-world

K. Naidoo (✉)
Fossil Fuel Non-Proliferation Treaty, Berkeley, CA, USA

P. Huntjens et al. (eds.), *Eco-Social Contracts for Sustainable and Just Futures*,
https://doi.org/10.1007/978-3-031-99109-7_13

initiatives are already demonstrating the potential of community-driven, eco-logically harmonious economic models.

Our interaction with the environment requires a paradigm shift from exploitative practices to a harmonious coexistence, recognising the inherent Rights of Nature. This shift is vital for our survival and necessitates embracing indigenous wisdom long marginalised by colonial legacies.

In this reimagined future, our identity transcends narrow nationalist senti-ments into an expansive global citizenship. The equitable sharing of resources, coupled with respect for diverse cultural tapestries, must lay the foundation for a new eco-social contract. Businesses must transform from competitive entities to cooperative agents of global good, aligned with the vision that no profits can be sustained on a dying planet.

Renewed labour rights also form a cornerstone of this transformation. The pandemic underscored the essential nature of undervalued workers; our new social contract must therefore prioritise fair labour standards in our quest for decent work.

To achieve such profound changes, state responsibility must shift from reactive measures to proactive investments in public well-being, fostering environments that nurture sustainable coexistence: between both humanity and nature and between humans themselves. Human beings in particular should seek greater social cohesion, integration and solidarity.

Ultimately, our path forward necessitates unprecedented mobilisation. Success hinges on fostering open participation and leveraging the power inherent within people—our autonomy, creative expression, collective wealth, and conscientious consumption. Each of these serves as pillars underpinning a movement driven not solely by leaders but by the collective participation of humanity.

As we strive towards this envisioned future, we need to summon the cour-age to embrace new dimensions of activism, recognising that conventional approaches will not suffice. Albert Einstein once suggested that insanity is doing the same thing repeatedly while expecting different results. To forge a truly sustainable, equitable future, we must reimagine our activism—a new paradigm for a new eco-social contract born from the grassroots and cemented by the collective will of humanity.

14

Eco-social Contracts as a Pathway Towards Inclusive and Sustainable Futures: Opportunities, Challenges and Lessons Learned

Katja Hujo

14.1 Introduction[1]

The twentieth-century social contract—an implicit bargain aiming to combine growth and productivity with redistribution and social protection—has broken down and cannot sustain the transformative vision of the 2030 Agenda. Unravelling under the pressure of neoliberal globalisation and failing to be fully inclusive and environmentally sustainable, the breakdown of the social contract now manifests itself in multiple global crises and the deep divisions in our societies (UNRISD, 2022a). Inequalities in many dimensions have grown, particularly in the last 40 years, and many people feel left out and left behind (Chancel et al., 2022). The failure of our economic model to account for the natural boundaries of our planet has led to environmental destruction and a climate crisis. And despite considerable progress in human development for more than half a century, this progress has been uneven and volatile, while recent gains have been partially reversed as a result of the COVID-19 crisis and global economic shocks related to an increasingly

---

[1] This chapter draws on Chap. 4 of the UNRISD flagship report (UNRISD, 2022c). The author acknowledges the permission of UNRISD to reproduce parts of the chapter here.

K. Hujo (✉)
UNRISD, Bonn, Germany
e-mail: katja.hujo@un.org

© The Author(s) 2025
P. Huntjens et al. (eds.), *Eco-Social Contracts for Sustainable and Just Futures*,
https://doi.org/10.1007/978-3-031-99109-7_14

239

uncertain geopolitical context (UN, 2021; World Bank, 2020; Brignone et al., 2024; Yilmazkuday, 2024). At the current juncture, too many people are living in or have been pushed back into poverty and hunger, struggling with multiple deprivations, vulnerabilities and insecurities, while often lacking the power and means to make their voices heard (UN, 2025). As a result, many citizens around the world have lost their trust in governments (Justino & Samarin, 2024; UNDP, 2022).

In this challenging context, as emphasised in the introduction to this edited volume, residual reforms of the social contract that fail to address root causes of problems will not be enough to achieve the necessary long-term transformations. The social contract needs a fundamental overhaul to achieve sustainable development for all; it must become an eco-social contract, incorporating the ecological dimension and creating a new contract that respects the natural environment and the needs of future generations and unleashes regenerative forces for eco-social change. This new eco-social contract needs to be grounded in a broad consensus between different stakeholders, embarking on a democratic, inclusive and participatory decision-making process at multiple levels, to arrive at a shared vision, concrete objectives and commitments, and accountability mechanisms.

This chapter aims to contribute to a better understanding of the origins and plurality of social contracts from a conceptual and empirical perspective. It provides insights into how social contracts are shaped by global trends and dominant economic paradigms, and presents case studies that show how social contracts are changing in response to new challenges and power reconfigurations. It discusses a set of normative principles developed by UNRISD (2021, 2022a) that could be used as benchmark criteria to assess the transformative potential of changes in social contracts and global social pacts, and concludes by summarising lessons learned on how to support eco-social contracts in the making.

14.2 Understanding the Social Contract: Normative and Empirical Approaches

In a world of multiple crises where many previous certainties have been shattered, large numbers of people are beginning to question the principles, values and public institutions our societies are founded upon, what philosophers such as Thomas Hobbes, John Locke and Jean-Jacques Rousseau have called the social contract. In present times, social contracts are understood to reflect basic societal decisions regarding the division of labour between states, markets, communities, families and individuals, on what is provided collectively and by whom in view of building a just society where equal opportunities exist for people to flourish and progress (Shafik, 2021). Social contracts are based on philosophical or normative frameworks and imaginaries and are implemented through concrete policies and institutions.

For an analysis of social contracts, it is useful to identify their scope (involved parties, application), their temporal dimension and their substantive content (Loewe et al., 2021). Based on human rights frameworks, it is also common to distinguish between the procedural (enforcement of rights), the distributive (access to resources and rights), the participatory (participation in decision-making) and the recognition function (promotion of dignity and respect) of social contracts (Plagerson et al., 2022).

Historically, theoretical or normative approaches and real-world examples of social contracts have differed according to how much weight they have given to social order (for example, protection of private property rights) versus social justice (for example, income redistribution) and regarding the balance between individual rights and responsibilities versus state regulations and provisioning (Hickey, 2011). More recent debates unfolding since the 1980s have taken an even wider view and an explicit critical stance, aiming to uncover how empirical social contracts deviate from the normative notion of mutual benefits based on cooperation among independent and roughly equal persons (Ulriksen & Plagerson, 2014).

Indeed, real-world social contracts tend to be far removed from the notion of free and equal persons creating a society based upon rules to which all agree (Sen & Durano, 2014:5). Rather, social contracts reflect existing power structures and inequalities at multiple levels and in varied forms, often creating de facto contracts of domination (Mills, 2007). They often do not grant meaningful political participation to non-elite groups, focusing in the best case on other legitimising factors such as security or welfare provision (Desai, 2022; Loewe et al., 2021). They are often the result of elite bargains and market power (Therborn, 2014). Critical scholars and activists have highlighted the racialist and patriarchal nature of existing social contracts (Mills, 1997; Pateman, 1988),[2] our missing contract with nature (Desai, 2022; Gough, 2021; Hopkins et al., 2020; Huntjens, 2021; Huntjens & Kemp, 2022; Willis, 2020), and problems of elite capture, corruption and lack of

---

[2] The attempt to deracialise social contracts was at the heart of several post-colonial nation-building social pacts in Africa, "often tolerating emerging vertical inequality along class lines," see Mkandawire (2012:10).

accountability undermining political institutions. More recently, problems of democratic backsliding, backlash against human rights and gender equality, as well as a rise in political polarisation, nationalism and authoritarianism, are putting pressure on social contracts, leading to their further demise (UNRISD, 2022b). It is therefore crucial to differentiate between ideal understandings of a social contract (the norms and values underpinning its vision and objectives, which vary according to different worldviews and ideologies) and real-world experiences (the actual institutions and policies that are implemented and their effects).

Finally, important questions arise regarding transnational issues and how to overcome the limitations associated with national social contracts when it comes to building eco-social contracts that aim to promote global social and climate justice, peace and human rights, concerns that are at the centre of the UN Charter and the Sustainable Development Goals (SDGs).

14.2.1 Diversity of Social Contracts

The Twentieth-Century Welfare State Bargain

Social contracts can be found in any society. There is a large diversity among them, each emerging from different contexts: everywhere, social contracts are shaped by historical and contextual factors and change over time, in response to changing political constellations or socioeconomic conditions. The twentieth-century welfare state social contract, mostly associated with indus-trialised countries in Western Europe and the Nordic countries, has received much attention and analysis, not the least because it was a highly institution-alised process of consultation and cooperation on economic policy issues between organised interest groups as well as the state (and in some cases civil society actors). Their key objectives were more equalised capital-labour rela-tions, shared growth and greater economic predictability as well as stable live-lihoods in times of rapid structural change and cold-war systems competition (Beveridge, 1942; Galbraith, 2022). Several late-industrialising countries in the Global South pursued a similar model, with different degrees of success, including several countries in Latin America and East Asia (Martínez Franzoni & Sánchez-Ancochea, 2020; UNRISD, 2010). Over time, gradual shifts towards greater individual and family responsibility were introduced, and in the context of the neoliberal reforms of the 1980s, a partial privatisation of government roles and a trend towards commercialisation of government-provided services emerged, reducing the responsibilities of the state. On the other hand, civil society movements brought forth concerns of constituencies outside the organised trade union movement, such as gender justice or the rights of migrant workers and asylum seekers, or of informal sector workers (Hujo & Koehler, Forthcoming).

Agrarian Social Pacts

Another example of a social contract associated with a dominant economic sector is the agrarian or rural social contract or pact, which describes the particular political economy of agricultural economies in terms of the challenge of creating collective action among farmers, coordinating with the state and involving farmers in politics (Sheingate, 2015). These contracts are often marked by unequal land distribution as a legacy of feudalism as well as colonial and neocolonial practices. In some historical cases, for example in Japan, South Korea and Taiwan, renegotiations of agrarian pacts involved substantial land redistribution that contributed to more egalitarian and developmental social contracts (Grabowski, 2002). In some countries, new agrarian social pacts (sometimes considered a subset of broader social contracts) have been combined with a specific social policy design, universal and tax-financed benefits, better adapted to the realities of rural workers and producers, who tend not to be covered by the contributory social insurance typical of urban manufacturing or service workers (Palme & Kangas, 2005). They also included a range of other measures such as producer subsidies, price controls or rural development policies (UNRISD, 2010:49).

As is the case with other types of social contracts, agrarian social pacts did not necessarily lead to egalitarian outcomes: large commercial farmers would usually dominate negotiations. As a result, support policies often benefited larger farmers and capital-intensive producers to the detriment of smallholders and subsistence farmers (Sheingate, 2015).

More recently, agrarian pacts are renegotiated in a context of climate change and environmental crisis. Examples of changing contracts can be found in the Middle East and North Africa (MENA) region (Houdret & Amichi, 2020). In Morocco and Algeria, for example, regimes traditionally granted rural elites access to water and land in exchange for loyalty. However, the neoliberal turn in the 1980s that propelled the liberalisation of agricultural policies and regulatory reforms changed the rural social contract, empowering a new elite of agricultural entrepreneurs and leaving traditional allies aside. This new rural social contract, however, is considered highly unstable in a context of rising inequalities, difficulties to access scarce natural resources and climate change.

Social peace is mainly enforced through a combination of subsidies and repression, while environmental costs are high as producers often circumvent protective regulations (Houdret & Amichi, 2020).

Social Contracts in Mineral-Rich Countries

Social contracts in mineral-rich countries have often been unequal and unstable due to elite capture and distributional conflicts, as the case of Zimbabwe shows (Saunders, 2020), leading some scholars to argue that resource-rich countries are afflicted by a resource curse (Auty, 1993). However, there are also examples of governments which have included marginalised groups in social contracts in mineral-rich contexts by widely distributing the benefits of resource extraction, while also strengthening their developmentalist social contracts through taking greater control within the sector and setting up institutions to better manage the challenges associated with mineral-led development (Hujo, 2012, 2020; UNRISD, 2010, 2016). Bolivia, a mineral-rich country, is an example showing how a historically elite-dominated and exclusionary social contract can be renegotiated, as occurred during the government of Indigenous President Evo Morales in the early 2000s (Paz Arauco, 2020; Barié, 2014). However, while progress was made in terms of domestic resource mobilisation by renationalising a significant share of the gas sector and channelling fiscal revenue into social inclusion policies and poverty reduction, the environmental question remains a challenge, despite the fact that the social contract is underpinned by the Indigenous cosmovision of *Sumak Kawsay* that establishes a contract with nature (see Chaps. 2, 6 and next section).

Communitarian Approaches

Beyond the social contracts associated with dominant economic sectors such as industry, agriculture and mining, different types of social contracts and associated narratives or normative frameworks can be identified across the world. These value frameworks or imaginaries rarely use the terminology of the social contract and can even be critical of the notion of consensus and equal partners that is associated with contractual theory as well as of the separation between individuals and communities/societies engrained in Western liberal philosophy. They tend to make less reference to vertical state–citizen relations or corporatist arrangements based on bargains between powerful interest groups and the state and are more concerned with horizontal social relations or human–nature relations. Prominent examples are *Ubuntu*, *Sumak Kawsay* or *buen vivir*, *Eco-Swaraj* and faith-based contracts (see Chaps. 4, 6 and 10). In some cases they have been mobilised as instruments of moral persuasion and reflect governments' efforts to promote social responsibility in a context of weak regulatory state capacity vis-à-vis elites and companies (Mkandawire, 2012). In other cases, practical implementation is not keeping up with normative visions, as in the case of Bolivia and Ecuador, where interventions are often detached from Indigenous struggles for decolonisation and autonomy, whereas macro policies and development frameworks continue to rely on an extractivist model.

14.3 Social Contracts in a Changing Global Context

While social contracts are often deemed successful if they coincide with or contribute to periods of stability, for example, during the so-called golden age of coordinated capitalism between 1945 and 1973 (Marglin & Schor, 1992; Piketty, 2022), pressures to renegotiate social contracts can arise in times of crisis, in particular if the crisis is identified as a systemic one that would make a return to the status quo an undesirable and unstable option (UNRISD, 2022: Chap. 2). Periods of instability and transformation are associated with the breakdown of accustomed norms and beliefs, when people's lived realities conflict more and more with familiar practices and they become convinced that the contract is no longer working (Sen & Durano, 2014:6). Crises and national emergencies can provide incentives for concertation and cooperation to overcome multiple challenges across different policy areas, sometimes leading to substantial paradigm shifts in different directions. Examples are the post-war international order and the development of welfare states; the neoliberal turn of the 1980s that radically redefined the social contract in many countries as a response to the economic crises of the 1970s and early 1980s in the context of an ideological revolution (Nugent, 2010); the social turn which aimed to reinfuse social objectives into market-centred development strategies in the 1990s after the World Summit for Social Development that was convened in Copenhagen in 1995 (UNRISD, 2016); the stalled efforts to reform economic governance after the financial crisis of 2008, and most recently, the aborted attempt to strengthen public social services and access to social protection in post-pandemic times. These critical junctures, in combination with an accelerating climate and care crisis, demonstrate the importance of crisis for changing existing social contracts. However, it is important to note that new bargains might not lead automatically to political and economic stability and greater social justice, either because they are skewed towards particular interests, they lack enforcement capacity or they do not address root causes.

The transition from welfare or developmental state social contracts to neoliberal or adjustment contracts is of particular importance. Over the last three decades, many twentieth-century social contracts forged in the post-war/post-colonial era that had aimed at economic development, social inclusion and a stronger public sector began unravelling during the period of economic crises, neoliberal policies and accelerated globalisation starting in the 1980s. In this period, power was shifted towards capital, and state capacity to enforce contracts weakened, in particular in the Global South affected by stabilisation and structural adjustment policies. Welfare and developmental social contracts were increasingly replaced by new types of contracts that emphasised individual responsibilities to the detriment of solidarity, redistribution and public provision. These changes also affected more traditional social contracts based on communitarian values, as these communities were increasingly integrated into world market dynamics, while traditional informal institutions of mutual support instead of evolving into employment-based social security were replaced by residual social assistance schemes, for example, cash transfer programmes for the poor, affecting social relations (Moore & Seekings, 2019). Welfare state social contracts were also affected by ongoing processes of demographic change, rapid technological progress and multiple crises, including the triple crisis of climate change, biodiversity loss and pollution.

A common characteristic of most twentieth-century social contracts was the absence of rules to respect planetary boundaries, preserve biodiversity and promote the sustainable use of natural resources, ushering in a global environmental crisis (Huntjens, 2021; Kempf & Hujo, 2022). The consumption and production patterns associated with these contracts were not sustainable and resulted in the depletion of natural resources, climate change, pollution and environmental deterioration. A binding obligation for economic actors, including the state, to protect the environment was missing, while the right to extract resources and deposit waste and emissions, to use natural resources for profit making or to privatise global commons was taken for granted (Standing, 2019).

The lack of respect for nature and the commercialisation of natural resources had widespread negative effects on the environment, and on health and economic opportunities of all people, but in particular for less powerful groups, for example, those groups whose livelihoods are embedded in the natural environment. Traditional farmers, fishers and Indigenous communities with livelihoods based on sustainable use of forests, land and water resources were often deprived of land and resource rights by big corporations or predatory rulers, for example, through privatisation, commercialisation or land grabs, often with negative impacts on women (Tsikata & Eweh, 2017). They also saw their livelihoods based on natural resources destroyed as a result of pollution and commercialised resource exploitation, a tendency that continues in present times.

14.4 Renegotiating Social Contracts: Evidence from the Global South

Reforming or renegotiating social contracts can take different forms, and entail complex transformations of institutions and structures that shape horizontal and vertical relations within societies. Social contracting can lead to policy, legal or institutional reforms, including more fundamental ones such as constitutional reforms, which often occur at critical junctures such as post-conflict situations, decolonisation or democratisation, as well as during authoritarian backlash. In post-conflict scenarios, the need to address root causes of conflict (often related to real or perceived inequalities and exclusion), to strengthen social relations, to attend to grievances and injustices, and to rebuild a peaceful and cohesive society based on shared values, trust and solidarity are of paramount importance, if relapse into violent conflict is to be avoided (McCandless, 2018; United Nations and World Bank, 2018).

Social contracts have also been renegotiated in response to peaceful regime changes, or because of collective mobilisation and claims making. This has often involved inclusion of previously excluded groups, for example through extending cash transfers to low-income groups, improving access to social services and to social protection and labour rights (Haggard & Kaufman, 2004), granting reproductive rights to women and rights to sexual and gender minorities such as LGBTIQ+ groups, or extending labour rights and social protection to informal workers. These incremental changes are often accelerated in times of crisis: most countries, for example, strengthened social protection during the recent COVID-19 crisis, as part of government's efforts to shield vulnerable populations from the adverse impacts of the pandemic and the associated lockdown measures (Kempf & Dutta, 2021). However, economic crises often prompt governments with limited fiscal space and dependence on foreign investors to implement austerity policies and cut social spending (Hujo, 2021).

Whether changes are so fundamental that we would speak of a new social contract, or whether incremental reforms are still part of the original societal–political consensus, depends on a variety of factors and arguably opens space for different interpretations. Key factors indicating a change in the social contract are the scope, temporality and substance of reforms, who participates as contractual parties and the ideational and value frameworks contracts are built on.

14.4.1 Constitutional Reform

Some countries have created new social contracts through a process of constitutional reform. The constitutions of Kenya and Nepal were created in a highly participatory manner and have progressive articles on the inclusion of Indigenous peoples and pastoralists, as well as a quota for women in Parliament (Berry et al., 2021; see Chap. 20). National constitutions were rewritten in Brazil in 1988, formalising a process of democratic transition after 21 years of military rule (1964–1985), and in South Africa in 1994, when the African National Congress (ANC) party under President Nelson Mandela took power, upending the racist apartheid regime that had been institutionalised by the National Party in 1948.

Constitutional reform as an instrument to rewrite the social contract is not always a progressive and democratic move, however, as recent examples from Hungary, Libya, Russia and Turkey show (El Gomati, 2022; Oross & Tap, 2021). Here, constitutional reform has been used to consolidate authoritarian regimes (strengthening, for example, presidential rule or possibilities for re-election or lifetime rule and weakening checks and balances) or entrench elite interests. These regressive outcomes occurred in some cases despite citizen participation, national consultations and referendums, as these were instrumentalised to legitimise the process rather than to shape it (Oross & Tap, 2021). In addition, constitutional reforms often lag behind in terms of implementation, though they open the way for litigation processes that have sometimes proven successful, as the case of South Africa demonstrates (Stern Plaza et al., 2016). Finally, high-level reforms such as constitutional reform need to be accompanied by change processes from below, in particular regarding social norms: the example of gender backlash after introducing the quota system in Kenya's Parliament shows the "need for women's rights activists to prioritise a parallel bottom-up process of transforming gendered power relations alongside top-down institutional efforts" (Berry et al., 2021:2).

14.4.2 Expanding Social Rights

Social protection reform has been a key instrument to make social contracts more inclusive and to expand social protection in times of crisis, in particular to mitigate negative impacts on vulnerable groups in the context of the recent COVID-19 pandemic. In responding to the health and economic crises, countries in the Global South have relied on recently introduced cash transfer programmes, raising expectations of a possible longer-term revision of social contracts (Martínez Franzoni & Sánchez-Ancochea, 2020, 2022). However, in the context of subsequent economic and political shocks, these expectations have in most cases not materialised.

An analysis of social policy responses to the pandemic suggests that these have been shaped by existing systems and national solutions rather than international support, while also highlighting that the pandemic shed light on gaps in social contracts. Groups most affected by the crisis were workers in informal or precarious employment conditions without social insurance coverage, as well as undocumented migrants and care workers (Cook & Ulriksen, 2021). These groups, often marked by intersecting inequalities, continue to be vulnerable and have yet to be included permanently into existing social protection schemes. However, these prospects are currently overshadowed by emerging anti-migrant and anti-rights discourses that have already translated into policy responses in major Western democracies such as the European Union and the United States (Kapelner, 2024; Amnesty International, 2025; The New York Times, 2025).

14.4.3 Renegotiating Social Contracts in the Global South: Lessons Learned

Social contracts, though usually designed for the long term, are not static and are often reformed in times of crises or at critical junctures, for example in the context of peacebuilding, regime change or ideological revolutions such as the neoliberal turn in the 1980s. Regional trends in the evolution of social contracts reveal important historical and contextual factors that drive their reform, while also pointing towards the importance of political settlements, developmental visions, dominant economic sectors and associated interests, cultural factors and values, and historical legacies (UNRISD, 2022a). Constitutional reform and inclusive social policy reforms are measures that can usher in substantive changes in previous social arrangements, but much depends on the scope, temporal dimension and actual design of policies and institutions, as well as on the inclusiveness of the contracting process, whether it counts on the support of important elite groups, whether state capacity exists to enforce it and whether it provides viable solutions to key development challenges and existing conflict issues.

Social contracts have been more successful when they (a) guarantee national reach and buy-in of key organised interest groups, (b) are coordinated by states with sufficient capacity to implement policies and enforce compliance, (c) are led by state actors with a proactive and long-term development vision, and (d) create consensus on concrete substantive issues within elite factions and the broader citizenry. Contracts have been more inclusive and produced better social outcomes when they were based on values of participation, recognition, democracy, social justice and solidarity. A context of growth and stability is a further enabling factor as well as policy space to design social contracts in line with the opportunities and constraints of each country's context (Mkandawire, 2012; UNRISD, 2010).

Finally, empirical cases referred to in this chapter, though far from constituting an exhaustive analysis, have demonstrated that few social contracts have established clear guidance on relationships with nature and the impacts of social and economic arrangements on future generations. Where this has happened, for example in the case of Bolivia or Ecuador (Barié, 2014), or in Chile in the recent failed attempt at constitutional reform (Palanza & Sotomayor Valarezo, 2024), difficulties have arisen in translating these visions into practice in contexts of mineral-dependent economies, or to gain the necessary political support, as in the case of Chile. One lesson emerging is that communitarian visions and the associated Indigenous knowledge systems which promote a more holistic human–nature relationship based on emancipatory local struggles should play a stronger role in shaping new eco-social contracts, including legislation, policy, professional practices and values. This requires broader social change processes of deracialisation of social contracts as well as greater attention to social and ecological reproduction processes.

On the positive side, environmental issues have moved to the centre of contemporary political debates and new social movements, reflected in new narratives ranging from approaches such as the green economy to the green new deal (UNRISD, 2022a; Hujo & Koehler, Forthcoming,) to more transformative approaches such as just transition, post-growth alternatives or regenerative economic systems (Gibbons, 2020). The latter approach scales up ambition to "address the dysfunctional human–nature relationship by entering into a co-creative partnership with nature. It aims to restore and regenerate the global social-ecological system through a set of localised eco-logical design and engineering practices rooted in the context and its social-ecological narratives" (du Plessis, 2012:19). Concrete proposals are developed and partially implemented in various countries to embark on more sustainable and regenerative pathways supporting justice and fairness in all spheres, for example through eco-social policies, loss and damage funds, increased investments in climate adaptation and comprehensive just transition strategies that move beyond the focus on protecting workers in carbon-intensive industrial sectors (Krause et al., 2022; Hujo & Koehler, Forthcoming). However, these initiatives and movements are currently undermined by a political conjuncture that fuels conflict, nationalism and backlash against global agendas, human rights and rules-based multilateralism, threatening the implementation of the SDGs, the Paris agreement and other important elements of global environmental governance.

14.5   Seven Principles for Building a New Eco-social Contract

This chapter has introduced different models and historical experiences with social contracts, with a focus on the diversity of normative and real-world approaches as well as renegotiations of contracts at critical junctures and in times of major societal or political transformation.

Based on the evidence and analyses presented, this chapter argues that the vision of a new eco-social contract needs to differ fundamentally from the twentieth-century social contract. A new eco-social contract should be instrumental in reconfiguring a range of relationships that have become sharply imbalanced—those between state and citizen, between capital and labour, between the Global North and the Global South, and between humans and the natural environment. The following normative principles have been proposed by UNRISD (2021, 2022a) to open a discussion on how future eco-social contracts could be strengthened. Similar and different principles can be found in various other chapters in this volume, demonstrating the evolving nature of these discussions (see Chap. 2).

Human Rights for All: *A new eco-social contract must surpass the post-war welfare state settlements by ensuring human rights for all, including those excluded from previous social contracts or relegated to a secondary role such as women;*

informal workers; ethnic, racial and religious minorities; persons with disabilities; migrants; and LGBTIQ+ persons. This requires a human rights-based approach that goes beyond formal-employment-dependent social benefits.

Universal human rights and inclusion resonate with the SDGs in various ways. Consistent with its promise to address inequalities, the 2030 Agenda commits "to leave no one behind," to ensure "targets [are] met for all nations and peoples and for all segments of society" and "to reach the furthest behind first." The social turn which led to a revival of social policy in development approaches in the late 1990s has resulted in expansion of social insurance and social assistance programmes in a range of countries, while some governments have also scaled up investments in public social services and strengthened workers' rights. Social protection has also been greatly expanded during the COVID-19 pandemic. The challenge is now to institutionalise universal social programmes and to close coverage gaps (for example, for informal workers, migrants or persons engaged in unpaid or community work) while providing adequate benefits across the lifecycle and in times of shock.

Progressive Fiscal Contracts: *A new eco-social contract must go hand in hand with a new fiscal contract that raises sufficient resources for climate action and SDG implementation and fairly distributes the financing burden.*

The provision of universal social policies requires a strong fiscal base. For many low-income countries, this will not be possible without strong support from the international donor community (Cattaneo et al., 2024). However, domestic financing schemes are the better option in the long term, as progressive distributional impacts support social integration by creating a social contract and strengthening relations within society, between economic sectors, between rich and poor and between different social groups and between society and governments (Hujo, 2020). A fiscal contract for the SDGs should favour financial instruments which are supportive of environmental goals and the sustainability transition (UNRISD, 2016).

Transformed Economies and Societies: *A new eco-social contract must be based on the common understanding that we need to transform economies and societies to halt climate change and environmental destruction and promote social inclusion and equality.*

Transformative change in our societies and economies demands deep-seated structural changes in order to overcome long-term stratification patterns that impact on future generations, locking people into disadvantage and constraining their choices and agency. Such structural change can be catalysed through innovative social policies in areas such as pensions, education, health, care, employment and equal opportunity based on universal approaches that enhance the role of the state and community organisations, and with strong regulatory frameworks and monitoring by citizens (UNRISD, 2016). It is also highly compatible with alternative economic approaches such as social and solidarity economy, new sustainability metrics used by enterprises, as well as just transition strategies which create synergies between social and climate justice (Morena et al., 2019; Kothari et al., 2014).

A Contract for Nature: *A new eco-social contract must recognise that humans are part of a global ecosystem. It must protect essential ecological processes, life support systems and the diversity of life forms, and pursue harmony with nature.*

Establishing a contract for nature requires changing dominant growth strategies and decoupling them as much as possible from natural resource use and adverse environmental impacts (see Chap. 22). This includes changing consumption and production patterns to ensure climate and intergenerational justice, which illustrates the link between resource use and equity (Cook et al., 2012). Moreover, the Rights of Nature approach describes inherent rights of ecosystems and species, as living beings that need to be given a voice and protected by law (see Chap. 11). It is based on recognition that the development and survival of human beings depend on a healthy environment and biodiversity. Earth jurisprudence considers the governance and regulation of relations between all members of the earth community, not just between human beings, and is thus an important aspect of a new eco-social contract (Berry, 2002).

Historical Injustices: *A new eco-social contract must be decolonised and informed by Indigenous knowledge, social values and capacities from the Global South. It must remedy historical injustices and combat the climate crisis fairly through just transitions.*

Securing workers' rights and decent work while economies are shifting to sustainable production and lifestyles will be paramount, as is the principle of common but differentiated responsibilities for Global South countries in addressing climate change (Morena et al., 2019). A new eco-social contract needs to replace the colonial tradition of resource exploitation with participatory and sustainable use of natural resources, as well as benefit sharing. The historical injustice of colonialism created mistrust and discrimination and is still institutionalised. Self-determination and recognition of Indigenous peoples' rights will be a vital part of applying traditional knowledge in climate change adaptation (Kempf & Hujo, 2022).

Gender Justice: *A new eco-social contract must recognise that previous social contracts have been built upon an unequal gender arrangement. It must go hand in hand with a gender contract in which activities of production and reproduction are equally shared by women and men and different genders, and where sexual orientations and expressions of sexual identity are granted equal respect and rights.*

Establishing an eco-social contract in which gender identity and sexual orientation are not a basis for discrimination requires dismantling gendered power hierarchies that subordinate women; establish gender as a static, definite and binary category; and devalorise social reproduction. Such a new contract must dismiss the gendered division of labour and centralise the work of care, a function that is essential for the maintenance of our social, economic, political and cultural institutions, and for our continued existence.

Solidarity and Peace: *A new eco-social contract requires new bottom-up approaches to transformative change for development, bringing together social movements and progressive alliances between science, policymakers and activists. It must overcome the mindset of "us against them," fostering instead a spirit of "all united against" global challenges such as climate change, inequalities and social fractures.*

Forging a new eco-social contract requires a new process where everyone gets a seat at the table. Civil Society Organisations (CSOs) and social movements as well as scientists, private sector actors and policymakers need to come together and discuss a fair distribution of costs and benefits of reforms. Informal workers, unpaid carers and community volunteers have to be invited to participate in social dialogue processes to shape public policies in line with their needs and interests. Public policies and institutions should strengthen the solidarity principle and support poor and marginalised groups and share benefits and risks in a fair way. The multilateral system needs to be strengthened to promote sustainable development, peace and security and to foster social and climate justice at the global level.

14.6 Looking Ahead: Challenges and Opportunities for New Eco-social Contracts in an Uncertain World

The world has been in a state of acute polycrisis for several years, starting with the COVID-19 pandemic. But if we think about the destructive and exclusionary effects of our global economic model, the starting point for the multiple, interlocking crises we face goes back several decades. This crisis context is now worsening by the day, making the task of repairing our broken social contracts both more urgent and more difficult, as we see the established post-war international order crumbling and nationalism, violent conflict, backlash and polarisation intensifying.

What are the prospects for the eco-social transition and transformative social change? Research evidence helps us to identify challenges, for example the challenge to (a) find momentum for reforming social contracts; (b) have the necessary time for building consensus, overcome polarisation and changing social norms; (c) bring the ecological dimension and interests of future generations into short-term political thinking; (d) reduce entrenched inequalities, power asymmetries and elite influence; (e) address the lack of credibility, trust in and legitimacy of political system; (f) mobilise economic resources and state capacity to deliver on promises and (g) prevent that transformative concepts are mobilised instrumentally and rhetorically and delinked from local struggles.

In terms of opportunities, reforming social contracts at critical junctures can forge societies together behind shared values and goals and give a clear mandate to act; reforming decision-making processes and building social contracts bottom-up can strengthen and innovate democracy and rebalance power relations; aligning national systems with global challenges and multilateral commitments can support eco-social transitions; social contracts can become more inclusive and sustainable through policy and institutional reform; continuous pressure from social movements can help to hold governments to account and keep up the demand of progressive policy changes for inclusion and sustainability.

Some movements suggest that a new eco-social contract should be informed by citizens through citizen assemblies, bringing new perspectives and expertise to combat climate change and serving as a counterbalance to vested interests (Willis, 2020; Chap. 22). Others point to the importance of integrating those vested interests representing key economic actors into social contracts following the example of neo-corporatist welfare pacts, to ensure that groups committed to the contract have the actual power to follow up on agreed goals (Mkandawire, 2012).

Grassroots participation and the inclusion of previously excluded voices are especially necessary, making sure that resource- and time-poor persons participating in these processes are supported and empowered. Consensus also implies that not everyone will see his or her original preferences succeed. Compromise is warranted, without getting stuck in the status quo or the lowest common denominator. Contestation and bargaining, protests and collective action, and building of strategic alliances will be necessary to challenge and overcome the status quo. Southern voices and Indigenous peoples' traditional knowledge as well as communitarian visions have hitherto been neglected in this debate, yet much can be learned from them, in particular regarding the sustainable management of natural resources and how everyone is part of a network of social relations defining rights and responsibilities. Last but not least, the people have more power resources than they tend to be aware of to shape the social and ecological transition and make their voices heard. In the words of Kumi Naidoo (2022: 268): "As citizens and individuals, we have the only real power to safeguard our planet and our children's future. This brave and potentially beautiful new world will require the highest level of moral courage yet, from all of us."

References

Amnesty International. (2025). *EU: Return proposals a "new low" for Europe's treatment of migrants*. News item 11 March 2025. URL: EU: Return proposals a "new low" for Europe's treatment of migrants - Amnesty International.

Auty, R. (1993). *Sustaining development in mineral economies: The resource curse thesis*. Routledge.

Barié, C. G. (2014). Nuevas narrativas constitucionales en Bolivia y Ecuador: el buen vivir y los derechos de la naturaleza. *Latinoamérica: Revista de Estudios Latinoamericanos, 59*, 9–40.

Berry, M., Bouka, Y., & Kamuru, M. (2021). Implementing inclusion: Gender quotas, inequality, and backlash in Kenya. *Politics & Gender, 17*(4), 640–664.

Berry, T. (2002). Rights of the Earth: Recognising the rights of all living beings. *Resurgence, 214*, 28–29.

Beveridge, W. (1942). *Social insurance and allied services*. His Majesty's Stationary Office. http://pombo.free.fr/beveridge42.pdf

Brignone, D., Gambetti, L., & Ricci, M. (2024). *Geopolitical risk shocks: When the size matters*. ECB Working Paper No. 2024/2972, Available at SSRN: https://ssrn.com/abstract=4919668

Cattaneo, U., Schwarzer, H., Razavi, S., & Visentin, A. (2024). *Financing gap for universal social protection: Global, regional and national estimates and strategies for creating fiscal space*. ILO Working Paper 113. International Labour Office.

Chancel, L., Piketty, T., Saez, E., Zucman, G., et al. (2022). *World Inequality Report 2022*. World Inequality Lab. wir2022.wid.world.

Cook, S., & Ulriksen, M. (2021). Social policy responses to COVID-19: New issues, old solutions? *Global Social Policy, 21*(3), 381–395.

Cook, S., Smith, K., & Utting, P. (2012). *Green economy or green society? Contestation and policies for a fair transition*. Occasional Paper: Social Dimensions of Green Economy and Sustainable Development no. 10. United Nations Research Institute for Social Development.

Desai, M. (2022). *Communitarian imaginaries as inspirations for rethinking the eco-social contract?* Issue Brief no. 12. United Nations Research Institute for Social Development.

du Plessis, C. (2012). Towards a regenerative paradigm for the built environment. *Building Research & Information, 40*(1), 7–22. https://doi.org/10.1080/0961321 8.2012.628548

El Gomati, A. (2022). The weaponisation of Libya's elections. *IPS Journal*, 14.02.2022.

Galbraith, J. K. (2022). Resource limits to American Capitalism & the Predator State Today. *GPEnewsdocs.com*, February 7. https://gpenewsdocs.com/resource-limits-to-american-capitalism-the-predator-state-today/

Gibbons, L. V. (2020). Regenerative—The new sustainable? *Sustainability, 12*(13), 5483. https://doi.org/10.3390/su12135483

Gough, I. (2021). *Two scenarios for sustainable welfare: New ideas for an eco-social contract.* The European Trade Union Institute.

Grabowski, R. (2002). East Asia, land reform and economic development. *Canadian Journal of Development Studies, 23*(1), 105–126.

Haggard, S., & Kaufman, R. (2004). Revising social contracts: Social spending in Latin America, East Asia, and the Former Socialist Countries, 1980–2000. *Revista de ciencia política (Santiago), 24*(1), 3–37.

Hickey, S. (2011). The politics of social protection: What do we get from a 'social contract' approach? *Canadian Journal of Development Studies/Revue canadienne d'études du développement, 32*(4), 426–438.

Hopkins, C., Greenfield, O., & Mohamed, N. (2020). *Is the moment for a new social contract here?* Green Economy Coalition blog, September 27. https://www.greeneconomycoalition.org/news-analysis/is-the-moment-for-a-new-social-contract-here

Houdret, A., & Amichi, H. (2020). The rural social contract in Morocco and Algeria: Reshaping through economic liberalisation and new rules and practices. *The Journal of North African Studies, 1848560.*

Hujo, K. (Ed.). (2020). *The politics of domestic resource mobilization for social development.* Palgrave Macmillan and United Nations Research Institute for Social Development.

Hujo, K. (2021). Social Protection and Inequality in the Global South: Politics, Actors and Institutions. Editorial Introduction to Themed Section on Social Protection and Inequality in the Global South: Politics, Actors and Institutions. *Critical Social Policy, 41*(3), 343–363.

Hujo, K. (Ed.). (2012). *Mineral rents and the financing of social policy: Opportunities and challenges.* Palgrave Macmillan and United Nations Research Institute for Social Development.

Hujo, K., & Koehler, G. (Forthcoming). Eco-social policies in the Global South and North: Potential and challenges for creating new eco-social contracts. In K. Bell et al. (Eds.), *The Sage handbook of eco-social policy and politics.* Sage.

Huntjens, P. (2021). *Towards a natural social contract: Transformative social-ecological innovations for a sustainable, healthy and just society.* Springer Nature. https://www.springer.com/gp/book/9783030671297

Huntjens, P., & Kemp, R. (2022). The importance of a Natural Social Contract and co-evolutionary governance for sustainability transitions. *Sustainability, 14*(5), 2976. https://www.mdpi.com/2071-1050/14/5/2976

Justino, P., & Samarin, M. (2024). *Trust in a changing world: Social cohesion and the social contract in uncertain times.* World Social Report 2025 Thematic Paper No. 2. DESA and UNU-WIDER.

Kapelner, Z. (2024). Anti-immigrant backlash: The Democratic Dilemma for immigration policy. *Comparative Migration Studies, 12*(1), 12.

Kempf, I., & Dutta, P. (2021). Transformative social policies as an essential buffer during socio-economic crises. *Sustainable Development, 29*(3), 517–527.

Kempf, I., & Hujo, K. (2022). Why recent crises and SDG implementation demand a new eco-social contract. In A. Antoniades, A. S. Antonarakis, & I. Kempf (Eds.), *Financial crises, poverty and environmental sustainability in the context of the SDGs* (pp. 171–186). Springer Nature.

Kothari, A., Demaria, F., & Acosta, A. (2014). Buen Vivir, degrowth and ecological Swaraj: Alternatives to sustainable development and green economy. *Development, 57*(3/4), 362–375.

Krause, D., Stevis, D., Hujo, K., & Morena, E. (2022). Just transitions for a new eco-social contract: Analysing the relations between welfare regimes and transition pathways. *Transfer,* 1–16.

Loewe, M., Zintl, T., & Houdret, A. (2021). "The social contract as a tool of analysis." Introduction to the special issue on Framing the evolution of new social contracts in Middle Eastern and North African Countries. *World Development, 145,* 104982.

Marglin, S. A., & Schor, J. B. (1992). *Golden age of capitalism: Reinterpreting the postwar experience.* Clarendon Press.

Martínez Franzoni, J., & Sánchez-Ancochea, D. (2022). *A lost opportunity to build social protection for all? Scenarios following emergency cash transfers in Central America.* UNRISD Think Piece Series, The Time is Now! Why We Need a New Eco-Social Contract for a Just and Green World. UNRISD.

Martínez Franzoni, J., & Sánchez-Ancochea, D. (2020). *Pactos sociales al servicio del bienestar en América Latina y el Caribe: ¿qué son y qué papel tienen en tiempos de crisis?,* Documentos de Proyectos (LC/TS.2020/169). Comisión Económica para América Latina y el Caribe.

McCandless, E. (2018). *Forging resilient social contracts for peace: Towards a needed re-conceptualisation of the social contract.* Working Paper. Witwatersrand University. https://www.wits.ac.za/wsg/research/research-publications-/working-papers/

Mills, C. W. (2007). The domination contract. In C. Pateman & C. W. Mills (Eds.), *Contract and domination* (pp. 79–105). Polity Press.

Mills, C. W. (1997). *The racial contract.* Cornell University Press.

Mkandawire, T. (2012). *Building the African state in the age of globalization: The role of social compacts and lessons for South Africa*. Inaugural Annual Lecture, Mapungubwe Institute for Strategic Reflection. MISTRA.

Moore, E., & Seekings, J. (2019). Consequences of social protection on intergenerational relationships in South Africa: Introduction. *Critical Social Policy, 39*(4), 513–524.

Morena, E., Krause, D., & Stevis, D. (Eds.). (2019). *Just transitions: Social justice in the shift towards a low-carbon world*. Pluto Press.

Naidoo, K. (2022). People's pathways to climate justice. Spotlight. In UNRISD (2022a), *Crises of inequality: Shifting power for a new eco-social contract*. United Nations Research Institute for Social Development.

Nugent, P. (2010). States and social contracts in Africa. *New Left Review, 63*, 35–68.

Oross, D., & Tap, P. (2021). Using deliberation for partisan purposes: Evidence from the Hungarian National Consultation. *Innovation: The European Journal of Social Science Research, 34*(5), 803–820.

Palanza, V., & Sotomayor Valarezo, P. (2024). Chile's failed constitutional intent: Polarization, fragmentation, haste and delegitimization. *Global Constitutionalism, 13*(1), 200–209. https://doi.org/10.1017/S204538172300028X

Palme, J., & Kangas, O. (Eds.). (2005). *Social policy and economic development in the nordic countries*. Palgrave Macmillan and United Nations Research Institute for Social Development.

Pateman, C. (1988). *The sexual contract*. Stanford University Press.

Piketty, T. (2022). *A brief history of inequality*. Harvard University Press.

Paz Arauco, V. (2020). Domestic resource mobilization for social development in Bolivia (1985–2014): Protests, hydrocarbons and a new state project. In K. Hujo (Ed.), *The politics of domestic resource mobilization for social development* (pp. 269–303). Palgrave Macmillan and United Nations Research Institute for Social Development.

Plagerson, S., Alfers, L., & Chen, M. (2022). Introduction: Social contracts and informal workers in the global south. In L. Alfers, M. Chen, & S. Plagerson (Eds.), *Social contracts and informal workers in the global south* (pp. 1–30). Edward Elgar.

Saunders, R. (2020). The politics of resource bargaining, social relations and institutional development in Zimbabwe since independence. In K. Hujo (Ed.), *The politics of domestic resource mobilization for social development* (pp. 371–404). Palgrave Macmillan and United Nations Research Institute for Social Development.

Sen, G., & Durano, M. (2014). *The remaking of social contracts: Feminists in a fierce new world*. Zed Books.

Shafik, N. M. (2021). *What we owe each other: A new social contract for a better society*. Princeton University Press.

Sheingate, A. (2015). Agrarian social pacts and poverty reduction. In Y. Bangura (Ed.), *Developmental pathways to poverty reduction*. Palgrave Macmillan and United Nations Research Institute for Social Development. https://doi.org/10.1057/9781137482549_7

Standing, G. (2019). *Plunder of the commons: A manifesto for sharing public wealth*. Pelican.

Stern Plaza, M., Binette, G., Ortiz, I., Schmitt, V., Frota, L., & Kelobang, K. (2016). *South Africa: Extending social protection by anchoring rights in law*. International Labour Organization.

The New York Times. (2025). How Trump's crackdown is drastically driving down migration. *The New York Times*.

Therborn, G. (2014). Los pactos en la teoría y en la historia social en Europa y la política de los Estados de bienestar: algunas experiencias. In M. Hopenhayn, C. M. Valera, R. Martínez, M. N. Rico, & A. Sojo (Eds.), *Pactos sociales para una protección social más inclusiva. Experiencias, obstáculos y posibilidades en América Latina y Europa* (pp. 134–140). Serie Seminarios y Conferencias no. 76. Comisión Económica para América Latina y el Caribe.

Tsikata, D., & Eweh, P. (2017). *Land and agricultural commercialisation and gendered livelihoods: A synthesis of the qualitative study of 4 districts in Ghana*. DEMETER Working Paper. Institute of Statistical Social and Economic Research, University of Ghana.

Ulriksen, M., & Plagerson, S. (2014). Social protection: Rethinking rights and duties. *World Development, 64*, 755–765.

UN (United Nations). (2021). *Our Common Agenda: Report of the Secretary-General*. United Nations.

UN. (2025). *World Social Report 2025: A new policy consensus to advance social progress*. United Nations.

UN (United Nations) and World Bank. (2018). *Pathways for peace: Inclusive approaches to preventing violent conflict*. World Bank.

UNDP. (2022). *Human Development Report 2021–22: Uncertain times, unsettled lives: Shaping our future in a transforming world*. United Nations Development Programme.

UNRISD (United Nations Research Institute for Social Development). (2021). A New Eco-Social Contract: Vital to Deliver the SDGs. Issue Brief No. 11. UNRISD.

UNRISD (United Nations Research Institute for Social Development). (2022a). *Crises of inequality: Shifting power for a new eco-social contract*. UNRISD.

UNRISD (United Nations Research Institute for Social Development). (2022b). Chapter 2.5. Political crisis: Protest, mistrust and threats to democracy. In *Crises of inequality: Shifting power for a new eco-social contract*. UNRISD.

UNRISD (United Nations Research Institute for Social Development). (2022c). Chapter 4. Toward a new eco-social contract: Actors, alliances and strategies. In *Crises of inequality: Shifting power for a new eco-social contract*. UNRISD.

UNRISD (United Nations Research Institute for Social Development). (2016). *Policy innovations for transformative change: Implementing the 2030 agenda for sustainable development*. UNRISD.

UNRISD (United Nations Research Institute for Social Development). (2010). *Combating poverty and inequality: Structural change, social policy and politics.* UNRISD.

Willis, R. (2020). A social contract for the climate crisis. *IPPR Progressive Review, 27*(2), 156–164.

World Bank. (2020). *Poverty and shared prosperity: Reversals of fortune.* World Bank.

Yilmazkuday, H. (2024). Geopolitical risks and energy uncertainty: Implications for global and domestic energy prices. *Energy Economics, 140*(2024), 107985.

15

The Role of Philanthropy in Supporting Ideas and Action for Eco-Social Futures: Some Lessons and Insights from the African Climate Foundation

Saliem Fakir and Danielle Hersch-Castros

15.1 Introduction

Similarly, under the virtuous-sounding guises of charity and philanthropy, the rich and powerful can bestow kindness from on high, without feeling implicated in the systems that produce poverty, oppression, and environmental degradation in the first place—the very systems that also produce their own wealth and power. (Hunt-Hendrix & Taylor, 2024).

This chapter explores an African climate philanthropy's approach to grant-making and assesses how philanthropy might act as a catalyst in implementing eco-social contracts in practice. This discussion is framed from the perspective of the African Climate Foundation (ACF), founded in April 2020 as the first African-led and African-run re-granting organisation on the continent.

A re-granter is a pooled fund that is either thematically or geographically focused, or both. ACF is both and channels support from global endowments to advance climate and development goals across the African continent. The majority of ACF's funding is from endowments and it does not receive any

S. Fakir (✉) • D. Hersch-Castros
African Climate Foundation, Cape Town, South Africa
e-mail: saliem@africanclimatefoundation.org;
danielle@africanclimatefoundation.org

© The Author(s) 2025
P. Huntjens et al. (eds.), *Eco-Social Contracts for Sustainable and Just Futures*,
https://doi.org/10.1007/978-3-031-99109-7_15

money from high-net-worth individuals. It operates autonomously, with no external foundation influencing its strategic direction. While endowments provide the funding, all investment decisions are aligned with ACF's strategy that is determined and ratified by ACF's board.

Beyond distributing funds, a re-granter plays a strategic role in shaping the strategy, fostering partnerships and building an ecosystem capable of delivering impactful long-term outcomes.

The establishment of ACF was driven by the significant gap in philanthropic funding for climate initiatives in Africa, as well as the absence of a dedicated African-based climate philanthropy that integrates both developmental and climate priorities. ACF has, through its grant-making, led these types of initiatives in a unique way, tackling the dual challenge of mobilising new climate investments and managing climate risk impacts across the African continent.

The ACF's focus on pooling funding around key initiatives stems from the identification of finance as one of the key constraints to Africa meeting its climate and development objectives. The ACF is therefore actively working to significantly increase support into initiatives that will catalyse low-carbon and resilient development pathways in key sectors—supporting long-term socio-economic well-being and inclusive development on the continent.

A concept central to guiding ACF's work is the climate-development nexus (Fakir, 2022). This concept highlights the critical need to integrate and connect climate and development goals and policies to achieve sustainability and equitable progress (Gomez-Echeverri, 2018). For Africa to meet its development objectives, building resilience against climate change is imperative. At the same time, achieving development goals is equally necessary for enhancing climate resilience. African countries need to therefore identify opportunities to simultaneously pursue their developmental objectives while building resilience against climate change impacts.

The climate-development nexus is embedded in key international agreements, including the Paris Agreement and the Sustainable Development Goals (SDGs), and regionally, in Agenda 2063, developed by the African Union (AU). These frameworks represent transformative milestones in the global development agenda. The overarching objective of these agreements is to redefine the approach to tackle the challenges of both climate change and development, which emphasises achieving a balance between climate resilience, economic diversification and developmental needs of the Global South. However, even with these agreements in place, the success of building climate-resilient communities relies heavily on the ability of countries to design and implement cohesive, coordinated and comprehensive national and sub-national action plans that align climate and development objectives, and to fund these.

In this chapter, two case studies will illustrate ACF's work in testing various approaches to facilitating systemic change through the climate-development nexus: the Just Energy Transition Partnership (JET-P) and the Adaptation and Resilience Investment Platform (ARIP). Parallels can be drawn between the approaches taken to design these initiatives and the principles underlying eco-social contract thinking. Such as (1) using climate solutions to solve underlying economic issues, (2) scaling climate finance in line with interventions that represent local needs, (3) empowering communities and fostering inclusive decision-making through a bottom-up approach and (4) advancing context specific climate solutions that challenge dominant economic models.

15.2 Advancing Eco-social Futures Through Funding the Climate-Development Nexus in Africa

Eco-social contracts allude to forms of relationships and agreements that connect the political, economic, social and ecological spheres, centred around the principles of equality and justice. Huntjens & Kemp (2025) define eco-social contracts as "the collective power of societies, people and nature to deal with the polycrisis of the twenty-first century through collective agreements (at multiple governance levels) among members of a society. This entails cooperation and the recognition of norms targeted at sustainability, equity and justice, with the associated rights and duties of care for the environment and the well-being of others (including future generations and all life on this planet)."

There are essential factors that should be taken into consideration when addressing climate issues in Africa. Africa has abundant natural resources, many of which, like the Congo Basin, are pristine and hold global significance. Africa is the world's poorest and least developed continent on the planet, with significant development needs. And generally, the economic growth and development models on the continent often mirror those used by advanced economies and emerging regions like China, Europe, Asia and the United States.

The context of the African development needs is one of underdevelopment and addressing this is crucial to building climate resilience and to be able to use that to also foster more pro-social types of arrangements. So, there is an opportunity to introduce new models of economic thinking, planning and practice through rethinking and redesigning solutions that are tailor-made to Africa.

From a climate philanthropy's perspective, eco-social contracts suggest that any initiatives aiming to transition to new social and economic systems, for example, a civilisation less dependent on fossil fuel, must work towards fundamentally reorienting social behaviour, economic structures and political systems. More importantly, this framework must place ecological considerations at the forefront of any decisions about the future, given humanity's intrinsic connection to nature.

In the case of Africa, investments in new energy are key to broader economic transformation due to the lack of energy access on the continent. Electrification is a core pillar of broader development on the continent, and while reducing energy emissions is important, it's not just about decarbonisation and switching to renewable energy sources. The decarbonisation agenda requires a reorientation and shift in how we think about energy sources, how they are chosen, integrated and scaled to balance the needs for actual sustainable growth.

The successful implementation of climate philanthropy in the Global South requires a balancing act between scaling up green economy solutions and dealing with the socio-economic impacts that these kinds of surge climate and environmental investments would bring about, particularly in regions with strongly entrenched fossil fuel industries. Achieving economic resilience and reducing overreliance on natural resources are key to this balancing act. The efforts of ACF and others in the co-design and establishment of the Just Energy Transition Partnership (JET-P), or similar forms of country investment platforms, aim to address both climate and development considerations.

The climate-development nexus highlights the approach required by the philanthropic sector to establish relationships that bridge the political, economic, social and ecological realms. Eco-social contracts strongly allude to the latter forms of relationships, with a strong focus on principles such as the balance between humans and nature, ensuring a more holistic and sustainable approach to addressing climate change and development challenges.

The African continent has a low emissions profile and contributes minimally to global climate change, apart from South Africa, which ranks among the top twenty emitters due to its reliance on coal. However, Africa faces significant climate-related impacts. The World Meteorological Organization (2022) estimates that climate change is affecting the African continent disproportionately, partly because the average temperature and sea level on the continent is rising faster than global average and partly due to pre-existing economic vulnerabilities. As many African nations fall within the world's poorest, they experience some of the most severe consequences of climate change. Advanced economies, which have consumed a disproportionately high share of the global carbon budget, have reaped economic benefits at the expense of more vulnerable regions (Down to Earth, 2021).

As the only African climate re-granter on the continent, ACF has observed three types of philanthropic models that are currently shaping the climate and development landscape:

The first focuses on strengthening civic and environmental rights by building movements to defend these rights, thus strengthening democracy or at least building momentum behind political change that can increasingly adopt democratic and participatory approaches.

The second, inspired by thinkers like Leah Hunt, is the dominant approach in philanthropy that is very state-centric and relies on elite experts, academics, think tanks and other types of advocacy groups to bring about policy reform and push the state to protect and defend those rights. Here, the fundamental issue is not so much about whether philanthropy should support social movements and academics, or grassroots versus elite-driven approaches, but rather what the process of change is, and what ecosystem of actors needs to be supported or developed to bring about those changes.

A third approach, which is more emergent, is the idea of using philanthropic funding to create investment environments to foster investment flows in areas like climate action, often through country investment platforms such as the JET-P. These country investment platforms hold significant value for structural change, not only for specific sectors, but have a ripple effect on the economy and even the political economy of countries as they shift the dependency, particularly in Africa, on natural resources and the export of fossil fuels. Furthermore, it offers an opportunity to use investment platforms to also reconfigure political, economic and social relations at a local level, especially in towns and regions that are dependent on fossil fuels.

Eco-social futures require the renegotiation and reimagination of globally and nationally transformative social contracts that recognise the climate-development nexus and highlight the need for a whole-of-society commitment: from governments, businesses and civil society to collaborate effectively. Eco-social contracts offer a framework to promote more accountable, transparent, and sustainable approaches to addressing the pressing environmental and socio-economic challenges of our time. We have found that elements of an eco-social contract have the potential to be embedded in just transition processes or other initiatives. In this way, it can highlight and frame the practical shifts that need to happen in order to lay the foundation for a different kind of economy, one that is just and fair, and does not perpetuate the existing and failing development system that maintains entrenched inequalities.

Integrating eco-social contracts into a country's own climate and development practices and programmes can take various forms, including (1) delivering a socio-economic agenda that better integrates and reflects a shift to an ecological paradigm, (2) recognising that we are embedded in forms of political and economic relations, but this should not dissuade us from using approaches that seek to achieve better socio-economic and ecological outcomes, (3) drawing on diverse knowledge systems, indigenous, local and global, to enrich the body of experience and wisdom, and lastly, (4) through the idea of experimentation and learning by doing, nurturing a diverse ecosystem of players that collaboratively support socio-economic transformation (Fakir, 2024).

The ACF's work in supporting and developing climate investment and development platforms (like the JET-P and ARIP) aims to establish an alternative path dependency of sufficient scale and weight that can reorient economies towards sustainable goals, without undermining developmental priorities.

15.3 Case Study 1: South Africa's Just Energy Transition Partnership (JET-P)

South Africa is a country deeply reliant on its extractive coal mining practice, which powers 85% of electricity demand in the country (Department of Mineral Resources and Energy, 2019). Coal also plays a critical role in the South African economy, as a source of export revenue and as a key feedstock for industries such as the iron and steel sector (InfluenceMap, 2024).

Amid the global shift towards net-zero economies, dismantling the South African coal system which is a crucial source of revenue and employment, as well as a significant contributor to the country's gross domestic product (GDP), is no easy feat. The challenge of decarbonising South Africa's economy becomes highly complex and poses a significant risk to the country's economic sustainability as many livelihoods depend on this sector for survival. This is further exacerbated by the nation's long-standing historical challenges, including systemic corruption, severe poverty, high unemployment and increased vulnerability to the impacts of climate change.

The Just Energy Transition Partnership (JET-P) arises as a mechanism to help address South Africa's deeply entrenched and increasingly ineffective minerals and energy complex, which has led the country down a coal-dependent economic path. The transition process in this country will be particularly challenging because much of the JET-P funding is focused on decarbonisation, with solutions largely dependent on techno-economic shifts.

While South Africa's transition to renewables is highly investable—especially given its abundant alternative energy resources (PCC, 2023)—the most crucial aspect is ensuring complementary investments in an alternative economic framework that boosts economic diversification. The "just" element of this transition is not only about shifting energy technologies, but also about fostering a more inclusive economy to guide South Africa's energy transition. This is what the Just Transition Framework (JTF) aims to achieve.

The Just Transition Framework was established by the multi-stakeholder Presidential Climate Commission (PCC) and chaired by President Cyril Ramaphosa. The PCC is mandated to steer the country's decarbonisation pathways towards a low-emissions and climate-resilient economy (PCC, 2022). As a foundational framework that reflects a new eco-social contract governing the JET-P, the Just Transition Framework extends beyond actions to achieve emission reductions, to reshaping economic and social relations. The overlapping and interconnected social and economic issues that compound this transition is especially important in regions where local economies and communities have been entirely shaped around coal.

The Just Transition Framework is underpinned by three core mechanisms for securing "justice," acknowledging that if not implemented with a justice framework that considers the relationship between humans and nature, the transition could significantly worsen integrated socio-ecological-economic conditions:

- Distributive justice considers inequalities related to gender, race and class resulting from the transition. It emphasises the need to ensure that the costs associated with the transition do not further burden communities and workers but instead remain the responsibility of those who are historically responsible and therefore accountable (PCC, 2022).
- Restorative justice speaks to the historical damages inflicted on South African communities and regions impacted by coal and other fossil fuel industries. It highlights the urgent need to heal the land and its people (PCC, 2022).
- Procedural justice highlights the participatory model required for a truly "just" and inclusive transition to take place. This must ensure that all South

Africans, particularly marginalised voices, are involved in key decisions that foster an inclusive and equitable economy that benefits everyone (PCC, 2022).

The Just Transition Framework provides the foundation for South Africa's just energy transition, guiding decarbonisation efforts while ensuring economic inclusion and social justice. However, implementing these plans requires significant financial resources, which the Just Energy Transition Partnership (JET-P) aims to mobilise. The JET-P includes four key components: first, the decommissioning of old coal plants; second, the rapid expansion of renewable energy and other investment like green hydrogen; third, the development of grid infrastructure and finally, the creation of job opportunities through green industrial development.

There are complementary additions that require investment in an alternative economic framework, particularly in regions with a long history of coal mining, where the local economies and communities have been entirely shaped around this sector. What presents an opportunity for the philanthropic sector is the way the Just Transition Framework can transform economic and social relations—not driven solely by technocratic solutions or financial incentives, but through participatory policy- and decision-making approaches being co-designed and adopted at the community level. This will require working with labour, business, and government to develop the eco-social contracts needed to ensure an inclusive and equitable green transition.

It is often overlooked in public and philanthropic discussions, but the JET-P relies on existing public and private capital structures, which are governed by liberal, capitalist economic systems. The assumption that these existing models can effectively deliver a just transition is itself a contradiction, revealing tensions that are likely to surface over time.

The JET-P is still struggling to allocate funds towards the implementation of techno-economic solutions (Business Day, 2024), and it remains significantly behind in developing eco-social contracts, where the intended beneficiaries of the just transition are actively involved in a participatory model of economic and development planning. Attempts are being made towards advancing and deepening the programme as a priority of the partners involved.

While there are no easy answers to what this approach should be, the just transitions approach is a real-life attempt to design and implement eco-social contracts through intersectional alliances in practice. It actively involves the participation of diverse groups, including the private and public sector, Indigenous communities, social justice movements, labour unions and other civil society organisations—all working together to address interconnected social, economic and environmental issues that transcend the energy sector. Within the boundaries of the existing models of planning and economic development, true success would require much more consensus-building, necessitating a complete re-evaluation of all economic models.

For programmes like JET-Ps to succeed, it has to be government led and fall within government accountability. Philanthropy must critically examine the nature of Africa's economies and the prevailing development paradigms to reshape how economic and environmental challenges are tackled and help strengthen the institutional framework for fair JET-Ps. This requires envisioning an eco-social contract that moves beyond outdated models structured around coal and other extractive industries, and has social mandate, ownership and demand. They should reflect and drive just transitions that not only focus on shifting energy systems but also address broader economic and social transformations which are necessary for sustainable and equitable development. In South Africa, the most crucial part is breaking the dependency and the technological lock-in on coal. The only way to do that is at scale and with the social mandate to do so.

15.4 Case Study 2: Adaptation and Resilience Investment Platforms (ARIPs)

The economic impact of climate change is mounting and depending on the scale of the climate vulnerability and the cost to the national fiscus; climate change has a way of affecting the future growth prospects of a country. It has a direct impact on key economic sectors that are vital to the GDP and national income. The World Meteorological Organization (WMO) estimates an average loss of 2–5% of GDP for African countries and 9% of budgets being diverted to respond to climate change events (WMO, 2024).

For example, in the case of Malawi (Araya & Fakir, 2024), the impact of the scale of cyclone Freddy slows down economic development and hence slows down the GDP growth, the ability to strengthen government budgets and improve foreign reserves. It disadvantages future investment opportunities and the ability of the government to strengthen its fiscus. Climate risk and vulnerability therefore contributes to financial distress. A recessionary fact is that climate change is a growing source of new debt distress that increases the cost of borrowing, increases debt service cost and makes it hard for the government to have the fiscal space to stimulate new economic growth.

There is another dominant observation that many developing countries face a circular dilemma, described by the Independent Expert Group on Debt, Nature and Climate (2023) as "the triple crisis." This arises because most adaptation finance predominantly originates from the public sector in the form of debt, while the most climate-vulnerable nations are often already in or on the brink of debt distress. Additionally, extreme climate events force countries to borrow further to cover losses and damages, intensifying fiscal strain and reducing financial flexibility. High debt levels and limited fiscal capacity also undermine governments' ability to invest in other critical sectors, exacerbating climate vulnerabilities, deepening inequality and impeding progress towards social and environmental justice.

In response to African countries not having the necessary financial resources and capacity to address the consequences of climate change they are increasingly facing, the ACF developed the concept of an Adaptation and Resilience Investment Platform (ARIP). Adopting a similar approach to the JET-P, it is also a form of country-specific investment platform aimed at exploring how financial mechanisms can be structured for climate adaptation that does not increase a country's debt burden.

Adaptation and Resilience Investment Platforms (ARIPs) are a way to deal with the debt issues of a country, as well as improve fiscal resilience and provide additional resources at national, sub-national and household levels, for strengthening economies against climate vulnerabilities. ARIPs are also designed to take a more systemic approach as it deals with adaptation and resilience from a fiscal approach, and not just at project-level. The approach of ACF is to expand an investment platform to scale adaptation finance, connecting the climate-development nexus in countries, with a specific focus on technological solutions. In this way ARIPs seek to link fiscal resilience with investments in key economic sectors that can strengthen climate resilience.

This approach is at a very early stage and is experimental in nature, but it is mentioned here as a case study because it offers opportunities to insert fresh and new ideas which may not have gained traction in the early phase of the design and development of climate adaptation finance initiatives.

The ACF's ARIP is currently in the initial stage of design, but a core component is devising innovative solutions to mobilise adaptation finance that can deliver co-benefits, targeting both mitigation and adaptation objectives. These finance solutions must also address strategies to reduce sovereign debt and enhance fiscal resilience. The overarching aim is to bolster the climate resilience of African nations, lower their emissions profiles and simultaneously foster growth in critical economic sectors essential to the countries' development.

More importantly, ARIPs are based on country-specific economic and fiscal circumstances. These investment platforms will therefore need to focus on improving African countries' ability to absorb finance, which requires improving government capacity, project pipeline development, coordination and monitoring of the distribution mechanism, etc. Improving the climate resilience of critical sectors is also likely to have a positive knock-on effect on the entire economy. This may be due to their systemic role (such as provision of electricity, transport infrastructure, etc.), that directly affects the productivity of other industries and due to the impact they may have on the country's economic position, such as in terms of trade balance and tax revenues.

This endeavour demands substantial political will and far-reaching changes in the global financial and economic architecture—challenges that seem almost Sisyphean in scale. ARIPs also require a more systemic approach: to both reduce onerous debts and to fund climate finance solutions that lead to lower debt reduction, freeing up of fiscal space and creating a more enabling environment for further investments in key economic sectors.

As the ARIP model is multi-stakeholder and country-led, philanthropy can play an important role in setting them up and facilitating them. However, philanthropy faces a number of challenges in effectively undertaking this role which the concept of eco-social contracts seeks to address—such as a focus on addressing historical injustices, human rights, contracts with nature, and a more progressive fiscal contract (UNRISD, 2021). This is because philanthropy operates within a global economic structure still dominated by a capitalist model. This model varies in form, yet climate risks and decarbonisation transitions have not fully redefined the future economic systems needed to address this challenge.

The persistent adaptation finance gap (UNEP, 2024) stems, in part, from the lack of understanding of the investment case for the private sector, alongside developing countries' fiscal constraints and the limited flow of grants from developed countries, relative to the adaptation needs. While priority adaptation measures often seek to improve the resilience of critical public infrastructure, such as roads and power plants, on which the entire economy depends, the difficulty in attracting funding and structuring "bankable" projects for climate adaptation initiatives is a prime example of "the tragedy of the commons" under the capitalist model (Hardin, 1968).

Adaptation and Resilience Investment Platform (ARIPs) and other similar country investment platforms offer an opportunity for the philanthropic sector to step away from the traditional philanthropy models focusing on small-scale, project-by-project interventions, to instead leverage its expertise, networks and funding to enable better use of limited public resources to catalyse more participation by the private sector—in ways the current system does not automatically provide. This also includes facilitating governments in taking a more just or similarly an eco-social approach within their national and sub-national strategies, by using country investment platforms like JET-Ps and ARIPs for the implementation of the principles for bottom-up initiatives and for enabling better, two-sided meaningful engagement between national-level planning and community-level know-how and buy-in.

To realise this potential, climate investment and development platforms like JET-Ps (in South Africa and more recently in Senegal) and ARIPs (in Malawi)—need to find ways to deepen this thinking through real-life experimentation on what this could look like in reality. This is already happening on the continent and globally. For example, movements like "transition towns (TT)," which gained traction in 2008, offer valuable insights as one manifestation of a suite of locally led economic planning and design approaches (Hopkins, 2008). And where they have not succeeded, we need to understand why.

The South African JET-P and regional ARIPs are long-term initiatives that are catalysed by significant foundational capital from bilateral and multilateral sources to address a climate and development challenge in tandem. Moreover, they are both initiatives of sufficient scale that they can in the long run shape a path dependency around which a network of actors and other economic spin offs will emerge. For example, the case of South Africa that is transitioning from a deeply entrenched fossil fuel economy, necessitating the gradual closure of all its coal plants, could see cleaner energy solutions becoming the preferred path dependency.

As a result, this new form of energy will have a new techno-economic set of characteristics which can effectively lead to a different set of skills and jobs market. New path dependencies are described in that they have the potential to completely uproot the incumbent political economy; however, the new path dependency is a double-edged sword. If it is not imbued with and rooted in strong social justice, it too would show similarities of exclusivity that we see in existing incumbent path dependencies, like the fossil fuel economy in South Africa.

This offers an opportunity for governments, businesses and civil society to reimagine things that could not happen in the existing coal sectors because it has been mostly an extractive industry. Path dependencies that attempt to build strong social and ecological elements to it have a stronger chance to be different from the past.

What we can do as an African-based philanthropic organisation is to strongly encapsulate the African perspective to build an ecosystem that not only addresses climate challenges but also directs investments towards an inclusive development and economic framework. The climate investment and development platforms above present an opportunity to enhance local capacities and promote engagement with alternative economic models that challenge outdated paradigms, by using local realities to design climate solutions. Philanthropic entities on the continent must use this strategic advantage to advance an agenda that promotes African solutions to ensure more resilient economies through economic diversification.

15.5 Role and Critique of Philanthropy in Advancing Eco-social Futures

This chapter raises the question of how we might embed eco-social contracts as both a guiding framework and a practical approach, encouraging us to reconsider the role and purpose of philanthropy. It is essential to understand that philanthropy is not monolithic; it includes a wide array of approaches, practices and impacts. Climate philanthropy increasingly recognises that economic and social justice are vital components of the transition process, a complex process that plays out at individual, local, national and global scales. However, for a long time climate philanthropy gave little consideration to socio-economic issues, but that has changed significantly in the last few years (Climate Justice Resilience Fund, n.d.; IKEA Foundation, n.d.).

A core question within both the JET-P and ARIP initiatives is whether philanthropy can be an instigator of revolutionary movements that can challenge existing systems or whether it is an approach that provides support to actors that adopt an incrementalist approach to systems change. Much of its funding is drawn from a capitalist system that also presents contradictions to the desire and motivation of the founders of philanthropic organisations that want to support social justice, democracy and rights-based approaches.

Far more important is determining how to make the state more responsive to the premises of the paradigms that underscore eco-social contracts, which is effectively about protecting citizen rights, fostering economic rights, restoring harmony between people and nature, and safeguarding citizens against abuse of state power. Achieving this is a complex endeavour and depends significantly on the political, social and economic structures of individual countries.

Furthermore, the diversity of the continent's political, social and economic systems clearly illustrates that a one-size-fits-all approach to fostering change or structuring philanthropic support will not be effective.

Through ACF's work, there are several observations that philanthropic organisations can draw on to support pathways to eco-social futures, through incorporating a climate-development lens into their philanthropic initiatives.

First, the role of philanthropy in supporting ideas and actions for eco-social contracts must go beyond achieving climate targets. It must actively work to reshape political and economic cultures to align with eco-social principles and ethos. This cultural shift needs to move from an individualistic expression towards collective action and reciprocity (with people and with nature), especially when all the evidence shows that we are moving towards a 2-degree world (Barton et al., 2024). In economic, social and political contexts, we must interrogate the antisocial enabling conditions that drive behaviours and that appear to be prioritising self-interest and exploitation over pro-social, collective goals. Only by addressing these foundational issues can philanthropy fulfil its potential to advance both climate and social justice.

Second, climate philanthropy has largely seen the solution as financing techno-economic outcomes, but over time it is being realised that this has to be balanced with a stronger embeddedness of social justice and other social welfare outcomes. Techno-economic shifts inevitably bring to the fore the challenges of mobilising the whole of society behind a transition process that is difficult and fraught with many trade-offs, where there will be winners and losers. Lindsey McGoey highlights a significant critique of philanthropy, noting that it often operates within the existing form of modern capitalism while simultaneously using its altruistic efforts as a facade to sustain forms of generational wealth (McGoey, 2012). This practice inherently contradicts the very goals and principles of philanthropy, as much of this wealth is derived from investments rooted in the capitalist system.

Third, there is a critical need for wider citizen engagement in economic governance, in which diverse models of economic thinking and experiments can be shared more widely (Mohamed, 2023). Similarities in this thinking have been adopted in the energy transition process through South Africa's Presidential Climate Commission's development of the Just Transition Framework. This is also happening through efforts like the JET-P being implemented in other countries on the continent, for example in Senegal. The ACF, in partnership with the non-governmental organisation Centre de Recherche et d'Action sur les Droits Économique Sociaux et Culturels (CRADESC), has initiated the work of implementing the JET-P through a community capacity-building workshop to discuss the methodology for implementing the JET-P in Senegal (Torche du Monde, 2025). However, this too needs to go beyond engaging people in the design and planning process and into the implementation and monitoring of initiatives to be truly successful.

Fourth, climate philanthropy has not entirely resolved whether effective climate solutions can be driven by the private sector or are best delivered through the public sector. This debate reflects broader critiques of market-based approaches, with concerns about whether profit-driven private firms can prioritise the creation of public goods (Christophers, 2022). Scepticism also arises over who bears the risks associated with scaling up clean technologies, as private firms often prioritise profitability over equitable and sustainable outcomes.

Fifth, there is a broader critique of the resurgence of "philanthrocapitalism," which promotes the idea that philanthropy can adopt the practices and principles of a capital market per se (Sklair & Glucksberg, 2021). This approach assumes that leveraging market power will inherently improve societal welfare of a wider community. However, this form of altruism diverges from the forms of altruism determined by notions of reciprocity, solidarity and community. Instead, it relies on market mechanisms that generated wealth in the first place, falsely presuming they can translate into widespread social benefit.

Finally, a key issue for climate philanthropy is that the political economy has to reorient towards a new cultural paradigm and value system that redefines value creation over value extraction, in order for the eco-social contract model to really work in reality.

In addressing climate change, philanthropy must support deeper transitions that address social justice and inequality that may arise from climate risk as there has been a tendency to focus predominantly on technological solutions, overlooking the need for a new eco-social contract for society—philanthropy needs to tackle both. There is some hope, as conversations around the just transition are gaining momentum, though they remain insufficiently critical of the underlying economic system that philanthropy itself depends on.

What the eco-social contract paradigm highlights is that there are elements of this that require fleshing out, especially in how philanthropic organisations are governed, which inherently impacts the socialisation of its respective programmes. The first process of change is to understand how eco-social contracts fit into philanthropic initiatives, as seen through the process of setting up the JET-P and ARIP. Then follows the process of reflection and embedding this into the programmatic and governance structures of philanthropies.

Finally, working with grantees to collectively catalyse experiments that seek to put this eco-social contract into practice.

Addressing the urgency of the climate adaptation and mitigation challenge and the risks posed by extreme climate events requires a shift in focus for climate philanthropy. Where neglect has occurred, greater attention must be given to strengthening economic and social resilience. Equally important, significant investments in new clean infrastructure for climate adaptation and mitigation must be increasingly aligned with the principles of an eco-social contract. This calls for philanthropies to engage in a more systematic and strategic manner than it has before, as climate adaptation and mitigation efforts that lack a social justice lens risk losing legitimacy and undermining their effectiveness.

The evolution of climate philanthropy is starting to reflect this shift. While early climate philanthropic efforts were technocratic, the growing recognition of social justice and equity, including ACF's work on just energy transition plans is already bringing this eco-social contract ethos to the forefront of philanthropic discourse, promoting inclusive economies and climate solutions that align with broader social and economic justice goals.

15.6 Conclusion

Africa has fifty-five countries with distinct political, economic and social systems, with a colonial legacy still looming as a spectre of the past. The challenges faced by the continent are, in many respects, unique.

The challenge for all of us is to embed eco-social contract thinking into climate and development practices. There is an opportunity for philanthropy to instil more inclusive approaches that the eco-social contract encourages, especially given the dependence of African economies on non-inclusive extractive models of economic development, which are not resilient, not sustainable and certainly have not benefited the broader populace. Philanthropy should prioritise pragmatic, country-led and locally owned initiatives that are shaped by the ecosystems and players within the African continent.

As we have highlighted in the chapter, supporting eco-social futures requires nurturing localised experiments and initiatives, such as JET-Ps and ARIPs, which provide real-life insights into implementing eco-social contracts in practice. Contextualising and applying the principles and precepts of the eco-social contract model as a practical framework for transforming political and economic culture within local spheres necessitates moving away from neoliberal capitalist thinking with its entrenched competitive values towards systems grounded in mutualism and solidarity.

Eco-social contract approaches, which ensure investments in climate solutions also advance justice and equality, provide a promising framework for fostering both economic diversification and strengthening social and political organisation.

At its core, philanthropy in Africa should adopt a long-term perspective, recognising that shifting the fundamentals of political, economic and social change is a slow-moving process. For the ACF, it involves increasing sustainable and inclusive investments that can bring about economic diversification and strengthening social organisations to build capabilities that can exert constructive pressure on the system. In this way, it can ensure that the process of increasing climate investments—often vulnerable to poor political behaviour and actors—is accompanied by the influence of robust social organisations and accountability, driving the emergence of a new kind of political and social contract and ultimately an eco-social future.

References

Araya, M., & Fakir, S. (2024). *The climate crisis is also an inflation crisis*. Project Syndicate. https://www.project-syndicate.org/commentary/climate-risks-must-be-integrated-into-economic-policymaking-by-monica-araya-and-saliem-fakir-2024-12

Barton, A., et al. (2024). The role of ecosystems in mitigating climate change: Towards a more integrated approach. *Bioscience*. https://doi.org/10.1093/biosci/biae087

Business Day. (2024). *SA's international finance pledges for its energy transition now over $13bn*. https://www.businesslive.co.za/bd/national/2024-11-21-sas-international-finance-pledges-for-its-energy-transition-now-over-13bn/

Christophers, B. (2022). *The price is wrong: Why capitalism won't save the planet*. Verso Books.

Climate Justice Resilience Fund. (n.d.). *Climate rights funder collaborative*. Retrieved November 25, 2024, from https://www.cjrfund.org/climate-rights-funder-collaborative

Department of Mineral Resources and Energy. (2019). *Integrated Resource Plan 2019 (IRP2019)*. Department of Mineral Resources and Energy, South Africa. https://www.dmre.gov.za/Portals/0/Energy_Website/IRP/2019/IRP-2019.pdf

Down To Earth. (2021, October 5). *Carbon budget: Unfair share has been the way of the world*. https://www.downtoearth.org.in/climate-change/carbon-budget-unfair-share-has-been-the-way-of-the-world-79872

Expert Review on Debt, Nature and Climate. (2023). *Tackling the vicious circle: The interim report of the Expert Review on Debt, Nature and Climate*. United Nations Economic Commission for Latin America and the Caribbean (ECLAC). https://www.cepal.org/sites/default/files/document/files/tackling_the_vicious_circle_-_the_interim_report_of_the_expert_review_on_debt_nature_and_climate.pdf

Fakir, S. (2022). *Climate and development nexus in Africa: Additional observations*. Polity. https://www.polity.org.za/article/climate-and-development-nexus-in-africa%2D%2Dadditional-observations-2022-03-11

Fakir, S. (2024). Climate and development investment platforms as tools for economic transformation. *Engineering News*. https://www.engineeringnews.co.za/article/climate-and-development-investment-platforms-as-tools-for-economic-transformation-2024-05-17

Gomez-Echeverri, L. (2018). Climate and development: enhancing impact through stronger linkages in the implementation of the Paris Agreement and the Sustainable Development Goals (SDGs). *Philosophical Transactions of the Royal Society A: Mathematical, Physical and Engineering Sciences, 376*(2119), 20160444. https://doi.org/10.1098/rsta.2016.0444

Hardin, G. (1968). The tragedy of the commons. *Science, 162*(3859), 1243–1248. https://doi.org/10.1126/science.162.3859.1243

Hopkins, R. (2008). *The transition handbook: From oil dependency to local resilience*. Green Books.

Hunt-Hendrix, L., & Taylor, A. (2024). *Solidarity: The past, present, and future of a world-changing idea*. Pantheon Books.

Huntjens, P., & Kemp, R. (2025). The transformation flower approach for eco-social contracting: Comparative insights from eight case studies in the Global South and North. In P. Huntjens, N. Mohamed, K. Hujo, & M. Desai (Eds.), *Eco-social contracts for sustainable and just futures* (Chapter 16, pp. 283–312). Springer Nature.

IKEA Foundation. (n.d.). *Global South just transition fund*. Retrieved November 25, 2024, from https://ikeafoundation.org/grants/global-south-just-transition-fund/

InfluenceMap. (2024, April). *The carbon majors database: Launch report*. InfluenceMap. https://influencemap.org/briefing/The-Carbon-Majors-Database-26913

McGoey, L. (2012). Philanthrocapitalism and its critics. *Poetics, 40*(2), 185–199. https://doi.org/10.1016/j.poetic.2012.02.001

Mohamed, N. (2023). Building new social contracts: An overview of participatory mechanisms for economic governance. *Green Economy Coalition*. GEC. https://www.greeneconomycoalition.org/assets/reports/GEC-Reports/Building-new-social-contracts-mechanisms-paper-Najma-Mohamed-FINAL.pdf

Presidential Climate Commission (PCC). (2022). *A framework for a just transition in South Africa*. https://pccommissionflo.imgix.net/uploads/images/22_PAPER_Framework-for-a-Just-Transition_revised_242.pdf

Presidential Climate Commission (PCC). (2023). *A critical appraisal of South Africa's Just Energy Transition Investment Plan*. https://pccommissionflow.imgix.net/uploads/images/PCC-analysis-and-recommenations-on-the-JET-IP-May-2023.pdf

Sklair, J., & Glucksberg, L. (2021). Philanthrocapitalism as wealth management strategy: Philanthropy, inheritance, and succession planning among the global elite. *The Sociological Review, 69*(2), 314–329. https://doi.org/10.1177/0038026120963479

Torche du Monde. (2025). *Énergie: JETP et renforcement des capacités des communautés, l'ONG CRADESC engage le processus.* Retrieved January 29, 2025, from https://torchedumonde.com/energie-jetp-et-renforcement-des-capacites-des-communautes-long-cradesc-engage-le-processus/

United Nations Environment Programme (UNEP). (2024). *Adaptation Gap Report 2024: Come hell and high water.* United Nations Environment Programme. https://www.unep.org/gan/resources/report/adaptation-gap-report-2024

United Nations Research Institute for Social Development (UNRISD). (2021). *A new eco-social contract: Vital to deliver the 2030 Agenda for Sustainable Development.* UNRISD. https://sdgs.un.org/sites/default/files/2021-07/UNRISD%20-%20A%20New%20Eco-Social%20Contract.pdf

World Meteorological Organization (WMO). (2022). *State of climate Africa highlights water stress and hazards.* World Meteorological Organization. https://wmo.int/news/media-centre/state-of-climate-africa-highlights-water-stress-and-hazards

World Meteorological Organization (WMO). (2024). *Africa faces disproportionate burden from climate change and adaptation costs.* World Meteorological Organization. https://wmo.int/news/media-centre/africa-faces-disproportionate-burden-from-climate-change-and-adaptation-costs

16

The Transformation Flower Approach for Eco-Social Contracting: Comparative Insights from Eight Case Studies in the Global South and North

Patrick Huntjens and René Kemp

16.1 Introduction

The urgency of tackling converging crises of climate change, biodiversity loss, widening inequalities, and democratic erosion demands a profound societal transformation. This transformation calls not only for technological innovation and new policies but also for a rethinking of the underlying social contracts that govern relationships between citizens, the state, markets, and nature. At its heart lies a fundamental realisation that can no longer be ignored: the social contract has been broken for billions of people, and with it, the bonds between people, planet, and power must be rewoven. A top-down approach based on specific regulations to safeguard the natural environment and social protection is often too crude and weak for creating alternative configurations.

In this context, the concept of *eco-social contracts* offers a bold and actionable vision: a shared foundation for just and sustainable futures rooted in restored relationships between people, planet, and power in configurations

P. Huntjens (✉)
Inholland University of Applied Sciences, Delft, The Netherlands
e-mail: patrick.huntjens@inholland.nl

R. Kemp
Maastricht University, Maastricht, Netherlands

© The Author(s) 2025
P. Huntjens et al. (eds.), *Eco-Social Contracts for Sustainable and Just Futures*,
https://doi.org/10.1007/978-3-031-99109-7_16

which are mutually agreed upon and elaborated over time. We define eco-social contracts in an ideal-typical sense as: *Implicit or explicit collective agreements across multiple levels of governance, among members of society, aimed at addressing the interconnected polycrisis of the twenty-first century, including inequalities, injustices, climate and ecological breakdown, and faltering trust in institutions. These agreements are rooted in cooperation and the recognition of shared norms and values oriented towards sustainability, equity, and justice. Importantly, eco-social contracts encompass social, environmental, economic, cultural, and institutional dimensions, and articulate the corresponding rights and duties of care for the environment and the well-being of others, including future generations and all forms of life on Earth* (Definition by authors). The configurations created are societal innovations with an important role for values that are integrated in economic practices that go beyond philanthropy and doing good (Diepenmaat et al., 2020).

Eco-social contracts build on long-standing traditions of environmental justice, human rights, and social solidarity, while also challenging dominant paradigms of growth, extractivism, and anthropocentrism (Huntjens, 2019, 2021; UNRISD, 2022; IPBES, 2024). They reflect a shift from current social contracts, largely centred on economic security and individual rights, to more regenerative, inclusive, and pluralistic models of governance and coexistence. Table 16.1 below captures this shift in broad strokes.

This chapter introduces the Transformation Flower Approach (TFA) as a structured yet adaptive approach to support eco-social contracting, enabling multi-stakeholder governance transformations for a just and sustainable

Table 16.1 From current to new eco-social contracts. Source: Authors

From...	To...
Anthropocentrism	Recognition of human–nature interdependence
State-centric governance	Polycentric, multi-level governance
Economic growth as the goal	Regenerative and well-being-oriented economies (economies in the service of life)
Short-term market logic	Long-term planetary stewardship and care-based economies
Privileges for a few	Dignity, security and justice for all, incl. non-human life
Passive citizen roles	Active, empowered participation
Homo Economicus, a rational person pursuing wealth and self-interest	*Homo Ecologicus*, a person connected with and caring for the well-being of others, incl. non-human life
Nature as resource	Nature as co-inhabitant and rights-holder
Short-term planning	Intergenerational responsibility and resilience
Fragmented, sectoral approaches	Integrated, systemic approaches
One-size-fits-all solutions	Context-sensitive and plural pathways

future. The TFA provides a step-by-step approach to connect vision-building, institutional change, and value-driven governance, addressing multi-stakeholder complexities and resistance to change (Huntjens & Kemp, 2022; Huntjens et al., 2024). The TFA is oriented at concrete visions, ascertained in a stakeholder process, followed by a back-casting to identify the steps and policies needed to achieve it. A concrete vision is what motivates people to engage with each other, much more than abstract principles. Achieving concrete results is an important motivator, and absence of results will undermine the process.

Eco-social contracting refers to the deliberate process of establishing collective agreements across multiple governance levels, ranging from local communities to global institutions, to align social, ecological, and economic objectives in order to create just and sustainable futures (Huntjens & Kemp, 2022; UNRISD, 2022; IPBES, 2024). The word eco-social contract is not even needed in an eco-social contracting process. Even if participants never explicitly use the term, the process itself of building cooperation, shared responsibility, and care across people and nature is already an act of eco-social contracting. This concept emerges from the recognition that traditional market mechanisms and top-down policy interventions are often inadequate to address systemic challenges such as environmental degradation, social inequity, and economic instability. Instead, eco-social contracting emphasises participatory, collaborative frameworks that integrate ecological sustainability with social justice and economic resilience (Huntjens, 2021; Mohamed & Huntjens, 2023). In this context, the Transformation Flower Approach (TFA) serves as a governance and facilitation approach to co-design and implement such eco-social contracts in practice, enabling systemic change through participatory, actor-driven pathways (Huntjens & Kemp, 2022; IPBES, 2024). In being oriented towards alternative views, structures and practices, the TFA goes beyond a principle-based approach, by investigating how sustainability problems can be addressed via transformative ways of production and consumption. Such initiatives demonstrate how eco-social contracts can take shape in different domains, reconnecting citizens, producers, and public institutions through shared responsibility and cooperation. Short food supply chains for organic produce, for example, illustrate how consumers may take on multiple roles, such as harvesters, ambassadors, co-owners of land, and volunteer-workers in eco-supermarkets. These arrangements redefine consumption as participation and care, creating spaces where people co-produce food and trust.

Systemic transformation requires a shift from sectoral, technocratic policy-making to integrative, participatory governance models that reflect the complexity of socio-ecological-economic systems (Chaffin & Gunderson, 2016; Koontz et al., 2015). This means that governance strategies must:

- Address multi-level governance challenges, from local to global scales (Gupta, 2007; Newig & Fritsch, 2009; Bache et al., 2016; Di Gregorio et al., 2019; Hooghe & Marks, 2021).
- Support cross-sectoral integration to align environmental, social, and economic policies (Boas et al., 2016; IPBES, 2024).
- Facilitate inclusive decision-making through participatory and deliberative processes (Armitage et al., 2020; Huntjens, 2021).
- Incorporate adaptive governance mechanisms that allow for experimentation and learning over time (Sharma-Wallace et al., 2018).

Current governance structures often rely on top-down policy interventions, which may be ineffective in addressing the deep-rooted causes of unsustainability (IPBES, 2024). The TFA offers a novel governance approach that operationalises transformative change through stakeholder engagement, negotiated eco-social contracts, and institutional work (Huntjens & Kemp, 2022; IPBES, 2024).

16.1.1 Comparative Case-Study Methodology

This chapter presents a comparative analysis of eight case studies that have applied the Transformation Flower Approach (TFA) in diverse geographical, ecological, and governance contexts. The aim of the analysis is to identify patterns, challenges, and enabling conditions across the four phases of the TFA: (1) Vision and Values, (2) Leverage Points, (3) Coalitions, and (4) Governance Pathways.

The selected case studies were drawn from recent applied research projects and master's theses, including both European and non-European contexts. These include four cases from the Netherlands (Maastricht, Amsterdam, Midden-Delfland, and the Meuse River Basin), as well as single cases from Milan (Italy), Istanbul (Turkey), Patagonia (Chile), and the Brazilian Amazon. All cases explicitly applied the TFA in the analysis of sustainability transition processes in domains such as food systems, agriculture, energy, climate, water management, and ecosystem governance.[1]

---

[1] The TFA was not identically applied. This is also not possible: no stakeholder process can be done on the basis of a prescriptive model. Even in the cases where there was a mutually agreed plan (because of a funded proposal which stipulates actions and tasks), there was a great deal of improvisation. Each project involved a unique process and setting, a unique history of pre-existing relations and collaborative action, a governance system with institutional arrangements, certain problems and opportunities for dealing the problem. These projects involved discursive work, intermediation work and institutional works, details of which are described in the reports. For those details we refer to the reports, noting that the reports contained certain omissions because of the specificness of the research. In several reports, the TFA is used as a scheme for retrospective analysis.

Table 16.2 Overview of case studies. Source: Authors

Country	Location	Topic/Sector	Source
Brazil	Amazon Rainforest	Indigenous Rights and Eco-social Governance	Daenen (2023)
Chile	Patagonia	Energy Transition	Alves (2023)
Italy	Milan	Short Food Supply Chains (SFSC)	Cappellini (2022)
Netherlands	Maastricht	Short Food Supply Chains (SFSC)	Cappellini (2022)
Netherlands	Amsterdam	Short Food Supply Chains (SFSC)	Zorer (2022)
Netherlands	Midden-Delfland	Nature-inclusive Circular Agriculture	Huntjens et al. (2024)
Netherlands	Meuse River Basin (Limburg)	Nature-based River Basin Management	Hebben (2024)
Turkey	Istanbul	Short Food Supply Chains (SFSC)	Zorer (2022)

Data for each case were extracted from original research reports and theses (see Table 16.2), which include in-depth qualitative interviews, document analysis, and participatory observations. A structured template was developed to extract key insights aligned with each TFA phase. Specifically, we coded for:

- Phase 1: Vision and Values—Dominant narratives, shared visions, and guiding values identified by local actors.
- Phase 2: Leverage Points—Key systemic interventions, tipping points, and transformation drivers.
- Phase 3: Coalitions—Actor configurations, partnerships, and relational dynamics.
- Phase 4: Governance Pathways—Institutional changes, new governance practices, and learning processes.

The analysis focused on interpretative, pattern-oriented synthesis, identifying both converging and diverging themes across the cases. The qualitative findings were systematically compared and synthesised in a cross-case matrix (see Table 16.3 in the results section), allowing for an integrated overview of how the TFA phases manifested across contexts. Instead of seeking generalisable outcomes, the QCA aimed to reveal context-sensitive patterns and recurring elements that support transformative change.

This analysis is interpretative and exploratory in nature. While the diversity of case contexts adds richness, it also limits direct comparability. The cases also vary in research depth, scale, and data availability. Nevertheless, the QCA provides meaningful insights into how the TFA can support transformation processes across diverse socio-ecological systems.

16.2 Conceptual Foundations of Transformative Change

16.2.1 Theories of Transformation and Governance

While the terms *transition* and *transformation* are sometimes used interchangeably in sustainability literature, this chapter distinguishes between them. *Transition* typically refers to sectoral or technological shifts that unfold over time (e.g., energy or agriculture transitions), often within existing institutional frameworks. *Transformation*, by contrast, implies a deeper, systemic reconfiguration of structures, values, power relations, and governance models. As such, the Transformation Flower Approach (TFA) is primarily concerned with transformation, focusing on long-term, multi-actor, value-driven change that addresses the root causes of unsustainability.

The study of transformative change is an interdisciplinary field drawing from sustainability transitions, institutional theory, political ecology, and systems thinking. Various governance models have been proposed to facilitate systemic shifts. The following four are particularly relevant:

Transition Management (TM) is a widely used governance model that fosters long-term sustainability transitions through experimentation, participatory governance, and system innovation (Kemp et al., 2007; Loorbach, 2010). It emphasises adaptive learning processes and niche innovation to promote systemic shifts. However, critics argue that TM underestimates power dynamics, political resistance, and vested interests in maintaining the status quo (Lawhon & Murphy, 2011).

Transformative Governance (TG) extends TM by incorporating multi-level governance strategies, empowering marginalised groups, and enhancing institutional adaptability (Visseren-Hamakers et al., 2021). This approach acknowledges that transformative governance requires justice-oriented policies and the recognition of diverse knowledge systems. It aims to balance top-down policy interventions with bottom-up social movements (Huntjens et al., 2012; Homsy et al., 2019).

Leverage points for change highlight different levels of intervention in complex systems (Meadows, 1997; Murphy, 2022). Some leverage points address policy instruments and regulations, while deeper interventions require shifts in worldviews, paradigms, and institutional cultures. Scholars argue that governance models should focus on higher-order leverage points, such as transforming economic structures and social values, rather than surface-level policy adjustments (O'Brien & Sygna, 2013).

Co-evolutionary governance posits that institutions, actors, and technological systems co-evolve over time (van Assche et al., 2014). It emphasises that governance structures must remain adaptive to changing socio-political and ecological realities, ensuring that policy frameworks are reflexive, learning-oriented, and inclusive (Kooiman, 2003).

16.2.2 Power, Institutions, and Resistance to Change

A significant challenge in transformations lies in overcoming the power of existing institutions, through the creation of alternative ones and adapting current arrangements. Established power structures often reinforce economic models and technological dependencies that hinder systemic shifts (Seto et al., 2016; Aarts & Leeuwis, 2023). Power asymmetries manifest in multiple ways:

- *Structural power* is embedded in legal, financial, and bureaucratic institutions, perpetuating unsustainable practices and limiting transformative agency (Meadowcroft, 2011).
- *Discursive power* influences how sustainability is framed in policy debates, shaping what solutions are considered legitimate (Rosenbloom & Rinscheid, 2020).
- *Instrumental power* refers to the direct influence of economic and political elites in resisting transformative policies through lobbying, financial incentives, and regulatory control (Markard et al., 2012).

Overcoming these barriers requires governance models that explicitly engage with power dynamics, stakeholder negotiations, and institutional adaptation (Rinscheid et al., 2021). The Transformation Flower Approach (TFA) provides mechanisms for multi-actor deliberation, participatory learning, and leverage-point strategies, ensuring that governance transformations are inclusive and equitable. A change in power relations often depends on alternative views, structures and practices becoming more powerful. It is not just a political battle but a political and economic battle.

16.2.3 Multi-actor Engagement Approaches to Transformative Change

Transformative change is inherently multi-actor and cross-sectoral, requiring collaboration among governments, businesses, civil society, and communities to ensure effective and equitable outcomes (Avelino & Wittmayer, 2016; Allen et al., 2023). Governance for sustainability must move beyond single-actor leadership to embrace diverse stakeholder interactions across scales. The Just Transition Framework emphasises social equity in transformation, ensuring that economic and environmental shifts do not deepen inequality but instead foster inclusion and opportunity, especially for marginalised groups (McCauley & Heffron, 2018; Mohamed & Huntjens, 2023).

Polycentric governance advances this by promoting overlapping, distributed decision-making structures that encourage local innovation while aligning with broader sustainability goals (Ostrom, 2010; Huntjens et al., 2012; Termeer et al., 2013). These systems foster adaptive learning, experimentation, and resilience by empowering institutions at multiple levels.

Networked Governance complements this approach by highlighting the role of horizontal collaboration, alliances, and co-production of knowledge (Fligstein & McAdam, 2012). Rather than top-down control, transformation is facilitated through dynamic networks of policymakers, civil society, private actors, and researchers. These networks enable cross-sectoral learning and co-create solutions that are context-sensitive and widely supported, improving the legitimacy and long-term impact of sustainability interventions.

16.2.4 The Role of Values in Transformative Change

Values play a critical yet often underutilised role in transformative governance (Rinscheid et al., 2025). While personal, relational, and systemic values shape sustainability transitions, governance systems frequently prioritise technical and economic solutions, overlooking ethical and cultural dimensions (IPCC, 2022; IPBES, 2022).

At the personal level, values such as justice, autonomy, and ecological responsibility shape behaviour and influence societal norms (Schwartz, 1992). Relational values, which emphasise reciprocity and care in human-environment relationships, are reflected in Indigenous knowledge and community-based governance (Chan et al., 2016). Systemic values embedded in institutions influence how ecological, social, and economic priorities are balanced. Unchallenged, dominant systemic values tend to reinforce extractive systems and inequality (McGreevy et al., 2022). Governance that supports long-term value alignment is needed to shift towards regenerative approaches.

Value misalignment across governance levels remains a key challenge. Policies may reflect incorrect assumptions about public preferences, causing resistance (Bouman & Steg, 2023). Plural valuation, which includes intrinsic, instrumental, and relational perspectives, helps reconcile differing values in governance processes (IPBES, 2022).

Deliberative dialogue and participatory approaches help build consensus and bridge value divides. However, institutional adaptation is necessary to ensure that value change leads to practical implementation (Bouman et al., 2024). Co-evolutionary governance supports ongoing learning and alignment between values and institutional arrangements (Van Assche et al., 2014), while recognising that co-evolution is a path-dependent process with inherent limits.

16.3   The Transformation Flower Approach (TFA) for Eco-social Contracting

The Transformation Flower Approach (TFA) is a transformative governance and change facilitation approach designed to support multi-stakeholder decision-making in addressing complex socio-ecological-economic challenges. Rooted in systems thinking, sustainability transitions, and institutional adaptation, the TFA provides a structured yet adaptable approach for fostering eco-social contracting (Huntjens & Kemp, 2022). Unlike traditional governance models that rely on top-down interventions, the TFA emphasises participatory, iterative approaches that engage multiple actors across governance levels and sectors, co-creating solutions through collaborative decision-making (IPBES, 2024).

16.3.1   Core Components of the TFA

The TFA offers a systemic approach for transformative change, identifying key actors, leverage points, and institutional mechanisms to create multiple pathways for transformation (Fig. 16.1). Drawing on theories of transition pathways (Loorbach, 2010; Geels et al., 2016), transformative pathways (Westley et al., 2011), leverage points (Meadows, 1997), and transformative governance (Visseren-Hamakers et al., 2021), it integrates conceptual insights with practical implementation.

The TFA consists of four interlinked phases that structure multi-stakeholder collaboration and systemic transformation. These phases guide multiple actors in aligning institutional mechanisms and values to foster systemic change and resilience.

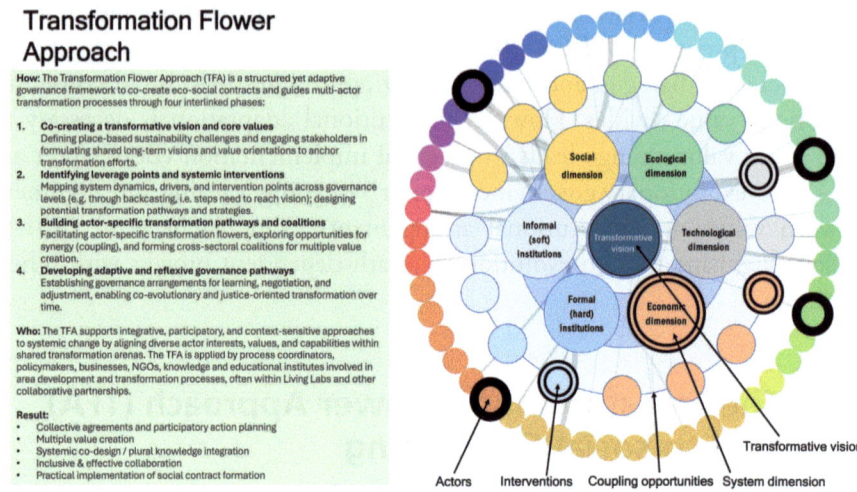

Transformation Flower Approach

How: The Transformation Flower Approach (TFA) is a structured yet adaptive governance framework to co-create eco-social contracts and guides multi-actor transformation processes through four interlinked phases:

1. **Co-creating a transformative vision and core values**
   Defining place-based sustainability challenges and engaging stakeholders in formulating shared long-term visions and value orientations to anchor transformation efforts.
2. **Identifying leverage points and systemic interventions**
   Mapping system dynamics, drivers, and intervention points across governance levels (e.g. through backcasting, i.e. steps need to reach vision); designing potential transformation pathways and strategies.
3. **Building actor-specific transformation pathways and coalitions**
   Facilitating actor-specific transformation flowers, exploring opportunities for synergy (coupling), and forming cross-sectoral coalitions for multiple value creation.
4. **Developing adaptive and reflexive governance pathways**
   Establishing governance arrangements for learning, negotiation, and adjustment, enabling co-evolutionary and justice-oriented transformation over time.

Who: The TFA supports integrative, participatory, and context-sensitive approaches to systemic change by aligning diverse actor interests, values, and capabilities within shared transformation arenas. The TFA is applied by process coordinators, policymakers, businesses, NGOs, knowledge and educational institutes involved in area development and transformation processes, often within Living Labs and other collaborative partnerships.

Result:
- Collective agreements and participatory action planning
- Multiple value creation
- Systemic co-design / plural knowledge integration
- Inclusive & effective collaboration
- Practical implementation of social contract formation

Fig. 16.1 Summarised overview of Transformation Flower Approach. Source: Authors

Phase 1: Vision and Core Values

The first phase of the TFA involves conducting a comprehensive multi-scalar analysis to define key sustainability challenges and stakeholder priorities (Huntjens et al., 2024). This step ensures that transformation efforts are tailored to the specific socio-economic and environmental conditions of the target area (e.g., neighbourhood, city, region, state, etc.) and the priorities of stakeholders (in particular the need to make money from alternative practices). By engaging stakeholders early, a shared vision emerges that integrates eco-social values and sustainability principles while respecting local priorities (Visseren-Hamakers et al., 2021; Garritzmann et al., 2024). Utilising systemic mapping tools like the X-curve model (Loorbach, 2014) helps visualise the transformation from unsustainable views, structures and practices to emerging sustainable and just alternatives that are part of an alternative eco-social contract (Fig. 16.2). This phase establishes a foundation for long-term collaboration and commitment by aligning diverse interests around a common goal. Such a transformation cannot be simply managed but requires enabling circumstances and active participation. This dual nature, as both a normative goal and a participatory process, defines the transformative power of eco-social contracts.

Phase 2: Leverage Points for Change

Once the transformation arena is defined, the second phase focuses on identifying the key drivers and governance levers that can accelerate systemic

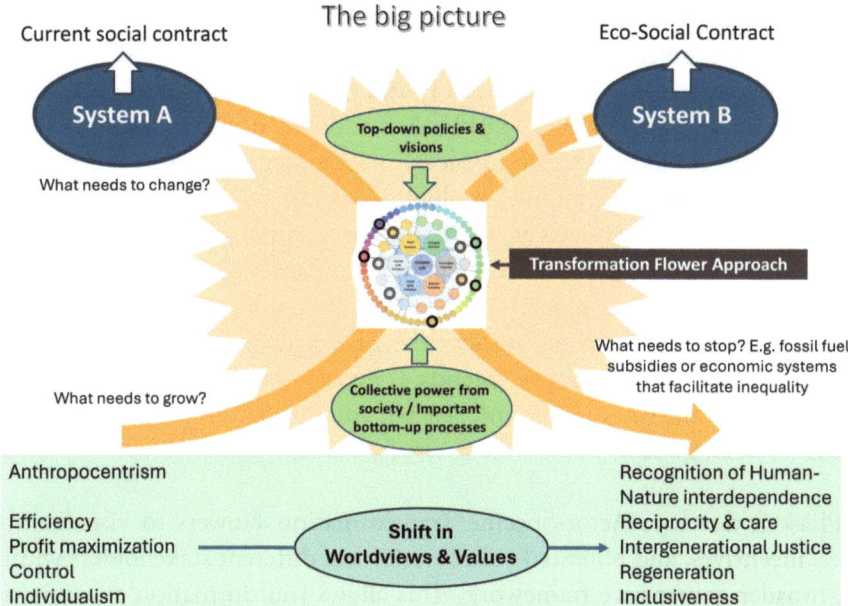

Fig. 16.2 The X-curve model (Loorbach, 2014; modified by authors) as a heuristic device to define the transformation arena, corresponding to the first analytical phase in the TFA. Source: Authors

transformation (Meadows, 1997; Murphy, 2022). This phase involves mapping transformation drivers or "key enablers" and understanding how they interact across different levels of governance. Scenario-based pathways are developed to align strategic policy interventions with practical, area-based solutions (Kemp et al., 2007). Strengthening the connection between governance levers and sector-specific strategies, such as regenerative agriculture and circular economy models (Godfray et al., 2010), ensures that interventions are both impactful and feasible. By integrating diverse actors into this phase, the TFA fosters an inclusive governance structure that promotes effective collaboration and systemic change. Coherent packages of measures are needed. Since high environmental taxes and tight regulations will be heavily resisted, they can only be gradually introduced and tightened. Subsidy schemes for innovators will run into problems of their own (of budgets getting cut or being exhausted). The choice of leverage actions is thus subject to constraints.

Phase 3: Actor Coalitions

This phase recognises that governance transformations cannot follow a one-size-fits-all model (Meadowcroft, 2011) and must be adapted to specific contexts and actors (Westley et al., 2011). Actors differ in their roles, power, and influence within networks. A multi-level stakeholder assessment ensures that interventions address economic, ecological, social, and institutional dimensions comprehensively (Geels et al., 2016). For example, civil society or labour unions can pressure governments to adopt standards that affect resource use beyond their immediate scope (Ponte, 2019). By tailoring roadmaps for policymakers, businesses, and civil society, the TFA ensures that diverse values are reflected in the transformation agenda (Huntjens & Kemp, 2022). This is especially relevant where economic, social, and ecological goals must be balanced (IPBES, 2022).

Phase 3 develops Actor-Specific Transformation Flowers to visualise the roles, incentives, and potential collaborations of different stakeholders within the broader governance framework. This allows transformation strategies to remain flexible and responsive to each actor's influence and constraints (Aarts & Leeuwis, 2023).

At its core, Phase 3 aligns with eco-social contracting principles by fostering negotiated, multi-stakeholder agreements that balance priorities within long-term governance structures. It supports the broader call for new eco-social contracts (Huntjens, 2021; UNRISD, 2022; IPBES, 2024), shifting from anthropocentric to inclusive, regenerative, and participatory governance models.

By aligning actor-specific pathways with regional governance initiatives, Phase 3 enhances policy coherence and stakeholder ownership. It encourages cross-sectoral collaboration and co-creation, ensuring that transformation is co-designed, co-owned, and co-implemented. This approach supports institutional adaptation while minimising resistance to change.

Phase 4: Pathways for Co-evolutionary Governance

The fourth phase of the TFA acknowledges that transformation is a continuous and iterative process. Governance literature highlights the necessity of reflexivity in policymaking (Jessop, 2003) and the role of deliberative democracy in conflict resolution (Dryzek, 2002). Learning mechanisms, such as experimental governance (Sabel & Zeitlin, 2008) and co-creation processes (Ansell & Gash, 2008), enable strategic adjustments based on new information and stakeholder feedback.

By embedding participatory learning loops, the TFA ensures that governance processes remain adaptive, context-sensitive, and inclusive over time. Systemic change is not a one-time intervention but an evolving process that requires ongoing negotiation, learning, and adaptation. This phase institutionalises mechanisms for stakeholder engagement and policy refinement, ensuring governance remains flexible and responsive to emerging challenges and opportunities.

Transformative governance requires continuous learning, negotiation, and reflexivity. Phase 4 of the TFA institutionalises adaptive governance mechanisms, ensuring policies and interventions evolve based on stakeholder feedback and real-world outcomes (Visseren-Hamakers et al., 2021). Defining coherent packages of policies is an important task for co-evolutionary governance, requiring policy evaluation and barrier-based analysis. Independent institutes (such as PBL in the Netherlands which is specialised in policy evaluation and policy advise) could play an important role here.

Multi-actor deliberation platforms provide spaces for ongoing negotiation and knowledge exchange, while transformative learning cycles ensure policy experimentation and feedback loops refine interventions. Additionally, conflict resolution strategies help address resistance from incumbent actors and ensure inclusive governance (Rosenbloom, 2020; Meadowcroft, 2011).

16.4 Comparison of Case Studies

To support a comparative analysis across the eight case studies, Table 16.3 provides a synthesised overview of how the Transformation Flower Approach (TFA) was applied across its four phases in each context.

Phase 1: Vision and Core Values
The foundation of any eco-social contract lies in a shared vision anchored in core values. Across the eight case studies, visioning processes allowed for recognition of place-based needs and aspirations while reflecting broader systemic concerns such as environmental degradation, social inequality, and fragmented governance. Most visions articulated a shift towards more localised, regenerative, and just systems of food production (NL, Italy, Turkey, Brazil), land and water stewardship (Chile, Netherlands river basin), and democratic, post-growth economies.

Table 16.3 Comparative overview of case applications across TFA phases. Source: Authors

Case location	Phase 1: Vision & Values	Phase 2: Leverage Points	Phase 3: Coalitions	Phase 4: Governance Pathways
Amsterdam (NL)	Shared values on food sovereignty, urban–rural solidarity, youth involvement	Short supply chains, community-supported agriculture (CSA), participatory logistics	Young activists, local food collectives, small-scale farmers, civil society actors	Bottom-up experimentation, embedded learning, collaborative infrastructure
Maastricht (NL)	Focus on local embeddedness, sustainable diets, trust and transparency	Transparent pricing, local procurement, citizen–producer agreements	Local governments, consumers, organic cooperatives	Pilots with city support, shifting narratives on food resilience
Midden-Delfland (NL)	Vision of regenerative rural–urban landscapes with multifunctional land use	Nutrient cycling, multifunctionality, education for transition	Farmers, regional planning authorities, educators	Integration into regional policy, long-term collaboration frameworks
Istanbul (Turkey)	Urban–periurban food resilience and democratic food governance	Municipal procurement reform, fair farmer–consumer pricing	Food assemblies, urban farmers, progressive city administration	Municipal-led innovation, institutional alignment with grassroots
Milan (Italy)	Reviving traditional food cultures, climate resilience, ecological production	Institutional food policy, food education in schools, waste valorisation	Milan Urban Food Policy Pact, food banks, NGOs	Cross-sector alliances, replication in other cities
Patagonia (Chile)	Energy justice, autonomy, recognition of Indigenous cosmologies	Off-grid renewables, cooperatives, energy commons	Mapuche communities, NGOs, local energy engineers	Legal empowerment, communal ownership models
Amazon (Brazil)	Stewardship, Rights of Nature, Indigenous territorial sovereignty	Forest-based economy, territorial governance, eco-education	Indigenous leaders, environmental justice movements, academic partners	Policy advocacy, biocultural restoration frameworks
Meuse River Basin (NL)	Nature-based development, multiple value creation (flood safety, biodiversity, recreation and regional development)	Blue–green infrastructure, buffer zones, adaptive practices	Waterboards, environmental NGOs, local stakeholders	Dynamic coalitions, polycentric governance

Despite contextual differences, the cases converged on core relational and regenerative values: care, solidarity, interdependence, and sustainability. The Midden-Delfland case emphasised heritage, circular agriculture, and community bonds (Huntjens et al., 2024), while the Maastricht and Amsterdam cases promoted community-driven food systems rooted in fairness, proximity, and ecological connection (Cappellini, 2022; Zorer, 2022). The Patagonia (Chile) energy transition focused on energy sovereignty and Indigenous autonomy (Alves, 2023), and the Brazilian Amazon case prioritised climate justice, Indigenous rights, and biocultural integrity (Daenen, 2023). In both Latin American cases, Indigenous cosmovisions, such as *buen vivir*, were central to imagining just and sustainable futures.

The Meuse River Basin case (Hebben, 2024) emphasised Nature-based River Management, working with natural processes and landscape DNA, and multi-value creation combining flood safety, biodiversity, recreation, and regional development. Core values included ecological recovery, stakeholder participation, trust, communication, and transboundary cooperation, underpinned by long-term institutional learning. A reframing of risk and safety, including acceptance of hybrid solutions, was also seen as necessary for long-term resilience. However, the process lacked full engagement of civil society, including youth and farmers, limiting the inclusiveness of its visioning process.

The use of the Transformation Flower Approach (TFA) proved helpful in making normative orientations explicit and in translating values into operational entry points. Yet in some contexts, such as Milan and Istanbul, values remained more aspirational than embedded in governance, reflecting the early-stage or informal nature of these transitions (Zorer, 2022; Cappellini, 2022).

Across all cases, co-creation methods such as backcasting, narrative visioning, and value mapping were applied to support participatory dialogue. While these helped crystallise shared aspirations, a recurring challenge was aligning visionary ambition with the institutional, political, and cultural readiness for transformative change.

Phase 2: Leverage Points for Change

Phase 2 focuses on identifying and amplifying strategic leverage points, i.e., high-impact interventions and system dynamics that can enable structural change. Across the eight cases, the Transformation Flower Approach (TFA) supported actors in diagnosing systemic barriers and exploring actionable pathways for transformation. While the specificity of leverage points varied, recurring patterns emerged across diverse contexts.

In the Dutch short food supply chain (SFSC) cases (Maastricht, Amsterdam, Midden-Delfland), leverage points included relational governance, trust-based collaboration, and the revaluation of local food economies. These were often framed as alternatives to centralised, efficiency-driven food systems. In Amsterdam, for example, institutional recognition of food councils enabled better integration of citizen voices (Zorer, 2022). In Maastricht, strengthening informal networks and regional identity served as a key lever (Cappellini, 2022). Midden-Delfland emphasised short value chains and nature-inclusive circular farming as strategic levers for maintaining landscape and livelihoods (Huntjens et al., 2024).

In the Latin American cases, more radical leverage points were emphasised. In Chile (Patagonia), the strategic focus lay on building energy democracy and recognising Indigenous territorial rights. Leveraging traditional knowledge and community-based planning enabled the formation of hybrid governance models (Alves, 2023). In Brazil, transformation was framed through a post-extractivist lens, with degrowth-compatible levers such as land redistribution, Indigenous governance, and climate litigation strategies (Daenen, 2023). In both cases, contesting dominant power structures and epistemologies was central.

The Istanbul and Milan cases similarly identified key leverage points in narrative change, food justice advocacy, and the promotion of cooperative platforms. However, these levers were often fragile due to limited institutional anchoring and fragmented coalitions (Zorer, 2022; Cappellini, 2022).

The Meuse River Basin case (Hebben, 2024) identified leverage points such as the adoption of nature-based solutions (NbS), adaptive river management, and transboundary collaboration. Yet the absence of strong bottom-up engagement or shared ownership limited the transformative potential.

Across all cases, the identification of leverage points required both systemic analysis and a strong understanding of local dynamics. The TFA enabled actors to link small-scale actions to wider structural shifts, including changing narratives, policy frameworks, and actor roles, offering a more integrated view of transformation beyond linear cause-effect models.

Phase 3: Actor Coalitions

Phase 3 centres on building inclusive coalitions and fostering synergies between diverse values, interests, and actor groups. Across the case studies, coalitions ranged from highly organised institutional partnerships to loosely connected grassroots networks. The Transformation Flower Approach (TFA) facilitated a shift from fragmented stakeholder engagement to the co-creation of coalitions committed to transformative goals, with varying degrees of success.

In the Amsterdam and Maastricht SFSC cases, coalitions were grounded in civil society leadership and multi-level coordination. In Amsterdam, the Food Council functioned as a hub connecting youth organisations, urban farmers, and policymakers, promoting food democracy and ecological justice (Zorer, 2022). Maastricht emphasised horizontal alliances among consumers, farmers, and cooperatives, enabling trust-based collaboration and regional anchoring of food initiatives (Cappellini, 2022).

Midden-Delfland showcased a territorially grounded coalition, including farmers, municipalities, nature organisations, and cultural heritage actors. A strong emphasis was placed on value coupling, linking agriculture, biodiversity, landscape, and tourism, to strengthen both legitimacy and continuity (Huntjens et al., 2024). This configuration of actors, leverage points, and guiding principles for transformation in Midden-Delfland is summarised in Figure 16.3.

In Istanbul and Milan, coalitions were more emergent and less formalised. While activist networks and food justice platforms played a key role, fragmentation and lack of institutional support limited long-term impact. Still, these cases demonstrated the potential of shared narratives and community spaces to foster alliances across class and cultural divides (Zorer, 2022; Cappellini, 2022).

In Patagonia (Chile), the coalition-building process brought together Indigenous communities, local governments, and energy experts around a shared vision for energy sovereignty. However, power asymmetries and historical distrust required deliberate efforts at intercultural dialogue and rights recognition (Alves, 2023). In the Brazilian Amazon, coalition-building centred on mobilising Indigenous organisations, environmental justice networks, and academic allies. Legal advocacy and biocultural claims were key strategies, with strong emphasis on intergenerational and multispecies justice (Daenen, 2023).

The Meuse River Basin case revealed the limitations of technocratic partnerships lacking strong societal ownership. While collaboration between Dutch and Belgian water authorities improved flood management and ecological resilience, engagement of non-institutional actors (e.g., youth, farmers) remained minimal (Hebben, 2024).

Overall, successful coalitions were those that integrated diverse knowledge systems, engaged underrepresented actors, and created space for multi-value synergies, linking ecological, social, and cultural goals. While government

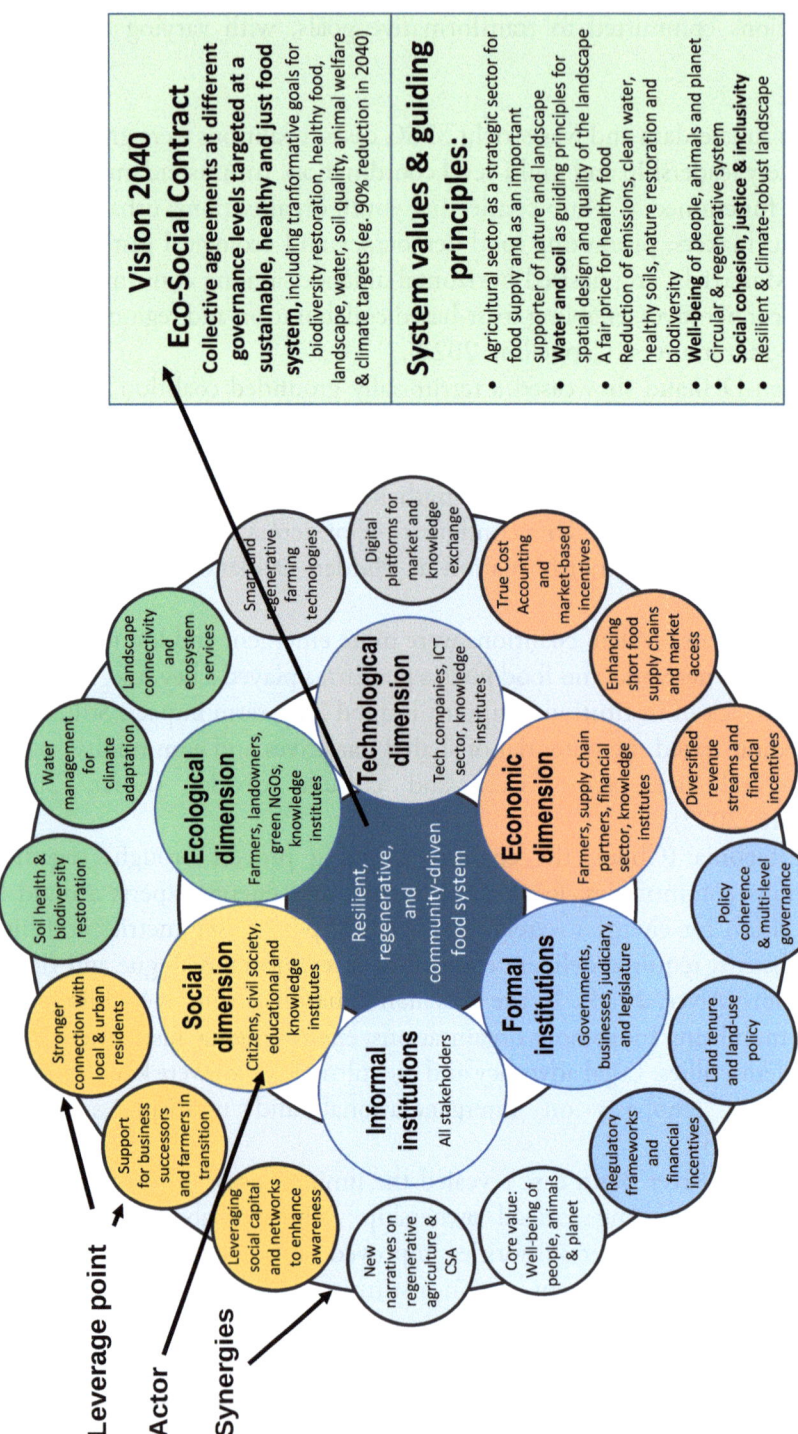

**Vision 2040
Eco-Social Contract**

Collective agreements at different governance levels targeted at a sustainable, healthy and just food system, including tranformative goals for biodiversity restoration, healthy food, landscape, water, soil quality, animal welfare & climate targets (eg. 90% reduction in 2040)

System values & guiding principles:

- Agricultural sector as a strategic sector for food supply and as an important supporter of nature and landscape
- **Water and soil** as guiding principles for spatial design and quality of the landscape
- A fair price for healthy food
- Reduction of emissions, clean water, healthy soils, nature restoration and biodiversity
- **Well-being** of people, animals and planet
- Circular & regenerative system
- **Social cohesion, justice & inclusivity**
- Resilient & climate-robust landscape

Leverage point

Actor

Synergies

Fig. 16.3   Overview of vision, leverage points, and actors in the Midden-Delfland case. Source: Authors

involvement often unlocked significant resources, it also provided constraints in the form of bureaucratic requirements, alignment with policy priorities, short-term funding cycles, and a shift in power balances that could limit local ownership and transformative ambition.

Phase 4: Pathways for Co-evolutionary and Transformative Governance
The fourth phase of the Transformation Flower Approach emphasises the development of inclusive, adaptive, and learning-oriented governance arrangements that support systemic change. This includes experimentation, feedback loops, and institutional adjustments based on emergent insights, especially in response to complexity, uncertainty, and contestation.

Across the eight case studies, efforts towards governance innovation were highly uneven. Some contexts exhibited substantial progress in embedding learning and reflexivity, while others remained at the margins of institutional change.

In Amsterdam and Maastricht, local food governance initiatives began integrating reflexive tools such as participatory evaluation, scenario planning, and policy dialogues (Zorer, 2022; Cappellini, 2022). Amsterdam's Food Council piloted inclusive deliberation mechanisms that involved youth, migrants, and informal actors in shaping food strategies. However, challenges persisted in scaling these initiatives and ensuring institutional continuity.

The Midden-Delfland case stood out for its long-term place-based approach, characterised by institutional memory, cross-sectoral dialogue, and multi-level collaboration. Stakeholders engaged in iterative learning through pilots and monitoring tools that aligned agriculture, water, and biodiversity goals (Huntjens et al., 2024). Yet, the approach remained partially vulnerable to shifting political agendas and funding structures. Building on these actor configurations and leverage points, Figure 16.4 illustrates potential transformative pathways towards an Eco-Social Contract in Midden-Delfland.

The Meuse River Basin (Hebben, 2024) demonstrated strong integration of nature-based solutions (NbS) and transboundary coordination in formal governance frameworks. Still, its highly technocratic character limited the inclusion of diverse societal perspectives. A more robust governance innovation would require deeper public engagement, especially from youth and rural communities affected by flood risks.

In Patagonia (Chile), governance learning was driven by bottom-up efforts to restore trust in institutions and create spaces for intercultural dialogue. Community energy cooperatives emerged as adaptive governance models,

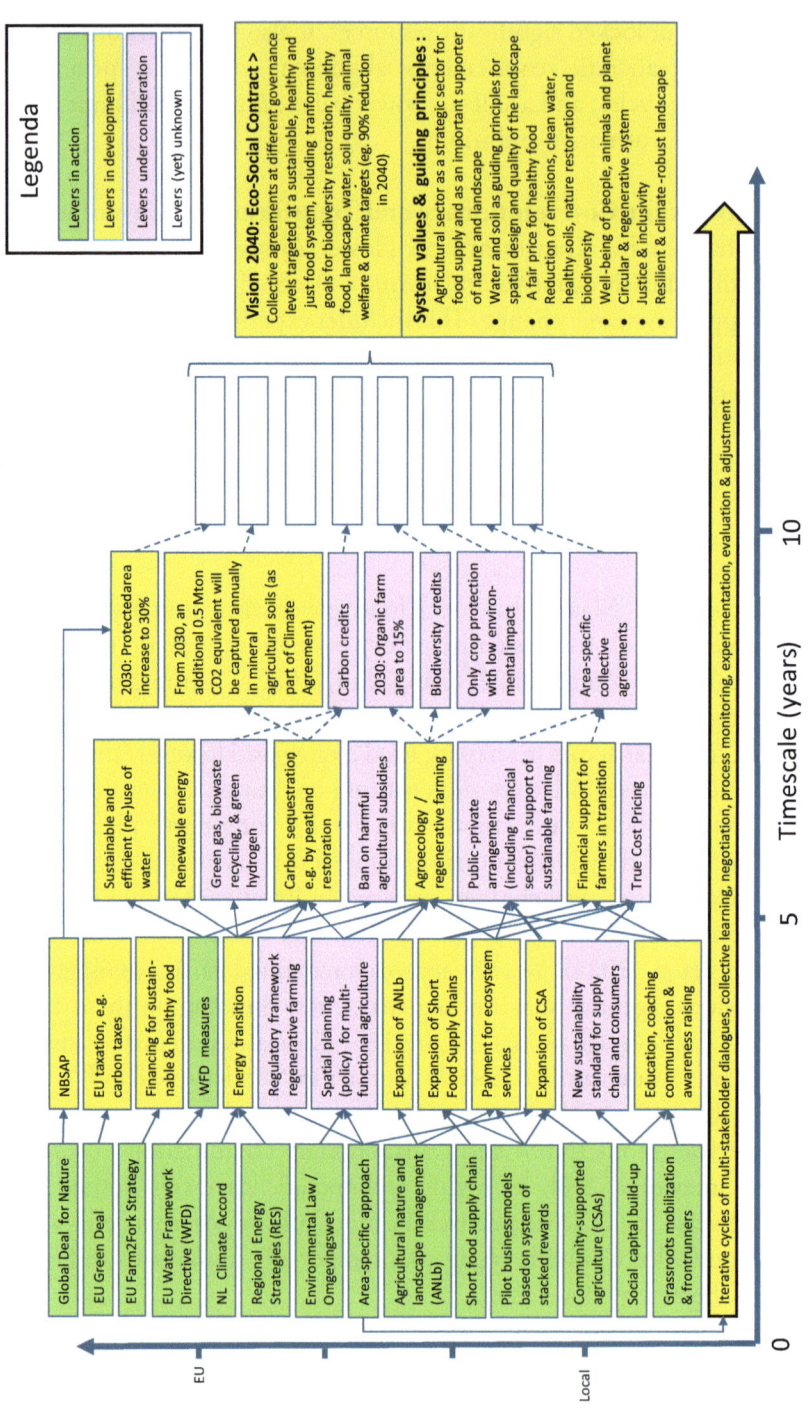

Fig. 16.4 Potential transformative pathways towards an Eco-Social Contract in Midden-Delfland. Source: Authors

supported by NGOs and progressive municipal actors (Alves, 2023). However, national policy alignment and long-term support remained limited.

In Brazil, governance innovation centred on legal frameworks and the defence of Indigenous rights, often in adversarial and high-stakes political environments. Networks of Indigenous leaders, scientists, and activists used tools such as community protocols and rights-based monitoring to document environmental harm and promote collective governance (Daenen, 2023). This demonstrates governance innovation rooted in biocultural resilience and resistance.

In Milan and Istanbul, governance experimentation was constrained by limited political will and policy continuity. However, civil society-driven models, such as urban food labs and community-supported agriculture, hinted at new pathways for institutional learning and experimentation.

Across cases, transformative governance required a shift from hierarchical control to reflexive, inclusive, and relational forms of stewardship, a core tenet of eco-social contracting. Such changes in governance take time to achieve. The creation of an eco-social contract is highly political and actively resisted by incumbents who are helped by experts favouring technical fixes and cost-benefit analysis in which relational values and fairness do not count. An important task for co-evolutionary governance is to define policies that alter economic decision-making. The instrumentation of transformative change is a journey in itself. The stakeholder process itself is inadequate for determining coherent packages and altering existing institutions but leverage points for action can be determined.

16.5 Patterns and Learnings Across Case Studies

The Transformation Flower Approach (TFA) has shown its versatility and adaptability in a wide range of settings, from the Brazilian Amazon and the Patagonian Andes in Chile to metropolitan food systems in Amsterdam and Milan. This section synthesises insights across the eight case studies to identify key principles, enabling conditions, and recurring tensions in the application of eco-social contracting. In this section we offer a cross-case analysis informed by all four TFA phases.

16.5.1 Guiding Principles

Across the cases, several guiding principles emerged as central to eco-social contracting. One recurring principle was the importance of relational values such as care, reciprocity, solidarity, and stewardship. These values were reflected in both the visions for transformation and the processes of collaboration and learning. They signify a departure from transactional and extractive logics and offer a foundation for collective responsibility and long-term resilience (IPBES, 2022; Raworth, 2018; Huntjens, 2021).

Another key principle was the emphasis on regenerative development. In most cases, transformation was understood not only as a technical or policy-oriented shift but also as a systemic process of restoring ecological integrity, social cohesion, and institutional trust. The Midden-Delfland case, for example, combined circular agriculture with cultural heritage and health concerns, while the Chilean case emphasised energy sovereignty and Indigenous autonomy (Huntjens et al., 2024; Alves, 2023).

The third guiding principle was the co-creation of desirable futures. Across the cases, participatory methodologies such as narrative inquiry, values mapping, and backcasting enabled communities and stakeholders to imagine and articulate shared visions. These futures were not treated as fixed endpoints but as evolving eco-social imaginaries that require continuous negotiation and learning (Wiek & Iwaniec, 2014; Hajer & Pelzer, 2018; Waddock, 2024).

A fourth guiding principle relates to the emphasis on multiple value creation, i.e., ensuring that transformations generate simultaneous ecological, social, and economic benefits. This focus allows the Transformation Flower Approach to transcend conventional sectoral silos and foster integrated solutions that are not only economically viable, but also environmentally sustainable and socially equitable (Huntjens, 2017, 2021; IPBES, 2022).

In relying on bottom-up initiatives, in which multiple stakeholders' concerns are addressed in a dialogical process which is results-oriented, the costs of imposed transformations are avoided. In "Seeing Like a States" James C. Scott (1998) highlights the high costs of imposed transformations through examples like collectivised agriculture, forced villagisation, and the creation of Brasilia. These projects, driven by modernist ideals of simplification and control, often disregarded local knowledge and resulted in unintended negative consequences for the populations involved. Transformative change aimed at restoring ecosystems and responsible production and consumption is thus best achieved through co-evolutionary governance, which relies on bottom-up change and the use of coherent packages based on analysis of barriers to systemic change (as a top-down element). The results-orientation is important for achieving buy-in from different stakeholders, especially for economic actors.

16.5.2 Enabling Conditions

Several conditions enabled the effective use of the TFA across different contexts. Leadership played a critical role, especially when boundary-spanning actors were able to connect different sectors and communities. In Amsterdam and Istanbul, for instance, food councils and advocacy networks were instrumental in initiating collective visioning and building multi-actor coalitions (Zorer, 2022; Cappellini, 2022).

The availability of a shared language and framework also helped to align diverse stakeholders. The TFA provided a structured but flexible tool for connecting normative values to concrete interventions. In more mature contexts, such as the Meuse River Basin, institutional memory and policy learning proved important. Long-term collaboration between governments, civil society, and experts facilitated the implementation of nature-based solutions and the incorporation of climate adaptation goals (Hebben, 2024).

Another enabling factor was the ability to link local initiatives with broader governance levels. Several cases demonstrated the importance of multi-scalar governance, in which local actions were informed by and contributed to regional, national, or even global agendas. This was particularly visible in the Latin American cases, where Indigenous worldviews and global climate justice narratives converged (Daenen, 2023; Alves, 2023).

16.5.3 Barriers and Tensions

At the same time, the case studies revealed a number of persistent barriers and tensions. Fragmented governance and lack of coordination between sectors often prevented integrated approaches. For example, both Milan and Istanbul faced difficulties in aligning food, water, and climate agendas across institutional boundaries.

Power asymmetries and political instability emerged as another major challenge. In Chile and Brazil, efforts towards eco-social contracting were constrained by entrenched economic interests and shifting political environments, even as grassroots networks mobilised for change.

In many cases, limited time and resources hindered deep engagement with diverse stakeholders. Early-stage initiatives, such as those in Amsterdam and Istanbul, showed strong ambition but were not always matched by institutional capacity or policy support.

Finally, there was often a gap between visionary ambition and the development of viable operational pathways. Translating shared values into actionable strategies proved difficult, especially when existing governance structures were not receptive to systemic change. This finding echoes broader critiques in the sustainability transitions literature, which highlight the need for both structural reform and cultural readiness (Geels et al., 2016; Abson et al., 2017).

16.6 Conclusion

This chapter has explored how the Transformation Flower Approach (TFA) can be applied across diverse socio-ecological contexts to co-create eco-social contracts. Building on eight illustrative cases from Europe and Latin America, ranging from food system transitions in Maastricht, Milan, Amsterdam, and Istanbul to landscape stewardship in Midden-Delfland, river basin governance in the Meuse, and socio-environmental justice in Patagonia, Chile, and the Brazilian Amazon, the chapter demonstrates how the TFA supports place-based, value-driven, and systemic transformation.

In all cases, the TFA proved effective in guiding actors through the complexities of transformation by structuring the process in four distinct yet interconnected phases: (1) Vision and Values, (2) Leverage Points, (3) Coalition Building and Multi-value Creation, and (4) Governance Learning and Adaptation. While the method facilitated local ownership and translation of broad visions into situated strategies, it also revealed recurring challenges, such as asymmetries in power and participation, institutional inertia, and tensions between short-term pragmatism and long-term ambition.

Despite these challenges, comparative insights from the cases highlight a number of promising patterns. Most notably, relational and regenerative values such as care, interdependence, justice, and sustainability formed a consistent backbone across the cases, though their operationalisation varied. Furthermore, the identification of leverage points often centred on cultural norms, policy frameworks, and institutional narratives, aligning with findings by Abson et al. (2017) on deep leverage for sustainability transformations. In coalition-building, the approach fostered cross-sectoral dialogue and narrative alignment, although the inclusiveness and durability of these coalitions remained context-dependent.

The TFA thus provides not only a practical toolkit for eco-social contracting but also a conceptual bridge between transformative visioning, systemic design, and adaptive governance. Its application in both European and Global South contexts illustrates its flexibility and cross-cultural resonance. Yet the chapter also calls for ongoing refinement of the approach to better integrate justice-based perspectives, engage marginalised groups, and institutionalise transformative learning.

Tilting the playing field is a key challenge for any transformation. It depends on political support for coalitions of change. Interdependencies that currently favour the status quo must be changed into interdependencies that underpin and sustain a transformation. This can be approached in a technocratic way and through an approach which is value based.

Ultimately, eco-social contracting requires more than the use of methodological tools. It demands a fundamental reorientation of societal values, governance norms, and economic models. As this chapter has shown, the TFA offers a hopeful yet grounded pathway to support that reorientation in diverse real-world settings. Future research could explore its scalability, integration with other participatory frameworks, and long-term transformative impacts.

References

Aarts, N., & Leeuwis, C. (2023). The politics of changing the Dutch agri-food system. *Journal of Political Sociology, 1*(1).

Abson, D. J., Fischer, J., Leventon, J., Newig, J., Schomerus, T., Vilsmaier, U., Von Wehrden, H., et al. (2017). Leverage points for sustainability transformation. *Ambio, 46*(1), 30–39.

Allen, C., Malekpour, S., & Mintrom, M. (2023). Cross-scale, cross-level and multi-actor governance of transformations toward the sustainable development goals: A review of common challenges and solutions. *Sustainable Development, 31*(3), 1250–1267.

Alves, I. (2023). *Developing a natural social contract – A case-study of the energy transition in Chile and the protection of Patagonia's environment and people.* SSP3021 Master Thesis, Master Programme Sustainability Science, Policy and Society, Maastricht Sustainability Institute, Maastricht University.

Ansell, C., & Gash, A. (2008). Collaborative governance in theory and practice. *Journal of Public Administration Research and Theory, 18*(4), 543–571.

Armitage, D., Mbatha, P., Muhl, E. K., Rice, W., & Sowman, M. (2020). Governance principles for community-centered conservation in the post-2020 global biodiversity framework. *Conservation Science and Practice, 2*(2), e160.

Assche, V., Kristof, R. B., & Duineveld, M. (2014). *Evolutionary governance theory: An introduction.* Springer.

Avelino, F., & Wittmayer, J. M. (2016). Shifting power relations in sustainability transitions: A multi-actor perspective. *Journal of Environmental Policy & Planning, 18*(5), 628–649.

Bache, I., Bartle, I., & Flinders, M. (2016). Multi-level governance. In C. Ansell & J. Torfing (Eds.), *Handbook on theories of governance*. Edward Elgar.

Boas, I., Biermann, F., & Kanie, N. (2016). Cross-sectoral strategies in global sustainability governance: Towards a nexus approach. *International Environmental Agreements: Politics, Law and Economics, 16*, 449–464.

Bouman, T., & Steg, L. (2023). The role of values in sustainable decision-making. *Current Opinion in Environmental Sustainability, 57*, 101206?.

Bouman, T., Steg, L., & Dietz, T. (2024). Value-based approaches for sustainability transitions. *Sustainability Science, 19*(1), 23–45.

Cappellini, F. (2022). *Transition to a sustainable food system in urban areas: An assessment of leverage points and options for co-evolutionary steering for Short Food Supply Chains in Milan and Maastricht.* SSP3021 Master Thesis, Master Programme Sustainability Science, Policy and Society, Maastricht Sustainability Institute, Maastricht University.

Chaffin, B. C., & Gunderson, L. H. (2016). Emergence, institutionalization and renewal: Rhythms of adaptive governance in complex social-ecological systems. *Journal of Environmental Management, 165*, 81–87.

Chan, K. M. A., Satterfield, T., & Goldstein, J. (2016). Rethinking ecosystem services to better address and navigate cultural values. *Ecological Economics, 120*, 8–18.

Daenen, S. (2023). *Towards a natural social contract in the Brazilian Amazon Rainforest.* SSP3021 Master Thesis, Master Programme Sustainability Science, Policy and Society, Maastricht Sustainability Institute, Maastricht University.

Diepenmaat, H., Kemp, R., & Velter, M. (2020). Why sustainable development requires societal innovation and cannot be achieved without this. *Sustainability, 12*(3), 1270.

Dryzek, J. S. (2002). *Deliberative democracy and beyond: Liberals, critics, contestations.* Oxford University Press.

Fligstein, N., & McAdam, D. (2012). *A theory of fields.* Oxford University Press.

Garritzmann, J. L., Häusermann, S., Palier, B., & Zollinger, C. (2024). *A model to follow? The EU and global eco-social policy.* Global Social Policy.

Geels, F. W., Kern, F., Fuchs, G., Hinderer, N., Kungl, G., Mylan, J., Neukirch, M., & Wassermann, S. (2016). The enactment of socio-technical transition pathways: A reformulated typology and a comparative multi-level analysis of the German and UK low-carbon electricity transitions (1990–2014). *Research Policy, 45*(4), 896–913.

Godfray, H., Charles, J., Beddington, J. R., Crute, I. R., Haddad, L., Lawrence, D., Muir, J. F., et al. (2010). Food security: The challenge of feeding 9 billion people. *Science, 327*(5967), 812–818.

Gregorio, D., Monica, L. F., Paavola, J., Locatelli, B., Pramova, E., Nurrochmat, D. R., et al. (2019). Multi-level governance and power in climate change policy networks. *Global Environmental Change, 54*, 64–77.

Gupta, J. (2007). The multi-level governance challenge of climate change. *Environmental Sciences, 4*(3), 131–137. https://doi.org/10.1080/15693430701742669

Hajer, M. A., & Pelzer, P. (2018). 2050—An energetic Odyssey: Understanding 'Techniques of Futuring' in the transition towards renewable energy. *Energy Research & Social Science, 44*(2018), 222–231.

Hebben, B. (2024). *Transformative change in river basin management: Towards natural river management.* SSP3021 Master Thesis, Master Programme Sustainability Science, Policy and Society, Maastricht Sustainability Institute, Maastricht University.

Homsy, G. C., Liu, Z., & Warner, M. E. (2019). Multilevel governance: Framing the integration of top-down and bottom-up policymaking. *International Journal of Public Administration, 42*(7), 572?–5582.

Hooghe, L., & Marks, G. (2021). Multilevel governance and the coordination dilemma. In *A research agenda for multilevel governance* (pp. 19–36). Edward Elgar.

Huntjens, P. (2017). Mediation in the Israeli-Palestinian water conflict: A practitioner's view. Water diplomacy in action: Contingent approaches to managing complex water problems 1 (2017): 203.

Huntjens, P. (2019). *Sociale innovatie voor een duurzame samenleving: Op weg naar een natuurlijk sociaal contract.* Lectorale boek. IMPACT Lectoraat Sociale Innovatie in het Groene Domein, Hogeschool Inholland, juni 2019. https://doi.org/10.48544/90f2b3be-ab4d-4acf-a9c3-13effea07bcf

Huntjens, P. (2021). *Towards a natural social contract: Transformative social-ecological innovation for a sustainable, healthy and just society.* Springer Nature. https://www.springer.com/gp/book/9783030671297

Huntjens, P., Herkrath, M., & Brouwer, D. (2024). *Gebiedsgericht Werken op Basis van de Transitiebloem-aanpak (TBA) – Casus: Landschap-inclusieve Kringlooplandbouw Midden-Delfland.* Hogeschool Inholland, Delft, May 3, 2024.

Huntjens, P., & Kemp, R. (2022). The importance of a natural social contract and co-evolutionary governance for sustainability transitions. *Sustainability, 14*(5), 2976. https://www.mdpi.com/2071-1050/14/5/2976

Huntjens, P., Lebel, L., Pahl-Wostl, C., Camkin, J., Schulze, R., & Kranz, N. (2012). Institutional design propositions for the governance of adaptation to climate change in the water sector. *Global Environmental Change, 22*(1), 67?–681.

Intergovernmental Panel on Climate Change (IPCC). (2022). *Climate change 2022: Impacts, adaptation, and vulnerability.* Contribution of Working Group II to the Sixth Assessment Report of the Intergovernmental Panel on Climate Change. Cambridge University Press. https://doi.org/10.1017/9781009325844

Intergovernmental Science-Policy Platform on Biodiversity and Ecosystem Services (IPBES). (2022). Methodological assessment report on the diverse values and valuation of nature of the intergovernmental science-policy platform on biodiversity and ecosystem services. In Balvanera, P., Pascual, U., Christie, M., Baptiste, B., & González-Jiménez, D. (Eds.). IPBES Secretariat. https://doi.org/10.5281/zenodo.6522522

Intergovernmental Science-Policy Platform on Biodiversity and Ecosystem Services (IPBES). (2024). Thematic Assessment Report on the Underlying Causes of Biodiversity Loss and the Determinants of Transformative Change and Options for Achieving the 2050 Vision for Biodiversity. Edited by Karen O'Brien, Lucas Garibaldi, and Arun Agrawal. IPBES Secretariat. https://doi.org/10.5281/zenodo.11382215

Jessop, B. (2003). Governance and meta-governance: On reflexivity, requisite variety, and requisite irony. In *Governance as Social and Political Communication*, 101–116.

Kemp, R., Loorbach, D., & Rotmans, J. (2007). Transition management as a model for managing processes of co-evolution for sustainable development. *The International Journal of Sustainable Development and World Ecology, 14*, 78–91.

Kooiman, J. (2003). *Governing as governance*. Sage Publications.

Koontz, T. M., Gupta, D., Mudliar, P., & Ranjan, P. (2015). Adaptive institutions in social-ecological systems governance: A synthesis framework. *Environmental Science & Policy, 53*, 139–151.

Lawhon, M., & Murphy, J. T. (2011). Socio-technical regimes and sustainability transitions: Insights from political ecology. *Progress in Human Geography, 36*(3), 354–378.

Loorbach, D. (2010). Transition management for sustainable development: A prescriptive, complexity-based governance framework. *Governance, 23*(1), 161–183.

Loorbach, D. (2014). *To transition! Governance Panarchy in the new transformation.* Erasmus University Rotterdam.

Markard, J., Raven, R., & Truffer, B. (2012). Sustainability transitions: An emerging field of research and its prospects. *Research Policy, 41*(6), 955–967.

McCauley, D., & Heffron, R. (2018). Just transition: Integrating climate, energy and environmental justice. *Energy Policy, 119*, 1–7.

McGreevy, S. R., Rupprecht, C. D. D., Niles, D., Wiek, A., Carolan, M., Kallis, G., Kantamaturapoj, K., Mangnus, A., Jehlička, P., Taherzadeh, O. & Sahakian, M., (2022). Sustainable agrifood systems for a post-growth world. *Nature Sustainability* [online].

Meadowcroft, J. (2011). Engaging with the politics of sustainability transitions. *Environmental Innovation and Societal Transitions, 1*(1), 70–75.

Meadows, D. H. (1997). Places to intervene in a system. *Whole Earth, 91*, 78–84.

Mohamed, N., & Huntjens, P. (2023). *Dismantling the ecological divide: Toward a new eco-social contract.* United Nations Research Institute for Social Development (UNRISD) Issue Brief 15. UNRISD. https://cdn.unrisd.org/assets/library/briefs/pdf-files/2023/ib15-a-new-contract-with-nature.pdf

Murphy, R. J. A. (2022). Finding (a theory of) leverage for systemic change: A systemic design research agenda. *Systemic Design Association* 1, no. 10.58279, v1004.

Newig, J., & Fritsch, O. (2009). Environmental governance: Participatory, multi-level- and effective? *Environmental Policy and Governance, 19*(3), 197–214.

O'Brien, K., & Sygna, L. (2013). *Responding to climate change: The three spheres of transformation.* Proceedings of Transformation in a Changing Climate Conference, Oslo, Norway.

Ostrom, E. (2010). Polycentric systems for coping with collective action and global environmental change. *Global Environmental Change, 20*(4), 550–557.

Ponte, S. (2019). *Business, power and sustainability in a world of global value chains.* Bloomsbury Publishing.

Raworth, K. (2018). *Doughnut economics: Seven ways to think like a 21st-century economist.* Chelsea Green Publishing.

Rinscheid, A., Huntjens, P., & Aarts, N. (2025). Do stakeholders' values support transformative change in the food system? Evidence from the Netherlands. *Sustainability: Science, Practice, and Policy, 21*(1). https://doi.org/10.1080/15487733.2025.2549160

Rinscheid, A., Rosenbloom, D., Markard, J., & Turnheim, B. (2021). From terminating to transforming: The role of phase-out in sustainability transitions. *Environmental Innovation and Societal Transitions, 41*, 27–31.

Rosenbloom, D. (2020). Engaging with multi-dimensional dynamics in socio-technical transitions. *Environmental Innovation and Societal Transitions, 34*, 25–38.

Rosenbloom, D., & Rinscheid, A. (2020). Deliberate decline: An emerging research agenda for the study of industrial phase-outs. *Environmental Innovation and Societal Transitions, 35*, 269–282.

Sabel, C. F., & Zeitlin, J. (2008). Learning from difference: The new architecture of experimentalist governance in the EU. *European Law Journal, 14*(3), 271–327.

Schwartz, S. H. (1992). Universals in the content and structure of values: Theoretical advances and empirical tests in 20 countries. *Advances in Experimental Social Psychology, 25*, 1–65.

Scott, J. C. (1998). *Seeing like a state: How certain schemes to improve the human condition have failed.* Yale University Press.

Seto, K. C., Davis, S. J., Mitchell, R. B., Stokes, E. C., Unruh, G., & Ürge-Vorsatz, D. (2016). Carbon lock-in: Types, causes, and policy implications. *Annual Review of Environment and Resources, 41*, 425–452.

Sharma-Wallace, L., Velarde, S. J., & Wreford, A. (2018). Adaptive governance good practice: Show me the evidence! *Journal of Environmental Management, 222*, 174–184.

Termeer, C. J., Stuiver, M., Gerritsen, A., & Huntjens, P. (2013). Integrating self-governance in heavily regulated policy fields: Insights from a Dutch farmers' cooperative. *Journal of Environmental Policy & Planning, 15*(2), 285–302.

United Nations Research Institute for Social Development (UNRISD). (2022). *Overcoming Inequalities: Towards a new eco-social contract.* UNRISD. ISBN: 9789210021609. Permalink: http://digital.casalini.it/9789210021609

Visseren-Hamakers, I. J., Razzaque, J., McElwee, P., Turnhout, E., Kelemen, E., Rusch, G. M., Fernandez-Llamazares, Á., Chan, I., Lim, M., Islar, M., & Gautam, A. P. (2021). Transformative governance of biodiversity: Insights for sustainable development. *Current Opinion in Environmental Sustainability, 53*, 20–28.

Waddock, S. (2024). Holistic eco-social imaginaries for a life-centered future. *Sustainability Science, 19*(6), 2119–2134.

Westley, F., Olsson, P., Folke, C., Homer-Dixon, T., Vredenburg, H., Loorbach, D., Thompson, J., Nilsson, M., Lambin, E., Sendzimir, J., Banerjee, B., Galaz, V., & van der Leeuw, S. (2011). Tipping toward sustainability: Emerging pathways of transformation. *Ambio, 40*, 762–780.

Wiek, A., & Iwaniec, D. (2014). Quality criteria for visions and visioning in sustainability science. *Sustainability Science, 9*(4), 497–512.

Zorer, E. (2022). *Transition to a sustainable food system through co-evolutionary steering: Discovering leverage points in the short food supply chains of Istanbul and Amsterdam.* SSP3021 Master Thesis, Master Programme Sustainability Science, Policy and Society, Maastricht Sustainability Institute, Maastricht University.

17

The Climate Conference of the Parties (COP): Process and Pathways for Eco- and Peace-Promoting Social Contracts

Erin McCandless

17.1 Introduction

Social contracts, at their core, are dynamic agreements about how we live together in governed spaces, grappling in particular with questions of how we handle moral obligation and competing interests (McCandless, 2023). As we face profound and growing crises—including rising conflict, humanitarian disaster, environmental degradation, climate change, that are intersecting in ways that deepen inequality and vulnerability—these questions become increasingly paramount.

Existing social contracts are widely seen as exacerbating vulnerability and exclusion, undermining inclusive and sustainable development (UN Women, 2021; ILO, 2019). Scholars attribute this to a neoliberal, globalisation-driven economic model that neglects redistribution and social protection. The UN Research Institute for Social Development (UNRISD) (2022: 221) argues that social contract breakdown fuels multiple global crises and deepens societal divisions. This economic model has failed to respect planetary boundaries, accelerating environmental destruction and climate crises, with severe

E. McCandless (✉)
Qatar-South Africa Centre for Peace and Intercultural Understanding, (CPIU), Faculty of Humanities, University of Johannesburg, Johannesburg, South Africa

© The Author(s) 2025
P. Huntjens et al. (eds.), *Eco-Social Contracts for Sustainable and Just Futures*,
https://doi.org/10.1007/978-3-031-99109-7_17

313

consequences for future generations. Social contract erosion has also been linked to conflict, particularly economic grievances over poor living standards and opportunities (James, 2012; Addison & Murshed, 2001), resource-sharing disputes, and institutional failures to resolve conflicts peacefully (Murshed, 2011). The rise of non-state armed actors further complicates global stability (McCandless, 2024; Davis, 2013). At the international level, contestation over peace and security reflects legitimacy crises in the rules-based order, while critiques of global financial institutions highlight their role in perpetuating, rather than resolving, inequality (United Nations, 2023).

The pervasiveness and escalation of conflict and climate-related crises further underscore the urgent need for new social contracts that centrally engage these issues and include the stakeholders most affected. While beyond the scope of this chapter to engage the climate-conflict intersections in depth, this well-documented literature suggests that climate change acts as a "threat multiplier." While not explicitly causal, as threats multiply, they have social, political, and economic consequences that inevitably hit the vulnerable hardest (Busby, 2019; Huntjens & Nachbar, 2015), by exacerbating food insecurity, water scarcity, and displacement, which can amplify tensions and lead to conflict—particularly in contexts of conflict and fragility where needed institutions and infrastructure are weak or non-existent. Conversely, conflict undermines the capacity of governments and communities to manage climate-related risks, increasing vulnerabilities to natural disasters and compounding the humanitarian and development challenges in these areas (Schwoebel & McCandless, 2021; Busby, 2019).

As these crises compound and intersect, policymakers are increasingly calling for new social contracts and global governance architectures that better address contemporary challenges. The United Nations Secretary-General (UNSG) António Guterres' call for new and more inclusive social contracts at national and global levels, as articulated in his 2021 "Our Common Agenda," was catalytic in this regard. The 2024 United Nations Summit of the Future further harnessed this momentum, reflecting alarm at our social and political conditions, the inadequacies and incapacities of our existing global governance architecture, and a stark multilateral desire for new agreements.

This chapter examines social contract, environmental, and peace literatures, alongside the primary global policy process dealing with climate—the Conference of the Parties (COP) of the United Nations Framework Convention on Climate Change (UNFCCC)—to explore what a social contract sensitive to both climate crisis and conflict might encompass. It considers how these policy dialogues, agreements, and emerging practices—reflecting mechanisms of global social contracting—are factoring in the special needs and interests of countries affected by conflict and fragility, alongside future generations and nature, in an effort to map more inclusive, sustainable, and peace-promoting social contracts at global and national levels.

The need for such investigation is paramount when considering the disproportionate impacts on, and different resulting needs for countries affected by these intersecting crises (McCandless & Faus-Onbargi, 2023). Of the 1.3 billion people exposed to climate hazards globally, 40% are in conflict-affected and fragile states (Läderach et al., 2021). Extreme weather events affect three times as many people annually in these states compared to other countries, yet they receive significantly less climate finance—up to 80 times less—than non-fragile states (COP28 President and WFP Chief, 2023). This inequity is particularly unjust given that rich countries, accounting for just 12% of the global population, have been responsible for approximately 50% of greenhouse gas emissions over the past 170 years (Andrew & Peters, 2021).

Moreover, countries affected by conflict and fragility are least prepared to adapt; more than 50% of the 25 most vulnerable countries, least ready to adapt to climate change, are affected by conflict (ND-Gain, 2022). Their struggles are evidenced by the fact that they are disproportionately behind the curve in producing their National Adaptation Plans (UNEP, 2024: XII).[1] In short, they face a triple burden: greater exposure to climate risks, less capacity to adapt, and significantly less access to climate finance. This is occurring as they also face immediate and myriad forms of threats that accompany violence and instability.

Premised on the notion that new social contracts need to endeavour to address these crises in an integrated manner, this chapter seeks to develop a framework for analysis by addressing three key questions:

1. What conceptual and practical considerations are needed to frame thinking about eco- and peace-sensitive social contracts?
2. How (and how well) are key policy processes through the climate COP seeking to address the challenges faced by countries affected by conflict and fragility?
3. What conceptual, policy, and practice insights and implications can be drawn from the literature and these efforts to advance transformative eco- and peace-promoting social contracts?

---

[1] Seven of ten countries that show no indication of developing such an instrument rank highly on the Fragile States Index.

To pursue this inquiry, the chapter draws upon key informant interviews,[2] extensive prior research on different aspects of the topic, literature review, and evaluation of COP processes and outcomes. Case studies are undertaken to reflect on the second question, focusing on key processes of the Conference of the Parties of the United Nations Framework Convention on Climate Change. Specifically, these include the COP28 UAE Declaration on Climate, Relief, Recovery and Peace (the "Declaration"), and National Adaptation Plans (NAPs).

Conceptually, and as elaborated below, these processes are framed here as social contracting mechanisms, that is, the mechanisms (i.e., legal, political, policy, social) occurring at and through myriad levels (i.e., global, national, civil society) and spheres (i.e., political settlement processes, transitional, governance, everyday) through which a social contract is forged and enabled to sustain (McCandless, 2020). The COP itself (a governing body of the UNFCCC, an international convention made up of member state representatives and accredited observers) can also be considered a mechanism of the wider foundational Paris Agreement—the 2015 legally binding international treaty—one that seeks to ensure that climate commitments are accounted for and implemented. Reflecting scholarly debates on the social contract, the 2015 Paris Agreement is considered a global social contract here, for simplicity, although some would (rightly) argue that such a formal agreement is part and parcel of a wider social contract that includes informal agreements, norms, and practices. The very existence of a social contract is, of course, contested in today's highly polarised geopolitical landscape—representing both a limitation and a critical area of ongoing research for social contract analysis.

The next section examines relevant literature and is followed by case analysis of the COP Declaration and NAPs. Transformative principles and pathways needed at the core of eco-social contracts for peace and well-being within countries emerging from conflict, as well as the shifts and conditions needed to support this at the global level, are then proposed.

By examining these intersecting crises and the efforts to address them, the chapter aims to spur reflection and discussion on strategic elements necessary for crafting new, more inclusive, and sustainable social contracts that hold promise to address the complex challenges of our time.

---

[2] Key informant interviews were conducted with leadership in the NGN, the G7+, and civil society involved in the Peace@COP network.

17.2 The Social Contract in Environmental and Peace Literature

The social contract concept, while often associated with classical liberal thought, has deeper historical and geographical roots. Social contract theorists have long been concerned with both existing and potential social contracts, encompassing normative questions of how we could, might, or should live together—ideally without coercion (McCandless, 2024). Enduring themes of social contract theory, as identified by McCandless (2018), building on Lessenoff (1990) and Freeman (2013), have engaged leaders and communities since the ancient era (1000 BC–400 BC), demonstrating the concept's relevance across different eras and civilisations. These include questions about its purpose and who it is between, the mechanisms through which it is forged and that enable its sustainability, how moral obligations are addressed, and how competing interests are managed.

The following subsections examine how the social contract is addressed in environmental (including ecological and climate) and peace (and state-building) literatures, while this "enduring themes" framework guides analysis in the chapter's conclusions.

17.2.1 Eco-social Contracts

In environmental and ecological literature, the concept of eco-social contracts has emerged, representing an evolution of traditional social contract theory to incorporate nature. UNRISD (2022: 221) argues that an eco-social contract must incorporate the ecological dimension (the connections between living things and the environments in which they live) and be grounded in a broad consensus among diverse stakeholders, rooted in multi-levelled democratic, inclusive, and participatory decision-making processes. Such a contract must be practically oriented, reflecting a shared vision that incorporates the needs of the planet and future generations, with concrete objectives, commitments, and accountability mechanisms.

A central theme in this literature is the expansion of the social contract to include non-human entities and future generations. Scholars like Eckersley (2004) and Dryzek and Pickering (2018) argue for incorporating the Rights of Nature and future generations into a "green state" and "planetary justice," respectively. This expansion challenges traditional anthropocentric views and acknowledges the interconnectedness of human societies with nature.

Huntjens' (2021: 4) "Natural Social Contract" views people and communities as part of a natural ecosystem, and society as a social-ecological system.

A fundamental shift from individualistic to eco-centric consciousness is emerging in contemporary discourse. A variety of UNRISD publications (i.e., Kempf et al., 2023; Mohamed & Huntjens, 2023) advocate moving from strategic individualistic to relational ethical thinking that recognises human interdependence and natural commons. UNRISD (2022: 299) envisions "a reimagined global governance system" based on multilateralism and solidarity, acknowledging human-nature interconnections. This evolution requires viewing humans as part of an interdependent ecosystem, working within planetary boundaries (Mohamed & Huntjens, 2023). Such transformation demands profound cultural and psychological changes, breaking from anthropocentric economic and social systems driving ecological division (Kempf et al., 2023). This represents a fundamental shift in the social contract, prioritising ecological harmony over individual interests.

Economic transformation is another key aspect of eco-social contract theories. Huntjens (2021) advocates for a transition from linear to circular and regenerative economies, aligning with scholars like Raworth's "Doughnut Economics"—where the economy operates within social and planetary boundaries, creating a "safe and just space for humanity" (Raworth, 2018: 44). This requires redefining economic progress, redistributing wealth and opportunity, and regenerating natural systems.

The Green New Deal (GND) concept has gained traction as a comprehensive approach to economic transformation supportive of an eco-social contract. Pettifor (2019) argues that the GND demands both deep economic and ecological restructuring to address structural issues in the financialised, globalised economy. UNRISD critiques some GND and "green growth" approaches, arguing that they may not sufficiently address fundamental contradictions of current systems. Instead, alternative economic approaches centred on environmental and social justice are needed to rein in the power of the financial sector, stop the gross accumulation of wealth by the super-rich, and rebalance state-market-society-nature relations (UNRISD, 2022: 274). The United Nations Conference on Trade and Development (UNCTAD) (2019) similarly supports a GND approach that combines environmental recovery, financial stability, and economic justice through massive public investments in decarbonisation while guaranteeing jobs for displaced workers.

New forms of governance and decision-making are also emphasised. Dryzek and Pickering (2018) call for inclusive, reflexive, and anticipatory forms of collective decision-making in global governance. Huntjens (2021: 174) emphasises adaptive, reflexive, and deliberative approaches to governance that strive to address uncertainty and complexity. There is also emphasis on participatory decision-making and deliberative processes (Hayward, 2006; UNRISD, 2022: 221) as mechanisms to craft and maintain an eco-social contract.

A strength of the eco-social contract literature is its holistic approach, recognising the interconnectedness of social, economic, and ecological systems, and proposals for comprehensively addressing complex global challenges like climate change and biodiversity loss. The literature is also upfront about its normative foundations—which is required, as UNRISD argues (2022: 221), to bring about the transformative shifts needed to overcome our development challenges.

17.2.2 Peace-Promoting Social Contracts

The concept of social contracts has long been associated with fostering peaceful societies in political thought and scholarship. Classical political philosophers tied the social contract concept to issues of peace and stability. The Hobbesian social contract reflected an imposed peace—an agreement amongst all authorising one person to exercise "political powers necessary to enforce the articles of peace" (Estlund, 2012: 8). Later, Locke, followed by Rousseau, shifted conceptual thinking towards a democratic social contract, based on consent by and rights of citizens, upheld by institutions (Freeman, 2013: 2). For Rousseau, deeply concerned with inequality, these were not simply political rights; all people held inherent rights of freedom and equality (Rousseau & Cress, 1987: 49).

Critical perspectives emerged over the years with implications for the concept's relationship to peace, reflecting a tradition in social contract thought on questions of how to resist coercive rule and expand the political and moral space for inclusion and emancipation (McCandless, 2024)—core themes in peace studies. While Marx saw the concept's value in serving emancipatory (rather than capitalist) interests, later scholars like Pateman (1988) and Mills (2002) advocated for contracts inclusive of women and all races.

Recent scholarship examines social contracts in conflict-affected contexts. Following the New Deal for Engagement in Conflict Affected and Fragile States, policy actors have explored pathways from fragility through social contract perspectives, examining interplay between societal expectations, state capacities, elite will, and the political processes involved in forging and institutionalising bargains (Loewe et al., 2019; OECD, 2011; UNDP, 2012, 2016).

Scholars have examined these issues through case study research. A nine-country study examining social contracts in post-conflict contexts identified three drivers of resilient social contracts: inclusive political settlements addressing core conflict issues, institutions delivering fairly, and broadening social cohesion (McCandless et al., 2018). Findings highlighted their mutually reinforcing nature and challenges when political settlements fail to link the work of different social contracting mechanisms, do not effectively redistribute power and resources, and do not deliver inclusive outcomes.

Building on this work, scholars have explored inclusion as a specific driver of social contracts in sustaining peace, finding that the interaction of elite and societal inclusion across political settlement processes supports national actors' ability to navigate crises without reverting to violence (Zahar & McCandless, 2020). However, inclusion cannot indefinitely substitute for tangible progress in meeting citizens' expectations and delivering on agreements—as exemplified by South Africa, where persistent inequality undermines social cohesion despite apartheid's formal end (Ndinga-Kanga et al., 2020).

Critics like Richmond (2009: 565) argue that liberal social contracts serve state and market interests over societies. While compelling, this analysis is restricted to "a particular paradigmatic framing and historical use of the concept, rather than the essence of the concept and the diversity of perspectives that have shaped contractarian thinking" (McCandless, 2024). As Hickey (2011: 428) observes, social contract thinking encompasses diverse perspectives, from interest-based (Hobbesian) to rights-based (Rousseauian). Thus, social contracts need not be static or top-down, or reinforcing a neoliberal state, but can be reshaped by inclusive societal and grassroots movements, as seen in the Arab Spring and in South Africa (Ndinga-Kanga et al., 2020; McCandless & Schwoebel, 2021). While power dynamics remain a significant challenge across contexts, the social contract concept offers a heuristic tool for engaging with power structures and reimagining more inclusive national visions (McCandless, 2024).

17.2.3 Common Themes

Both environmental and peace literature emphasise the need for inclusivity, participation, adaptability, and a fundamental shift in how we conceptualise stakeholders in new social contracts—from generalised notions of the governed and governing, to a lens on those most affected, or left behind. They both emphasise the importance of participatory decision-making, addressing power imbalances and deep structural inequities, and reimagining governance structures and development models to meet contemporary challenges. They also highlight the tension between stability and flexibility in social contracts, recognising the need for both enduring agreements and the capacity to adapt to changing circumstances.

Several realisation challenges persist, however, across these theoretical frameworks and proposals. First, implementing expanded rights for non-human entities and future generations poses practical difficulties. Overcoming power dynamics and institutional inertia at global and national levels remains a significant obstacle to realising more inclusive and ecologically minded social contracts. Second and related, achieving meaningful participation from marginalised groups on the one hand and inclusive outcomes on the other are not guaranteed as highlighted earlier. Finally, addressing competing interests and ideas about moral obligation in an increasingly complex and polarised world with powerful political interests denying the climate crisis (or man-made contributions) and retreating from international commitments will continue to present profound obstacles.

The good news lies in the considerable traction amongst policy actors reflecting on these topics, for new social contracts that fundamentally align with what these eco- and peace literatures are calling for. Common themes across a growing body of policy reflection suggest their need to be more resilient, inclusive of all stakeholders, social- and economic-justice oriented, underpinned by human rights, with greater trust and solidarity to consolidate and sustain them (McCandless, 2023).[3] The UNSG has also importantly underscored the critical need to revitalise social contracts at all levels—required to address the complex crises that transcend state boundaries (UNSG, 2022: 14).

17.3 Social Contracting Through COP Policy Processes

The theoretical frameworks of eco-social and peace-promoting social contracts provide crucial insights for examining contemporary global governance mechanisms addressing climate change and conflict. As both environmental and peace literature emphasise the need for inclusive, participatory processes and recognition of complex stakeholder interdependencies, the COP process has emerged as a key arena for negotiating new forms of global social contracts.

---

[3] This UNICEF study included a review of 50 documents from 12 policy institutions—the UN and its agencies, the World Bank, and the Organisation for Economic Co-operation and Development (OECD).

Two mechanisms in particular—the COP28 UAE Declaration on Climate, Relief, Recovery and Peace ("Declaration") and National Adaptation Plans (NAPs)—demonstrate how theoretical imperatives for integrating environmental and peace considerations are being translated into practical governance frameworks. These mechanisms illustrate both the progress and challenges in implementing more inclusive, responsive approaches to the climate-conflict nexus.

The COP, as the supreme decision-making body of the UNFCCC, has driven global climate action since 1995 through annual meetings reviewing progress, negotiating agreements, and setting targets. The UNFCCC process has seen significant milestones over its decades-long history. The Kyoto Protocol at COP3 in 1997 marked the first legally binding commitments for developed nations to reduce greenhouse gas emissions. COP15 in 2009 introduced the Copenhagen Accord, calling for limiting global temperature increases to below 2 °C and mobilising USD 100 billion annually by 2020 for climate finance. The landmark Paris Agreement at COP21 in 2015 brought treaty status for climate change, with nearly every country joining a unified framework committing to limit warming to below 2 °C while striving for 1.5 °C. COP26 in Glasgow emphasised coal phase-down and accelerated climate action.

COP28 in Dubai (2023) marked a pivotal moment with the first Global Stocktake assessing collective progress towards Paris Agreement goals. Notable outcomes included establishing a loss and damage fund and unprecedented agreement on fossil fuel transition, despite industry resistance. However, adaptation targets and finance remained partially unresolved (Wascow et al., 2023). At COP29, developed nations agreed to mobilise USD 300 billion annually by 2035 to support climate initiatives in developing countries—a tripling of the previous USD 100 billion target. However, it falls short of the USD 500 billion per year that the G77 called for, reflecting the Independent High-Level Expert Group's report suggesting USD 1 trillion by 2030 is needed (Climate diplomacy, 2024).

COP28 also was the first COP to explicitly link climate action with addressing conflict, notably through the development of the UAE Declaration on Climate, Relief, Recovery and Peace ("Declaration"), which recognised the disproportionate effects of climate change on countries affected by conflict and fragility, necessitating targeted support (COP28 Official Website, COP28 Declaration). COP29 further advanced action in relation to this Declaration and more broadly in relation to the climate-peace nexus, as elaborated below.

These developments signal a growing awareness of the interconnected nature of climate change and global security challenges.

17.3.1 The COP28 UAE Declaration on Climate, Relief, Recovery, and Peace

The COP28 UAE Declaration on Climate, Relief, Recovery, and Peace, launched in December 2023, represents a significant shift in global climate policy by explicitly addressing the climate-conflict intersection. This Declaration, endorsed by over 90 countries and 40 organisations, aims to mobilise "bolder, collective action" to scale up climate finance and action in fragile and conflict-affected settings (Vasquez, 2023; COP 28 UAE, n.d.).

The Declaration's framework is built on three pillars: enhancing financial support for adaptation and resilience, improving good practices and programming, and strengthening coordination and partnerships across sectors (UNFCCC, 2023). It aims to increase climate action and investment in conflict-affected countries and communities, presenting a package of finance, policy, programmes, and practices to operationalise this commitment (COP28 President, 2023). The three pillars focus on the following:

First: scaling up financial adaptation and resilience, improving access and disbursement flexibility, and strengthening technical and institutional capacities. It emphasises local investing, ownership, and results, leveraging private sector involvement, and monitoring commitments and disbursements to fragile and conflict-affected states to help identify funding gaps.

Second: understanding and improving good practice and programming, targeting vulnerable populations through prevention, anticipatory action, and climate-smart infrastructure. It emphasises sustainable agriculture, resilient food systems, and social protection, while promoting gender sensitivity and affected groups' empowerment and leadership.

Third: strengthening coordination and partnerships across the humanitarian-development-peace nexus, and optimising mandate coherence and enhancing information exchange (UNFCCC, 2023). The declaration also set a roadmap for future action, with signatories committing to reconvene at COP29 to review progress and initiate potential additional actions (UNFCCC, 2023)—which indeed, took place, with commitments for ongoing engagement as discussed above.

While marking significant progress, the Declaration has both strengths and limitations. Its primary strength lies in its acknowledgement of the links between climate change, fragility, and conflict, and its emphasis on conflict-sensitive climate policies, funds, and programming. The declaration also highlights the need for granular and integrated risk assessments, multi-level approaches with a focus on local actors, and the importance of local ownership in climate adaptation and peacebuilding efforts (Toda Peace Institute, 2024; Saferworld, 2023).

Conversely, the Declaration's non-binding nature and lack of specific funding commitments limit immediate impact. Critics point out its silence on the climate impacts of military activities and wars, particularly relevant given their substantial contribution to greenhouse gas emissions (Toda Peace Institute, 2024). The devastating levels of emissions and wider environmental impacts in Gaza's conflict are illustrative (Otu-Larbi et al., 2024). The Declaration's implementation also faces challenges. Translating the increased focus on conflict sensitivity into best practices and policies will require further action and cross-sector collaboration and funding. This includes investing in understanding what conflict sensitivity[4] means in different contexts and how it can be effectively applied to climate action (Saferworld, 2023).

Despite these limitations, the Declaration is widely viewed as a crucial first step in establishing the climate-conflict-peace nexus as a priority in international climate policy. It sets the stage for future discussions and commitments, potentially leading to more concrete actions and resources directed towards climate adaptation and peacebuilding in fragile contexts (Toda Peace Institute, 2024)—so desperately needed given the disproportionate vulnerability, risk, and funding challenges these countries face.

In June 2024 a new coordination mechanism was set up by the UAE-based Anwar Gargash Diplomatic Academy (AGDA), the G7+, an intergovernmental organisation of 20 countries of self-proclaimed fragile and conflict-affected states, and the UK-based Overseas Development Institute (ODI), to drive climate action in fragile and conflict-affected areas (AGDA, G7+, ODI, 2024). Building on these efforts, COP29 dedicated a day to the topic of Peace, Relief, and Recovery with many related side events. A Baku Call on Climate Action for Peace, Relief, and Recovery was launched, collaboratively initiated by Egypt, Italy, Germany, Uganda, the UAE, and the United Kingdom. The Call aims to address the climate change, conflict, and humanitarian needs nexus. Simultaneously, a Baku Climate and Peace Action Hub was conceived to facilitate collaboration between national, regional, and international initiatives to ensure peace-sensitive climate action, while scaling up finance and wider support for the most climate-vulnerable countries also affected by conflict and humanitarian needs (COP29, 2024).

---

[4] As commonly understood, conflict sensitivity is the practice of understanding the context in which an intervention occurs, assessing the interaction between the conflict and the intervention, and taking action to minimise negative impacts and maximise contributions to peace.

17.3.2 National Adaptation Plans

National Adaptation Plans (NAPs) are crucial components of global climate response under the UNFCCC, helping countries identify and address climate adaptation needs. Adaptation, as defined by the UNFCCC, refers to "adjustments in ecological, social, or economic systems in response to actual or expected climatic stimuli and their effects," encompassing "changes in processes, practices, and structures to moderate potential damages or to benefit from opportunities associated with climate change" (UNFCCC, 2025a). The process is flexible and country-driven, following UNFCCC guidelines covering groundwork and addressing gaps, preparatory elements, implementation strategies, and monitoring. As of November 2023, 142 developing countries were undertaking NAP measures (UNFCCC-LEG, 2023). This widespread participation underscores the global recognition of the need for structured adaptation planning.

While the UNFCCC guidelines don't specifically address ensuring conflict sensitivity, a civic initiative, the NAP Global Network (NGN)[5] has developed such guidance, alongside other like-minded civic initiatives (ECCP, 2024). The guidance note emphasises the need to understand and factor local context and conflict dynamics in adaptation strategies. It suggests that NAPs can contribute to peacebuilding by addressing conflict root causes, such as resource scarcity or livelihood insecurity, often exacerbated by climate change (NGN, 2024). This guidance aims to help countries initiate, finance, implement, monitor, evaluate, and learn from their NAP process in a way that responds to peace and conflict dynamics (NGN, 2023).

Currently 77% of submitted NAPs mention conflict as part of their context, though only 30% address conflict in adaptation actions (NGN, 2024). The NGN is one of numerous organisations also assisting countries in their NAP processes, and where possible they are working to align NAPS with country peacebuilding strategies (Interview, Alec Crawford 2024). Particular champions in engaging this nexus are Somalia, South Sudan, and the Central African Republic. Somalia has taken significant steps in this direction, pursuing a joint climate change adaptation and peacebuilding agenda in their NAP (NGN, 2022). South Sudan and the Central African Republic (CAR) are noted for their comprehensive approach to conflict sensitivity—from conflict assessment in the context, to conflict-sensitive actions and adaptation priorities, and explicit alignments with their peace agreements and peacebuilding

---

[5] The NGN has assisted 62 countries, including 23 LDCs, in their NAP processes (UNFCCC-LEG, 2023).

strategies. South Sudan, Sudan, and CAR propose the strengthening of traditional or development of new land dispute mechanisms in their NAPs, where land and climate issues (i.e., around pastoralist migration routes, and property rights) intersect to fuel both crises (Remling & Meijer, 2024: 9–11).

Despite progress, Remling and Meijer (2024: 12) observe that many countries lack systematic integration of conflict considerations throughout their planning processes, which could lead to maladaptation, where adaptation efforts result in greater vulnerability (Lang, 2019), potentially exacerbating existing tensions in fragile contexts. While the UNFCCC's Least Developed Countries Expert Group (LEG) plays a crucial role in guiding NAP development, the LEG's technical guidelines could be enhanced to more explicitly address conflict dynamics in adaptation planning (Remling & Meijer, 2024; NGN, 2023).

In conclusion, while the NAPs represent a crucial tool for climate adaptation, their effectiveness in conflict-affected areas remains a challenge. The growing recognition of the climate-conflict nexus in NAPs is a positive step, but more systematic integration is needed. As the global community continues to grapple with the complex interplay of climate change and conflict, enhancing the conflict sensitivity of NAPs will be crucial for ensuring effective, sustainable, and peace-positive adaptation strategies.

17.4 Analysis

The Declaration and NAPs effectively constitute key social contracting mechanisms in the COP process—one at global and the other at national levels—that are uniting concerns of climate and peace in the crafting of an evolving global agreement on climate. While it remains too early to tell if these efforts will produce sufficiently impactful outcomes that meet objectives, what researchers can and need to be doing is garnering insights and evidence about the direction of needed new social contracts, monitoring and evaluating the processes by which they are being forged, and offering perspectives on how such processes might be improved to meet objectives. This chapter has sought precisely to do this.

This section considers the conceptual, policy, and practice insights; conclusions and implications that can be drawn from the literature; and by observing COP processes, to advance transformative eco- and peace-promoting social contracts.

17.4.1 Eco- and Peace-Promoting Social Contracts

Returning to the analytical framework presented at the outset, this synthetic, concluding analysis considers what eco- and peace-promoting social contracts might encompass and embrace.

Whom Is the Social Contract Between?

The eco-social contract literature and COP processes are expanding our understanding of who should be included in social contracts. In addition to the state, citizens, and communities and the private sector, this now encompasses current and future generations, nature, and importantly, those most affected by climate change and conflict. The conversations and policy actions reveal the deep intertwining of national and international drivers of these crises, and need for integrated pathways to address them, across levels. As with attention turning to the reality of and need for multi-levelled peace agreements, how we understand and facilitate the forging of eco- and peace-promoting social contracts will likely follow suit.

The COP process, representing a social contracting space for the evolution of an eco-social contract, reveals both the widening stakeholder inclusion, and the tensions that arise between a global agreement and national interests and needs. The UNFCCC and UAE presidency commitment to make the COP28 the most inclusive COP to date, and commitment to "underpinning everything with full inclusivity" (UNFCCC, 2023: 7) reflect facilitative leadership on these issues. Particular attention is also being paid to countries affected by conflict and fragility, recognising their special status and needs. The summary outcomes document, like most UNFCCC documents, underscores the disproportionate climate change impacts for communities and societies, and vulnerable communities and under-represented groups in particular (UNFCCC, 2023). Analysts attending COP29 however agreed that attention to the nexus is diversifying and expanding—with conversations on conflict and fragility within the climate space being led by development actors, private sector and finance actors, and member states, while reflecting more Global South participation, especially from Africa[6] (EPA, 2024).

At the same time, if inclusion is meaningful when tied to inclusive outcomes, the power dynamics that ensures negotiating power and positioning is stacked against more vulnerable countries (those most affected and least

---

[6] Speakers in this webinar generally agreed on wider diversity of engagement, though Celestine Procter from Devex believed that COP28 had stronger local representation.

responsible) that must be tackled. In the Convention on Biodiversity (COP16), a specific programme of work focusing on the needs of Indigenous peoples and local communities was agreed, including a permanent subsidiary body that will focus on elevating their participation in all convention processes.[7] This might prove a valuable pathway to be replicated in climate policy processes.

What Is Its Purpose?

The purpose of these new social contracts is to foster a just, sustainable, and peaceful world. An eco-social contract must enable economies and societies to thrive while ensuring inclusivity, protecting human rights, respecting planetary boundaries, and supporting new forms of solidarity within a strengthened multilateral system. A conflict- and crisis-sensitive approach demands that such a contract addresses the specific vulnerabilities that exacerbate climate-related crises, tackling the root causes of fragility and reshaping power dynamics, including humanity's relationship with nature.

As scholars and policy analysts argue, building resilience to climate change requires addressing the deep-seated structural inequalities that make certain populations disproportionately vulnerable (Vasquez, 2023). McCandless and Faus-Onbargi (2023) emphasise that in crisis-affected contexts, a transformative political economy approach is required—one that moves beyond adaptation by confronting systemic drivers of vulnerability. They stress that this includes addressing financial responsibility for climate-induced losses and framing action around measures that build resilience across multiple scales.

Root causes of vulnerability—including colonial legacies, governance failures, and unsustainable land and ocean use (Wascow et al., 2023: 50)—must be recognised and tackled to ensure a just transition. Many Global South actors, particularly in the peacebuilding sector, focus on adaptation and resilience-building, but this is not the root cause. Rather, our global addiction to fossil fuels is. As Vasquez (2023) highlights, the first priority must be reducing global heating to limit the frequency and severity of climate disasters. While adaptation is crucial, failing to decrease emissions released into the atmosphere and reducing carbon dioxide (CO_2) concentrations—or "mitigation" (UNFCCC, 2025b)—will deepen vulnerabilities, escalate the crisis, and ultimately prevent an eco- and peace-promoting social contract from emerging.

---

[7] https://www.cbd.int/article/agreement-reached-cop-16

Rapid decarbonisation is necessary to achieve 1.5 °C targets, and mitigation remains the most effective way to curb climate-induced suffering. While the Loss and Damage Fund is a step towards climate justice, ensuring its effectiveness requires sustained attention to equitable allocation, particularly for conflict-affected and fragile states (Hardaway, 2024). A reconfigured eco-social contract must not only protect the most vulnerable but also drive systemic transformation, ensuring that climate action is underpinned by justice, accountability, and long-term sustainability.

The Mechanisms Through Which It Is Forged

Both peace and environmental literature emphasise inclusive processes for forging new social contracts. This includes:

1. Inclusive political settlements, peace processes and political agreements at national, global, and transnational levels (addressing peace and environmental issues).
2. Inclusive, participatory processes enabling bottom-up visioning and collective decision-making to inform such agreements and wider governance systems, at all levels.
3. A conducive global policy framework and global governance architecture that serve to implement agreements through specific mechanisms (i.e., the Declaration at global level, and NAP processes at country levels), coherently, transparently, and inclusively.

The COP processes illuminated reflect progress on points two and three. However, challenges remain in connecting these mechanisms to address the root causes of the climate crisis, and conflict. These root causes are intricately connected to wider agreements, processes, and power dynamics in the international system, while also having national and sub-national sources. At national level, robust, inclusive peace agreements and political settlements are needed to ultimately contextualise the actual and potential pathways for eco- and peace-promoting social contracts. Further, without such agreements, reversion to war is more likely, with negative implications for the climate crisis.[8] The global governance architecture also needs to be conducive, yet current significant scrutiny (discussed above) suggests that serious reforms and far greater political commitment and action (i.e., on mitigation, and funding

---

[8] For extensive reviews of literature on this point, see McCandless (2018) and Zahar and McCandless (2020).

mechanisms) are needed to address the power and resource asymmetries that reflect and fuel root causes of this crisis.

Handling Moral Obligations and Competing Interests

An eco- and peace-promoting social contract framework calls for foregrounding principles of climate justice, "common but differentiated responsibilities," and human needs. This requires acknowledging the Global North's outsized role in driving climate change and the responsibility to finance loss and damage, adaptation, and just transitions to carbon neutral economies in the Global South. While the notion of just transitions generally refers to processes of shifting economies and societies towards sustainable and low-carbon futures ensuring equity, inclusivity, and justice for all stakeholders, some argue the need for more transformative notions—where transitions address the root causes of crises, reshaping political and economic systems, and fostering resilience through participatory and equity-driven solutions that prioritise marginalised communities and long-term systemic change (McCandless & Faus-Onbargi, 2023).

The COP process has made some progress in the face of pressure from oil and gas interests that reflect a competing social contract (one valuing markets and elite business interests, over the interests of societies, future generations, and sustainability). However, significant obstacles remain, particularly in addressing competing interests around fossil fuels and power dynamics that perpetuate inequities and inequalities. Fossil fuel companies, according to Greenpeace representative Tracy Carty, have made USD 1 trillion per year in profit annually for half a century—and need to be forced to contribute (Noor & Carrington, 2024).

Despite positive developments in the COP processes and outcomes, several challenges persist pointing to needed action, in the interests of addressing questions of competing interests and moral obligations, including the need to:

1. *Agree on, and tackle root causes, and prioritise attention to mitigation*: Addressing root causes requires bringing a peace and conflict sensitivity lens to mitigation, given that structural roots of conflict and crisis often lie in the abuse of power and resources.
2. *Finance the means of implementation for just climate action*: Obstacles to financing, including debt burdens in conflict-affected states, lower appetite for risk in international private finance, lacking political will by states and

societies to acknowledge the facts and address issues of moral obligation, i.e., of loss and damage arguments.

3. *Address flagrant ongoing violations of the principles the contract is aspiring to*: A comprehensive, immediate approach to addressing climate impacts of military activities and wars is needed.

Each of these areas demands greater attention to historical wrongs that have generated profound inequities and vulnerabilities impacted by, and driving, these crises. They demand addressing the deeply asymmetric power relations and underpinning economic and development models driving our rules-based order, and applying a transformative lens needed to drive systemic changes that, in theory, are agreed upon at the highest levels.

For the COP outcomes to constitute a coherent and winning roadmap with political solutions (agreed necessary by member states and parties to convention in addressing our collective crises) rather than a laundry list with many technical fixes, such a transformative lens needs to lie at the heart of these consensus-based efforts. This returns us to questions of addressing competing interests and moral responsibility at the heart of social contract thought.

17.5 Conclusions

The intersection of climate change, conflict, and peace presents one of the most complex challenges of our time. As evidenced in both academic literature and policy processes like the COP, frameworks that hold promise for eco- and peace-promoting social contracts are being forged, offering promising approaches to address these interlinked issues. By expanding our understanding of who should be included in social contracts, redefining their purpose to genuinely and boldly address root causes, and developing inclusive mechanisms for their formation, we can work towards more just, sustainable, and peaceful societies.

Significant challenges persist however, particularly in translating high-level commitments into concrete actions and in addressing the deep-rooted power dynamics that often perpetuate both conflict and environmental degradation. Moving forward, a concerted effort is needed to ensure that climate action across the board (i.e., mitigation and funding of these actions, not just adaptation) is conflict-sensitive, and that peacebuilding efforts are "climate-smart"—integrating climate change into planning and development of sustainable systems (Lipper et al., 2014). This will require unprecedented levels of collaboration across sectors, scales, and borders, and targeting and fostering political will to take the transformative actions needed.

As we navigate this complex landscape, a social contract framework provides a valuable heuristic tool for reimagining our collective futures. By centring the needs of the most vulnerable, including future generations and the natural world, and those affected by conflict and fragility, we can work towards social contracts that are truly transformative, fostering resilience, equity, and sustainability in the face of our interconnected global challenges.

Acknowledgements Special thanks to Tafadzwa Ndofirepi, for superb research support, and to the editors of this volume for their valuable editorial insights and support.

References

Addison, T., & Murshed, S. M. (2001). From conflict to reconstruction: Reviving the social contract. *Journal of Peace Research, 38*(2), 155–176.
AGDA, G7+, ODI. (2024). *Anwar Gargash diplomatic academy, g7+ and ODI establish a new mechanism to narrow the 'conflict blind spot' in climate action and finance.* https://www.agda.ac.ae/media-centre/agda-news/news-details/anwar-gargash-diplomatic-academy-g7-and-odi-establish-a-new-mechanism-to-narrow-the-conflict-blind-spot-in-climate-action-and-finance
Andrew, R. M., & Peters, G. P. (2021). *The Global Carbon Project's Fossil CO2 Emissions Dataset.* https://www.globalcarbonproject.org/carbonbudget/
Busby, J. (2019). The field of climate and security: A scan of the literature. *Social Science Research Council.* https://s3.amazonaws.com/ssrc-cdn1/crmuploads/new_publication_3/the-field-of-climate-and-security-a-scan-of-the-literature.pdf
Climate Diplomacy. (2024). At least $1 trillion in climate finance needed each year, report finds. *Climate Diplomacy.* https://unclimatesummit.org/at-least-1-trillion-in-climate-finance-needed-each-year-report-finds/?utm_source=chatgpt.com
COP 28 UAE. (n.d.). *COP28 declaration status report.* https://www.cop28.com/
COP 29 Baku Azerbaijan. (2024). *COP29 presidency launches Baku Call on climate action for peace, relief, and recovery.* https://cop29.az/en/media-hub/news/cop29-presidency-launches-baku-call-on-climate-action-for-peace-relief-and-recovery
COP28 UAE. (2023). *COP28 President and WFP Chief call for urgent climate action to reduce rising humanitarian needs.* https://www.cop28.com/en/news/2023/11/COP28-PRESIDENT-AND-WFP-CHIEF-CALL-FOR-URGENT-CLIMATE-ACTION-TO-REDUCE-RISING-HUMANITARIAN-NEEDS
Crawford, A. (2024). *International Institute for Sustainable Development* (IPPC). Interview with author. September 23.

Davis, D. E. (2013). Non-state armed actors, new imagined communities, and shifting patterns of sovereignty and insecurity in the modern world. In K. Krause (Ed.), *Armed groups and contemporary conflicts* (pp. 20–44). Routledge.

Dryzek, J. S., & Pickering, J. (2018). *The politics of the anthropocene.* Oxford University Press.

Eckersley, R. (2004). *The Green State: Rethinking democracy and sovereignty.* MIT Press.

Environment, Climate, Conflict and Peace Community of Practice (ECCP). (2024). *Conflict-sensitive climate action in practice.* https://static1.squarespace.com/static/61dc05c236d4333322aa36f4/t/6733bec5eba83429c270258a/1731444422072/Conflict-sensitive+climate+action+in+practice+-+November+2024.pdf

Environmental Peacebuilding Association (EPA). (2024). *Progress on peace and climate at COP29? Looking ahead to 2025.* Webinar. https://www.youtube.com/watch?v=54tIGGcdWa0

Estlund, D. (2012). The truth in political liberalism. In A. Norris & J. Elkins (Eds.), *Truth and democracy.* University of Pennsylvania Press.

Freeman, S. (2013). Social contract approaches. In D. Estlund (Ed.), *The Oxford handbook of political philosophy.* Oxford University Press.

Hardaway, A. (2024). The loss and damage fund must not leave fragile states behind. *Climate Home News.* https://www.climatechangenews.com/2024/07/10/the-loss-and-damage-fund-must-not-leave-fragile-states-behind/

Hayward, T. (2006). Ecological citizenship: Justice, rights and the virtue of resourcefulness. *Environmental Politics, 15*(3), 435–446. https://doi.org/10.1080/09644010600627741

Hickey, S. (2011). The politics of social protection: What do we get from a 'social contract' approach? *Canadian Journal of Development Studies, 32*(4), 426–438. https://doi.org/10.1080/02255189.2011.647447

Huntjens, P. (2021). *Towards a natural social contract.* Springer Nature.

Huntjens, P., & Nachbar, K. (2015). *Climate change as a threat multiplier for human disaster and conflict.* The Hague Institute for Global Justice. https://thehagueinstituteforglobaljustice.org/wp-content/uploads/2023/07/working-Paper-9-climate-change-threat-multiplier.pdf

International Labour Organization. (2019). *Work for a brighter future: Global commission on the future of work.* ILO.

James, A. (2012). *Fairness in practice: A social contract for a global economy.* Oxford University Press USA.

Kempf, I., Hujo, K., & Ponte, R. (Eds.). (2023). *Global study on new eco-social contracts.* United Nations Research Institute for Social Development.

Läderach, P., Ramirez-Villegas, J., Caroli, G., Sadoff, C., & Grazia, P. (2021). Climate finance and peace – Tackling the climate and humanitarian crisis. *Lancet Planet Health, 5*, 856–858. https://doi.org/10.1016/S2542-5196(21)00295-3

Lang, A. (2019). *Maladaptation: An introduction.* https://weadapt.org/knowledge-base/vulnerability/maladaptation-an-introduction/

Lessnoff, M. (1990). *Social contract theory.* New York University Press.

Lipper, L., Thornton, P., Campbell, B. M., et al. (2014). Climate-smart agriculture for food security. *Nature Climate Change, 4,* 1068–1072. https://doi.org/10.1038/nclimate2437

Loewe, M., Bernhard, T., & Tina, Z. (2019). *The social contract: An analytical tool for countries in the Middle East and North Africa (MENA) and beyond: Briefing paper.* IDOS. https://doi.org/10.23661/bp17.2019

McCandless, E. (2018). *Reconceptualizing the social contract in contexts of conflict, fragility and fraught transition.* Wits School of Governance. https://www.wits.ac.za/wsg/research/research-publications-/working-papers/

McCandless, E. (2020). Resilient social contracts and peace: Towards a needed re-conceptualization. *Journal of Intervention and Statebuilding (JISB), 14,* 1–21. https://www.tandfonline.com/doi/pdf/10.1080/17502977.2019.1682925?needAccess=true

McCandless, E. (2023). *Social contracts: Towards more child- and future-centred framings.* UNICEF. https://www.unicef.org/innocenti/reports/social-contracts-towards-more-child-and-future-centred-framings

McCandless, E. (2024). Social contracts and sustaining peace. In R. Mac Ginty (Ed.), *Routledge handbook of peacebuilding.* Routledge.

McCandless, E., & Faus-Onbargi, A. (2023). Just transitions and resilience in contexts of conflict and fragility: The need for a transformative approach. *Current Opinion in Environmental Sustainability, 65,* 101360. https://doi.org/10.1016/j.cosust.2023.101360

McCandless, E., Hollender, R., Zahar, M.-J., Schwoebel, M. H., Menocal, A. R., Lordos, A., and case study authors. (2018). *Forging resilient social contracts: Preventing violent conflict and sustaining peace.* UNDP. http://www.undp.org/content/undp/en/home/librarypage/democratic-governance/oslo_governance_centre/forging-resilient-social-contracts%2D%2Dpreventing-violent-conflict.html

McCandless, E., & Schwoebel, M.-H. (2021). Conflicts and natural disasters. In O. Richmond & G. Visoka (Eds.), *The Palgrave encyclopedia of peace and conflict studies.* Springer. https://doi.org/10.1007/978-3-030-11795-5_144-1

Mills, C. W. (2002). *The racial contract.* Cornell University Press.

Mohamed, N., & Huntjens, P. (2023). *Dismantling the ecological divide: Toward a new eco-social contract.* UNRISD Issue Brief, 15. UNRISD. https://cdn.unrisd.org/assets/library/briefs/pdf-files/2023/ib15-a-new-contract-with-nature.pdf

Murshed, S. M. (2011). Peace-building and the social contract. In Kozul-Wright & Fortunato (Eds.), *Securing peace state-building and economic development in post-conflict countries.* United Nations.

NAP Global Network (NGN). (2022). *Somalia's National Adaptation Plan (NAP) framework.* https://napglobalnetwork.org/wp-content/uploads/2022/11/napgn-en-2022-somalia-nap-framework.pdf

Ndinga-Kanga, M., van der Merwe, H., & Daniel, H. (2020). Forging a resilient social contract in South Africa: States and societies sustaining peace in the post-apartheid era. *Journal of Intervention and Statebuilding, 14,* 22–41. https://doi.org/10.1080/17502977.2019.1706436

NGN. (2023). *Peace, Conflict, and National Adaptation Plan (NAP) Processes: Guidance Note.* https://napglobalnetwork.org/wp-content/uploads/2023/12/napgn-en-2023-peace-conflict-nap-processes.pdf

NGN. (2024). *Trends in key themes: Conflict and peacebuilding.* Accessed October 17, 2024, from https://trends.napglobalnetwork.org/trend-in-key-themes/conflict-peace-building

Noor, D., & Carrington, D. (2024). Cop29 climate finance deal criticised as 'travesty of justice' and 'stage-managed'. *The Guardian.* Accessed January 26, 2024, from https://www.theguardian.com/environment/2024/nov/24/cop29-climate-finance-deal-criticised-travesty-justice-stage-managed?utm_source=chatgpt.com

Notre Dame Global Adaptation Initiative (ND-GAIN). (2022). *Country Index.* Accessed March 9, 2022, from https://gain.nd.edu/our-work/country-index/

Organisation for Economic Co-operation and Development (OECD). (2011). *Supporting statebuilding in situations of conflict and fragility: Policy guidance, DAC guidelines and reference series.* OECD.

Otu-Larbi, F., et al. (2024). *A multitemporal snapshot of greenhouse gas emissions from the Israel-Gaza conflict.* https://www.qmul.ac.uk/sbm/media/sbm/documents/Gaza_Carbon_Emissions.pdf

Pateman, C. (1988). *The sexual contract.* Stanford University Press.

Pettifor, A. (2019). *The case for the green new deal.* Verso.

Raworth, K. (2018). *Doughnut economics: Seven ways to think like a 21st-century economist.* Chelsea Green Publishing.

Remling, E., & Meijer, K. (2024). Conflict considerations in the United Nations Framework Convention on Climate Change's National Adaptation Plans. *Climate and Development,* 1–15. https://doi.org/10.1080/17565529.2024.2321156

Richmond, O. P. (2009). A post-liberal peace: Eirenism and the everyday. *Review of International Studies, 35,* 557–580.

Rousseau, J.-J., & Cress, D. A. (1987). *On the social contract.* Hackett Publishing.

Saferworld. (2023). *Peace@COP28: New commitments on climate, relief, recovery and peace.* Accessed October 17, 2024, from https://www.saferworld-global.org/resources/news-and-analysis/post/1026-peacecop28-new-commitments-on-climate-relief-recovery-and-peace

Schwoebel, M. H., & McCandless, E. (2021). Conflict-disaster nexus. In O. Richmond & G. Visoka (Eds.), *Palgrave encyclopedia of peace and conflict studies.* Palgrave.

Toda Peace Institute. (2024). *COP28 – Massive disappointments, slight glimmers of hope.* Accessed October 17, 2024, from https://toda.org/global-outlook/2024/cop28-massive-disappointments-slight-glimmers-of-hope.html

UN Women. (2021). *Beyond COVID-19: A feminist plan for sustainability and social justice*. UN Women. https://www.unwomen.org/sites/default/files/Headquarters/Attachments/Sections/Library/Publications/2021/Feminist-plan-for-sustainability-and-social-justice-en.pdf

UNDP. (2016). *Engaged societies, responsive states: The social contract in situations of conflict and fragility*. UNDP.

UNFCCC. (2023). *Summary of Global Climate Action at COP 28*. UNFCCC. https://unfccc.int/documents/636485?gad_source=1&gclid=CjwKCAjw6c63BhAiEiw AF0EH1DbtozXCwC3xNOzyMdLlp_e64Mvfl8Vd9UVodhrPBaeErNwcQYO-MJRoC8qEQAvD_BwE

UNFCCC. (2025a). *Introduction: Adaptation and resilience*. UNFCCC. https://unfccc.int/topics/adaptation-and-resilience/the-big-picture/introduction

UNFCCC. (2025b). *Introduction to mitigation*. UNFCCC. https://unfccc.int/topics/introduction-to-mitigation

UNFCCC-LEG. (2023). *National adaptation plans: Progress in the formulation and implementation of NAPs*. UNFCCC. https://unfccc.int/sites/default/files/resource/NAP-progress-publication-2023.pdf

United Nations. (2023). *Our Common Agenda Policy Brief 6 Reforms to the International Financial Architecture*. United Nations. https://www.un.org/sites/un2.un.org/files/our-common-agenda-policy-brief-international-finance-architecture-en.pdf

United Nations Conference on Trade and Development (UNCTAD). (2019). *Trade and Development Report 2019: Financing a global green new deal*. UNCTAD. https://unctad.org/system/files/official-document/tdr2019_en.pdf

United Nations Development Programme (UNDP). (2012). *Governance for peace: Securing the social contract*. UNDP.

United Nations Environment Programme (UNEP). (2024). *Adaptation Gap Report 2024: Come hell or high water*. Nairobi. https://www.unep.org/gan/resources/report/adaptation-gap-report-2024

United Nations Research Institute for Social Development (UNRISD). (2022). *Crises of inequality: Shifting power for a new eco-social contract*. UNRISD. https://cdn.unrisd.org/assets/library/reports/2022/full-report-crises-of-inequality-2022.pdf

United Nations Secretary-General. (2022). *Our common agenda: Report of the secretary-general*. United Nations.

Vasquez, M. (2023). Building climate resilience in conflict zones requires less emergency aid, not more. *The New Humanitarian*. https://www.thenewhumanitarian.org/opinion/2023/12/18/cop-28-climate-resilience-conflict-zones-less-emergency-aid-not-more

Wascow, D. et al. (2023). *Unpacking COP28: Key outcomes from the Dubai Climate Talks, and what comes next.* World Resources Institute. https://www.wri.org/insights/cop28-outcomes-next-steps

Zahar, M.-J., & McCandless, E. (2020). Sustaining peace one day at a time: Inclusion, transition crises, and the resilience of social contracts. *Journal of Intervention and Statebuilding, 14*, 119–138. https://doi.org/10.1080/17502977.2019.1673130

Part IV

**Implementing Eco-Social Contracts:
Prospects for Movements and
Intersectional Alliances**

18

The Terms of Our Existence: Can We Really Reimagine Our Future?

Lysa John

Reimagining our future requires us to renegotiate the terms of our collective existence, something societies do every few decades. Whether it's Gandhi's non-violence and civil disobedience, Mandela's fight against apartheid, Martin Luther King Jr.'s struggle against segregation, or mass movements like Black Lives Matter, the Arab Spring challenging dictators, or people resisting genocide and dispossession, all of these are acts of negotiating the terms of our collective existence. The contributors to this book are seeking to reimagine a just and sustainable world through the framework of new eco-social contracts.

In the face of political turmoil and the escalating climate crisis, it has become crucial to think about fundamental changes to how societies work. The exploitation of people and the disregard for the planet's natural limits demand new economic and social frameworks that inseparably integrate social and environmental justice.

The climate and environmental emergency has nudged governments to reduce emissions, aim for net-zero emissions and achieve a world living in harmony with nature. However, the burden of making this happen, as well as the adverse consequences of climate change and biodiversity loss, are unfairly shared by nations of the Global South. Similarly, political shifts, including the rise of authoritarianism, have deepened social and economic divides.

L. John (✉)
Atlantic Institute, Oxford, UK
e-mail: lysa.john@atlanticfellows.org

© The Author(s) 2025
P. Huntjens et al. (eds.), *Eco-Social Contracts for Sustainable and Just Futures*,
https://doi.org/10.1007/978-3-031-99109-7_18

Governments have used crises to justify limiting civil liberties and suppressing dissent. Additionally, cuts in foreign aid and weakened global cooperation are hindering efforts to tackle inequality and climate change worldwide. As a result, we see reinforced patterns of exclusion and marginalisation as well as environmental degradation.

In response to the climate and ecological crisis and the assault on rights and freedoms, ordinary people and grassroots movements worldwide are organising themselves to negotiate for their collective futures. They are increasingly recognising that climate justice is deeply intertwined with social and economic inequalities. Indigenous and Adivasi[1] movements have historically viewed land, water, and the environment as inseparable from their identity and culture, thereby imagining sustainability and social rights as one and the same. Similarly, many rural women's empowerment groups in the Global South link economic rights to access to resources for agriculture, pastoralism, and other land-based livelihoods.

These movements are diverse in where they are located, what they are demanding, how they came to exist and how they are run, but all of them are unified by a common goal: to build an equitable future for the people and the planet. From advocating for Indigenous rights to tax justice, from reclaiming urban spaces to resisting dispossession, they are driven by values of solidarity, cooperation, respect for the planet, and a commitment to human dignity and rights.

Our imagination of a new eco-social contract must be based on these very values. And for this new order to be sustainable, it must address the needs of the people as well as the planet. It must aspire as much for social and economic equity as it does for conservation. It should listen to the silenced voices of marginalised groups, women, minorities, refugees, migrants, and Indigenous peoples, because they know the planet and its problems through their everyday experiences.

In our struggle to reclaim our planet and make it equitable, we must remember that just like justice and sustainability, our freedoms and futures too are intertwined. These interconnections should be central to the terms of our collective existence.

---

[1] Tribal groups across the Indian subcontinent.

19

Between Resistance and Cooperation: A Balancing Act Towards New Eco-social Contracts in Latin America

Carlos Emiliano Villaseñor Moreno

19.1 Introduction

It soon shall come
My lucky day
I am sure that before my death
My luck is bound to change
—Willie Colón, Héctor Lavoe

Hence the poverty of the poor is not a call to
a generous relief action but a demand that we go
and build a different social order
—Gustavo Gutiérrez

In many of our modern societies, the language of dialogue, consensus, participation, community and understanding underpins a great deal of the formal conventions through which we attempt to establish arenas to renegotiate existing paradigms. The use of these specific terms and others like them can be at least partially attributed to the rapid expansion of both western representative democracy and liberal principles around the world during the second

C. E. V. Moreno (✉)
Ombudsman Energía México, Mexico City, Mexico
e-mail: carlos@oem.org.mx

345
P. Huntjens et al. (eds.), *Eco-Social Contracts for Sustainable and Just Futures*,
https://doi.org/10.1007/978-3-031-99109-7_19

half of the twentieth century. Under this form of governance, one of the main aims is to institutionalise disagreements and tensions in a way that renders conflict (understood as threat to the public order) unnecessary (Przeworski, 2018).

Even alternative proposals of democracy that aim to expand the reach of democratic government beyond just formal institutions, such as the deliberative model, has its roots in the resolution of moral conflicts in pluralistic societies (Gutmann & Thompson, 1997). It is therefore no surprise that it also dominates the way many eco-social contract theorists and practitioners envision the process of negotiation, design and implementation of new relationships with others and with nature (Huntjens, 2021; UNRISD, 2021, 2022; Norton & Greenfield, 2023).

However, since at least the nineteenth century the aim within liberal democracy to prevent conflict has been heavily underpinned by fear of the participation of the masses in politics due to the belief that they were irrational, ignorant, and prone to violence. This translated into a mistrust of democracy and the need to "tame democracy" in order to render it safe to use (Manin, 1997; Rosenblatt, 2018; Znoj, 2024). This is the start of a very common *modus operandi* within our political systems where under the argument of safeguarding liberal principles, democratic ones are downplayed (Mouffe, 2000, 2013).

More contemporary discussions don't speak in terms of "taming" or "subordinating" but of "balancing" or "reconciling" liberal and democratic principles. However, they do tend to consider, as the more pressing risk to both mature and developing democracies, a general leaning towards "too much democracy" or illiberal democracy, manifested mainly through the increasing establishment of authoritarian practices and actors through democratic means and the erosion of liberal principles (Mounk, 2018; Campati, 2021).

What this position doesn't consider, and this chapter will defend, is that in its efforts to eliminate or reduce conflict through reining in democracy, liberal democracy has eroded the foundations upon which it was built, creating a vacuum in the practice and defence of substantive democratic politics filled by new autocratic actors (Mouffe, 2000, 2013; Pabst, 2016, 2019). Additionally, it has left advocates for change to the existing social contract in a very weak bargaining position if they decide to act within the confines of the economic or political systems (Gourevitch, 2013) while more radical forms of action and dissent, like protests and activism, are criminalised (Vegh Weis, 2022; Selmini & Di Ronco, 2023).

Furthermore, I contend that under the previously described conditions agonistic democracy, meaning the recognition that conflict is not only inherent but has value in the democratic organising of social relations, can provide a helpful framework for dissent against the status quo and for building effective leverage for change. If the political reality worldwide is that a new eco-social contract will have to be negotiated in a very crooked bargaining table, then a theory whose main concern lies in the configuration of and challenge to existing power relations can offer valuable insights (Mouffe, 2000, 2013).

Finally, in an effort to contribute to the scant literature on how agonistic politics should play out in practice, this chapter will analyse five examples of grassroots organisations in Argentina, Brazil and Mexico. Latin America has a long history of strong social mobilisation and resistance aimed at systemic transformation (Eckstein, 2001; Petras & Veltmeyer, 2011). What's more, a new wave of mass political action is taking place in what some have dubbed, the "Latin American Spring" (Chen, 2023). However, neither past nor present movements have really been analysed through the lens of agonism as a way to further understand, explain or critique them or to inform and expand the theory.

19.2 The Problem of Insufficiently Democratic Democracies

In order to build a critique to liberal democracy as a way to structure not only governance but social relations more broadly, a brief review of the evolution of "crisis literature" over the last decade can be helpful to contextualise the state of affairs that informs my analysis. By "crisis literature" I am referring to the strand of research focused on either disproving or demonstrating a state of crisis within modern democracies. Through the early 2010s there seemed to be a push and pull between those who argued that a serious democratic rollback was on the horizon and those that, while recognising there was cause for concern, considered democratisation efforts as stable and enduring (Merkel, 2014; Levitsky & Way, 2015).

The arrival of Donald Trump to the White House in 2016 and the subsequent rise of similar figures and parties across consolidated democracies tipped the scales towards the view that there was indeed a democratic crisis. Not only that, but a picture was starting to form that showed a trend of democratic erosion through democratic means (Mounk, 2018; Campati, 2021). The effects are felt not only at the institutional level but also on more informal principles and values necessary to foster democracy leading to a state of "democracy without guardrails" (Levitsky & Ziblatt, 2018).

On the other hand, there are those whose position is that the assessment above is missing a key component, namely that liberalism has eroded its own foundations, playing a fundamental role in the decline of the political systems structured around its principles and institutions. The result is the establishment of the very type of social order that liberal democracy was trying to prevent (Deneen, 2018; Pabst, 2016, 2019). This is not a new criticism; deliberative democracy and later agonism would be born as political theories out of very similar misgivings regarding liberal democracy. It is in their common early warnings as well as agonism's subsequent critique of deliberative democracy that I believe some of the most valuable insights can be found to develop alternatives to the current political system.

Both deliberative and agonistic theories' starting point is the existence of a democratic deficit, meaning a lack of popular sovereignty, derived from its subordination to liberal principles and expressing the need for more radical forms of democracy (Habermas, 1996; Mouffe, 2000). Deliberative democracy's argument is that liberal institutions, such as rule of law, aren't there to limit or manage democracy but they are in fact formed in simultaneity; rule of law can't exist without substantive democratic politics (Habermas, 1996). Furthermore, it roots democratic participation in a discursive principle, meaning the legitimacy of a norm is based on an agreement built from a rational argumentative process informed by both formal and informal public spheres (Habermas, 1996; Gutmann & Thompson, 1997).

However, agonism will posit that deliberative theory is still too concerned with achieving neutrality and consensus, remaining as ill-equipped to deal with the reality of "the political". It is therefore prone to repeat liberalism's mistakes. The political in an agonistic context means accepting the inherent potential for antagonism that comes with any structuring of social relations. For proponents of this theory, any established order is a hegemonic order put in place through acts of power (Mouffe, 2000, 2013). Therefore, the role of "democratic politics then becomes [...] how to constitute forms of power which are compatible with democratic values" (Mouffe, 2000: 22). This entails moving from antagonism to agonism, from violent forms of conflict to democratic ones.

Subscribing to such a view of politics in general and democracy in particular would have strong implications to any attempt to define the path towards a new eco-social contract. The core argument advanced by proponents of this contract is that growing evidence has put the inadequacy of our current political and economic systems to deal with some of the most urgent problems afflicting modern societies on display. In order to course correct there must be a redefinition of major "societal fault lines" such as our anthropocentric relationship with nature, the prevalence of a capitalist-extractivist economic logic, etc. (Huntjens, 2021; UNRISD, 2021, 2022; Norton & Greenfield, 2023).

This narrative can be expressed in agonistic terms defining the efforts to shift from our current social contract to a new eco-social one as a contestatory process between the current hegemonic order and the one that aims to replace it. Under that framework the success of a new eco-social contract depends on it acquiring enough power of its own and using it to challenge and displace the existing system. This is consistent with empirical research on democratisation processes where formally disenfranchised groups are able to transform a temporary surge of power into lasting democratic institutional change (Acemoglu & Robinson, 2000, 2001, 2005; Boix, 2003). The old system will not go out willingly, it will be forced to leave.

Additionally, agonism addresses concerns from the perspective of actors from Global South countries or activist groups that a focus on formal expansion of political participation, like deliberation, as the main bargaining tool for change ignores how easily these tools can be rendered useless or co-opted in contexts of great power inequality (Young, 2001; Kapoor, 2002; Pansardi, 2016). Lastly, this political theory is by its very nature context-specific given that it relies on the specific configuration of power in the space being analysed, avoiding one-size-fits-all type of solutions and also acknowledging that whatever order is brought by the implementation of a new eco-social contract is a partial one that, just as its predecessor, can be challenged and changed (Mouffe, 2000, 2013).

Nonetheless, it is necessary to note that little effort has been put into operationalising agonism. More specifically, its adherents haven't developed a comprehensive theory of democratic conflict; meaning they are unable to offer a convincing explanation as to why different political actors would engage in democratic struggle recognising the opposing party as a legitimate adversary rather than opting for violent conflict against an enemy. The extended effects from this gap include a lack of consideration of how an agonistic model of democracy could look like at different points within conflict dynamics (August, 2022; August & Westphal, 2024), but also what potential institutions could be used to allow and mediate conflict in a democratic setting (Lowndes & Paxton, 2018).

Some progress has been made to address these shortcomings like the work of August and Westphal (2024) that proposes the identification of the different stages through which a conflict can shift such as escalation, de-escalation and reconciliation; understand the mechanisms that consolidate each stage or the shift from one to another; and what a democratic response to each would entail. Regarding institutionalisation more specifically, Lowndes and Paxton (2018) have suggested that rather than proposing specific fixes that allow for democratic agonism to take place, there is a need for a design theory for agonistic institutions whose foundation is rooted in being procedural, collective, contextual, contestable and provisional/temporary.

19.3 The Potential of Agonistic Politics: Examples from Latin America

The modern history of Latin America, from the twentieth century onwards, constitutes a strong revindication in practice of what agonism lays out in theory. During this period the driving force for change was the strong presence of social movements that ranged from activists to guerrilla movements. However, a very strong response emerged from the political system to halt their advancement. This took place not only through the use of force and violence, but by weakening or eliminating any avenue of political influence. Such a strategy entailed the "creation of a model for the destruction of social relations" (Modonesi, 2008, 119) that allowed for social movements to form. Under such conditions the transition to liberal democracy in Latin American countries was viewed not as the final triumph of the social movements in the region, but as the consolidation of a hegemonic order that subordinated and constrained any attempt to challenge the existing systems (Retamozo, 2005; Modonesi, 2008; Gutiérrez Aguilar, 2015).

Therefore, the region has been defined by the struggle to express what its societies will look like, with liberal democracy establishing itself as the current hegemonic order. Not only that, but just as agonistic theorists had warned liberalism has closed the door on further contestation of existing power relations making it harder to solve systemic problems of economic and political inequality (Mouffe, 2000, 2013). Recent empirical research seems to bear this out by identifying political inequality as one of the main drivers of economic inequality in Latin America which can be traced back to colonial times (Eslava & Valencia, 2023; Fergusson & Robinson, 2024). In addition, a reduction in political inequality, through democratisation, seems to have had little impact on reducing economic inequality. This has been due in part to the way it has been set up, with mechanisms that discourage the creation of strong redistributive policies from political actors (Fergusson & Robinson, 2024).

However, a new re-politicisation of social movements and the popularisation of radical forms of democratic action seems to be on the rise (Chen, 2023), alongside renewed efforts to suppress them (Human Rights Watch, 2022). The aim of this section is not to explain the way this new wave of mobilisation is inserted in the larger historical narrative of agonistic politics in the region. My objective is less ambitious, building on previous work analysing specific initiatives at the micro level to understand how they can constitute arenas for democratic expansion through the lens of agonistic theory (Merlinsky, 2021). To do so, a series of semi-structured interviews were carried out with representatives of three civil society organisations from Mexico and Argentina. A fourth case study from Brazil was also included, however, there wasn't a possibility to find a primary source and I rely on existing literature and interviews given by adherents of the movement to the media. All interviews were transcribed and then subjected to thematic analysis in an effort to identify the specific conflict dynamics that underpin the actions of the movement (August & Westphal, 2024) as well as a potential affinity with the constitutive elements of agonistic institutions from the way these movements have structured themselves (Lowndes & Paxton, 2018).

19.3.1 *Zines por morras*

When talking about the inherent presence of antagonism in political life, few examples represent this stance better than the evolution of the feminist movement in Mexico. A major driver for its actions has been the long-standing crisis in the country regarding gender-based violence (López Barajas & Guerra Favela, 2020) but the mobilisation around the issue has undergone radical changes. One of the most significant is the transition from more conventional forms of action like activism, peaceful protest, performance art, etc. to more "violent" means like iconoclastic violence, the destruction of public and private property, and confrontations with law enforcement.

It must be noted, however, that this is not a burst of wanton violence on the part of Mexican women. The success that these new strategies would have in achieving important legislative and public policy victories for the movement is in part due to feminism being able to turn pain, grief and rage into organised and targeted political action (Lamas, 2020). As a result, gender issues have slowly gone from a marginal or niche issue to being at the centre of many conversations at the national and local levels through what can be considered the escalation of conflict.

It is in this context that the Mexican collective *Zines por morras* was originally created to address a narrative void regarding sexual harassment towards women that belong to sex and gender dissidences. Nonetheless, the language its members use is a lot more conciliatory in nature than what my introduction to the current state of the feminist movement in Mexico would make you believe. They emphasise public discussion of an uncomfortable topic without leading the public to any premeditated conclusion. In the words of co-founder Ana C., they want their texts to be *manoseados*, to be used, underlined, crumpled, written on, etc.

This should not be interpreted as an attempt to distance themselves from more confrontational forms of action, given the admitted militancy of the members of the collective in the larger feminist movement. A more accurate description I could give of their actions is to inform the different sides involved in the conflict, a practice not uncommon in several strands of feminism such as the radical one in which revolutionary action was informed by collective spaces to share participants' lived experiences (Tong, 2009).

In the same vein the main publication of the organisation, the *Guía Rosi*, a pun on the *Guía Roji* mapbook used to move around Mexico City before the era of Google Maps, is the product of personal experience. More specifically that of Ana C. with sexual harassment and her effort to both share and allow others to share similar experiences in Mexico City, intersected by their sexual and gender identities. It is also the result of a dialogue with other founding members such as Poly E. who was the person that suggested to Ana to turn all the testimonials into a formal publication.

A case like this shows that an agonistic approach to democratic politics does not preclude the use of spaces for deliberation and direct participation, but it demands that these processes be integrated into a larger agonistic system that allows their use for either escalation or de-escalation of conflict. The work of *Zines por morras* more specifically seems to aim to build on the successful escalation of conflict by the feminist movement in Mexico to expand the discussion regarding sexual harassment, allowing for opportunities to explore other less confrontational methods that can lead to eventual de-escalation and reconciliation.

Regarding the internal structuring of *Zines por morras*, they admit to not really having any formal way of defining their roles and actions. However, they do consider that internal conflict among members regarding the way the project should operate and where it should be headed, became a source of intimacy with each other. This allows for a deeper understanding of who they were collaborating with and the emotional work that comes with any joint endeavour. The organisation has also been able to capitalise on the engagement of its members in other movements like the support of the former members of the now disintegrated *Wiki-Política*, a political activist organisation created by students as an offshoot of the *YoSoy132* student protest movement of 2012.

19.3.2 Youth for Climate Argentina

Contrary to the case of *Zines por morras* it is somewhat harder to succinctly summarise the dynamics that underpin the creation and operation of Youth for Climate Argentina (JOCA in Spanish). At the international level the moment was definitely one of escalation, of both size and tactics regarding climate activism. More specifically the Fridays for Future movement, one of, if not the largest example of youth-led mobilisation to protest against the climate change crisis demanding swifter and more decisive action from governments across the world, would be one of the catalysts for the creation of JOCA in 2019. This is framed in what some scholars have called the entrance of the climate change movement into a phase of direct radical action, stemming from repeated failures from formal climate governance (Maslin et al., 2023).

However, at the national level the first years of operation of the movement seemed to be characterised by the presence of fertile ground to put forward their own expression of environmentalism. Mercedes P. and Bruno S. described how their early years were marked by victories such as the support of figures within the Argentine government with a long trajectory in environmental issues such as the legislative representative Pino Solanas, the implementation of their Declaration of Climate Emergency by the Senate as well as the participation of members, such as Mercedes P, in the creation process of the first climate change legislation in the country. Their operation then was not drastically different from what we have come to expect from civil society engagement with decision-makers in a democracy.

Nevertheless, the most significant contribution came from Arturo C. who defined one of the reasons for the movement's success as the crisis of legitimacy of traditional political actors, political parties to be precise. The cracks in the hegemonic control of parties in the way politics is made, allowed space for contestation from movements with alternative narratives like JOCA to position themselves as the proponents of a new national project. This very same element contributed to the arrival of actors at the other side of the spectrum like Javier Milei, the self-defined anarcho-capitalist and current Argentinian president.

The scenario described above can be considered as an example of the warnings of agonistic theory coming true. The undermining of liberal democracy, through its democratic deficit, allowed figures like Milei to find electoral success with the promise of defending that which liberal democracy is perceived to have put at risk (Mouffe, 2000, 2013). However, the Argentinian case seems to also offer a more optimistic picture. Even though there is a vacuum left regarding democratic politics, it does not have to be filled by right-wing authoritarian figures, there is an opportunity for other movements to advance their own claims.

At the same time, the current political landscape of the country represents some of the complications of implementing agonistic democracy in practice. The central tenet of agonistic democracy is that despite there being agreement on shared values of liberty and equality, there will always be disagreement on how to interpret those values which give rise to a conflictual consensus (Mouffe, 2013). For this to happen, there must be a recognition of your adversary and the dissent in their position as legitimate. Legitimacy requires at least some degree of common ground. Otherwise, even if both sides claim to be on the side of democracy there is no credible reason why either should accept that claim.

It is then no surprise that both Milei and JOCA see each other closer to enemies than to adversaries. Each side considers the other as a potential threat to their vision for the social order. In these conditions, an adherence to democratic forms of conflict tends to come down more to personal conviction than to institutional restrictions. Despite this, JOCA seems to remain steadfast in their commitment to only challenge power through democratic means. They are seizing their current position at the fringes as an opportunity to take a step back, see where the affinity with existing political actors had made the discussion around climate change stale, define the new battlegrounds for political action, and simplify the agenda.

This is a testament to the strength and efficiency of the complex organisational structure that has been built to steer the movement. It was designed to be context-specific, focused on the Latin American experience and was informed by the long tradition of the normalisation of political action by Argentinian society. Additionally, despite being born as a middle class urban movement, it has aimed to integrate other sectors of the population in decision-making. Finally, due to previous experience in crisis management due to the impact of COVID-19 in social movements, JOCA has learned to redefine itself and the way it operates in a relatively quick manner. In sum, JOCA seems to be a strong contender to effectively implement institutional agonistic design (Lowndes & Paxton, 2018).

19.3.3 The MOCAF Network

Among the selected case studies, the Mexican Network of Forest Peasant Organizations (Red MOCAF in Spanish) is the largest running organisation since its inception in the early 1990s. What this long-standing history allows me to do is to be able to identify the presence of cycles of escalation and de-escalation in its conflictual relationship with the Mexican government. The creation and operation of the network is then an example of the nonlinearity of conflict and the need for an application of agonistic democracy to be able to respond to these back and forth scenarios that don't necessarily aim for a specific form of resolution. The aim is precisely the continuity of contestatory direct action against power, rooted in a historical mistrust of formal institutions (August & Westphal, 2024).

More specifically we can divide Red MOCAF's existence in three distinct moments. The first corresponds to its creation and comes at a moment of reinvigoration of direct and at times violent political action in Mexico with some notable examples such as the rise of the National Zapatista Liberation Army (EZLN). Red MOCAF is moved to action by the focus on market deregulation, economic competition and international trade that characterised the administration of Carlos Salinas de Gortari. Many rural and indigenous communities considered the policies that would come from this government as a direct threat to the more collective, sustainable and socially just modes of living they had been advocating for.

With similar sentiments echoing across the different states present during the early summits, it was natural for a sense of affinity to form between the organisations, leading to the desire to give a more permanent and purposeful shape to the collaboration. This culminated with the creation of a formal organisation in 1994 as a non-governmental organisation (NGO). Therefore, despite being founded in a context of violent conflict escalation in the country, Red MOCAF seems to focus its actions in challenging power through the weight of coordinated collective action through a mix of activism and formal participation in political institutions.

This led to a de-escalation period that coincided with the first years of democratic transition in the country where the network managed to position itself in multiple advisory bodies for the government such as the National Forestry Council or the National Technical Advisory Council for Sustainable Development. Through this, it provided assistance for communities looking to protect and manage wetlands, implement communitarian forestry practices or sustainable management, participate in the voluntary carbon market, etc.

Through these two first phases of their history, Red MOCAF would develop a sort of "federalised" structure that facilitated coordination between the different organisations spread across 11 states of the country and acting on behalf of over 120,000 beneficiaries. This allowed for collective decisions and the setting of a common agenda, while also lending resources and information on how to deal with more localised issues for each member. Nansedalia R. mentioned a successful case of mobilisation from the community of the Costa Grande region of the state of Guerrero, where she is from, in which the Forestry Decentralized Public Organism Vicente Guerrero, that overexploited the *ejidos* (communally owned territory), had its assets transferred to the local government with the mandate to create a new organism for social participation in forestry activities in the state.

Having a clear understanding of the internal structuring of Red MOCAF is crucial to understand how the third, and current, stage of the organisation's activities has developed. After 30 years the network finds itself facing a wide range of challenges like worsening economic conditions, pressures to sell their *ejidos* and a strong uptick in violence against rural and indigenous communities in the country. The network has seen its ability to act greatly diminished with only 25 out of 40 of the member organisations still able to offer support to their communities. This has been compounded with an almost total break of relationships and funding from the federal government, as well as fiscal reforms that limit the resources available to civil society.

In this new context of escalation of conflict, the network as a whole has seen a push for reducing dependency on traditional funding and looking for stronger relations of mutual aid, not only within the country but internationally as well. On the other hand, the recent tensions with the Mexican government have not led Red MOCAF to retreat from the political sphere. There are now efforts to move beyond an advisory role to campaigning for its members to occupy electoral positions in the legislature and the executive at the local levels. This is not without its tensions within the network given that there is no consensus on how to approach this goal while upholding the principles of non-partisanship, non-profit, plurality and self-sufficiency upon which it was created.

Therefore, faced with the reality of the precarious place they still occupy within formal places of power, the network has been able to use their internal structure to reinvent themselves through different moments within their struggle against the political system and doing so by mostly democratic means. Nonetheless, it is necessary to clarify that its "federalised" aspects mean that despite these decisions taken at the organisational level, members' actions didn't necessarily follow in the same direction at the local level. The most notable case of this being the municipality of Cheran that famously armed themselves, expelled both government and organised crime from the territory and have become an autonomous, self-ruled community. This speaks to the need to consider not only how conflict dynamics are shaped by the interactions between the different collectives that aim to reshape social order, but also by those interactions happening within the collective (August & Westphal, 2024).

19.3.4 The Articulation of Indigenous Peoples of Brazil

Similarly to Red MOCAF, the Articulation of Indigenous Peoples of Brazil (APIB in Portuguese) is born out of a desire to give permanence to the collective action capacities gained by previous social mobilisation among indigenous groups in the country. More specifically it is a result of the Camp Terra Livre, a yearly national mobilisation that started in 2004 aimed at making visible the state of indigenous rights as well as presenting their demands to the Brazilian State. The APIB was created the next year in 2005 with one of the main demands being the demarcation of indigenous land (APIB, 2023).

However, unlike the case of Red MOCAF I will not carry out an analysis of the overarching progression of the movement over time. Instead, I will focus on the way APIB has responded to the biggest threat to their demands for formal recognition and demarcation of indigenous land, the temporal framework argument. This reduction in scope will allow me to focus on a specific element highlighted in the previous case study in more detail. Namely, on the potential divergence in approaches to challenges to power within the same movement.

The temporal framework argument is the product of strong lobbying by the powerful agribusiness and mining companies in the country that have time and time again tried to encroach on indigenous territory. In a nutshell it consists of a legal argument that pushes the idea that the only lands that should be recognised as indigenous are those that were occupied by indigenous people at the moment of the publication of the Brazilian Constitution in October of 1988. What the argument purposefully ignores is the massive displacement suffered by the indigenous populations in the years prior to 1988, meaning only a small portion of the current 1393 indigenous territories in the country would be officially recognised. As a result, there could be a potential escalation in the displacement and violence against the communities that inhabit the spaces that would no longer be considered as legitimately indigenous territories (Alves, 2023; APIB, 2023).

The response of APIB to the issue has developed on two fronts, the first being the juridical one by presenting legal arguments of their own and the second through the activism that created the organisation in the first place, mainly the yearly Camp Terra Libre protests. It is then a mix that seems to be a very common method of action between our different case studies of direct action and participation through formal institutions whenever possible.

For a while it seems that coordinated action through the judicial system might be enough to neutralise the threat of the temporal framework argument with the legal team of the APIB managing to get the Supreme Court to rule against it, first in 2017 and then again, very recently, in 2023 (Business and Human Rights Resource Centre, 2023). In an interview to Folha de S. Paulo (2023), Mauricio Terena, legal coordinator of the APIB, presented succinctly the case that had been advanced by the organisation, based on the legal theory of *indigenato*, recognising indigenous people as the primary inhabitants of the territory and their congenital right to its possession, as well as climate change arguments that relied on the importance of indigenous stewardship of the land to prevent the destruction of the Amazonia.

However, the end result is a pretty clear-cut example of the limitations of formal political participation in the face of significant inequalities in the distribution of power (Pansardi, 2016). A couple of months after the ruling of the Supreme Court in 2023 a new bill was introduced to the legislature that effectively aimed to operationalise the temporal framework argument. Prominent actors like the UN Special Rapporteur on the rights of Indigenous Peoples, José Francisco Calí Tzay, expressed not only their concern but also their bewilderment as they couldn't understand "what could justify a re-discussion of the legal understanding already determined by the Supreme Court, given this short period of time" (OHCHR, 2024).

This is where the divergence in opinions among the indigenous population starts to happen. While there is a consensus to go back to focusing on direct political action, not everyone takes the same road to advance the cause for indigenous rights. Many important voices in the APIB have focused on strengthening the Camp Terra Livre mobilisations and have adopted a stronger discourse away from legal arguments. Simão Norivaldo, executive coordinator of APIB by the Aty Guasu Assembly, for example claimed during an interview at the 2024 mobilisation that "the time frame proposal for us means a genocidal landmark" (NIATERO, 2024).

Others have adopted more radical practices such as the *retomadas*, or the practice of retaking traditional territories by the local population, in direct opposition to the government and the private sector. In defence of this practice, Dinamam Tuxá, also a member of APIB, argues, "when we say retaken it means something has been taken from us. And that we are claiming and returning [to the land], taking back what was once ours and which, due to this very violent colonization process, was stolen from us". The complete loss of credibility of formal institutions and the frustration of the people that engage in these practices is also a common theme present in affirmations made by people like Laura, an indigenous leader from the Avá Guarani people that claims "We've been receiving empty promises from white people for a long time—and we're getting tired of it. That's why what we're doing isn't an invasion" (Moncau, 2024).

Therefore, a fundamental conversation that the APIB will have to develop internally is how legitimate it is for them to engage in actions that could be perceived as increasingly undemocratic, including the use of coercion and violence, in the pursuit of democratic goals. The development of a solid ideological framework for the instances where the use of these tactics is justified, might help prevent more excessive impulses, maintain the APIB as a legitimate political actor and reduce the probability of splinters in the movement due to increasingly irreconcilable positions (Gourevitch, 2018).

19.4 Conclusion

Over the course of this chapter, I have put forward a case that in the current global context it can be helpful to revisit early warnings made by deliberative and agonistic political theory regarding the erosion of the democratic foundation of liberal democracies by liberalism itself. In an effort to firmly establish liberal principles as universal and uncontestable, modern democracies have gradually diminished popular sovereignty and therefore the ways that most of the citizenry has to advocate for change. A reaction to this has been the development of "illiberal democracy" or the rise of authoritarian actors and practices through democratic means (Mouffe, 2000, 2013).

For alternative visions of the social order such as the one defended by proponents of a new eco-social contract, this means that the toolkit at their disposal to effectively challenge existing economic and political systems through democratic means is shrinking by the hour. Not only that, but increasing conditions of economic and political inequality mean that proposed solutions through the amplification of formal democratic participation, like the one suggested by deliberative democracy, might only be marginally effective (Young, 2001; Kapoor, 2002; Pansardi, 2016).

A successful bargaining process for a new eco-social contract must, then, be informed by both the current configuration of power relations that hold unequal, exploitative and undemocratic systems in place and engage more directly in a struggle to take that power away from them. This effort must be underpinned by the central question that agonistic theory sets for democratic politics: How do we constitute forms of power which are compatible with democratic values? (Mouffe, 2000). The answers to this question in a practical sense have been few. However, recent research on conflict dynamics for the creation of a cohesive theory of democratic conflict or the proposal of a design framework for agonistic institutions start to chart a path towards more frequent implementation of this vision for democratic politics. Further work needs to be done to develop these nascent efforts into fully realised theories for action (Lowndes & Paxton, 2018; August & Westphal, 2024).

The relevance of these conversations to continue can be seen over the four case studies of social movements in Argentina, Brazil and Mexico analysed in this chapter. A common trend in every single one of them is their lack of aversion to engaging in confrontation with the political system in order to change it, combined with the desire to do so as democratically as possible. An agonistic view of democracy seems to be present, even if not explicitly identified as such by the interviewees.

How this has translated in practice is usually the creation of mixed models that alternate between direct contestatory action and formal participation in existing decision-making spaces. They have also built internal structures that are designed to be able to switch between these strategies according to changes on the conditions imposed by the political system. There is always some level of political pragmatism away from permanent and universal answers to what constitutes legitimate democratic action. Although how much this pragmatism allows for more radical forms of action in defence of democratic principles varies from one movement to another.

A stress test to these forms of operation is also taking place right now for most cases given that they are dealing with increasingly hostile political contexts and have harder questions to answer regarding their commitment to struggle through democratic means. This has resulted in deeper conversations at the internal level on how to conceptually engage with the use of power, coercion, violence, domination and conflict as they become increasingly part of their political reality. Further research is then necessary to offer a clear picture on the way these conversations are taking place in different contexts and the paths forward that are being laid out.

References

Acemoglu, D., & Robinson, J. A. (2000). Why did the west extend the franchise? Democracy, inequality and growth in historical perspective. *Quarterly Journal of Economics, 4*, 1167–1199. https://doi.org/10.1162/003355300555042

Acemoglu, D., & Robinson, J. A. (2001). A theory of political transitions. *The American Economic Review, 4*, 938–963.

Acemoglu, D., & Robinson, J. A. (2005). *Economic origins of dictatorship and democracy*. Cambridge University Press.

Alves, S. (2023). 'We defended our right to the land': Brazil's Indigenous people hail supreme court victory. *The Guardian*. Accessed December 24, 2024, from https://www.theguardian.com/global-development/2023/sep/25/we-defended-our-right-to-the-land-brazils-indigenous-people-hail-supreme-court-victory

APIB. (2023). *The Brazil's indigenous people articulation – APIB is an agglutination instance and national reference of Brazil's indigenous movement*. APIB Official. Accessed September 30, 2024, from https://apiboficial.org/sobre/?lang=en

August, V. (2022). Understanding democratic conflicts: The failures of agonistic theory. *European Journal of Political Theory, 2*, 182–203. https://doi.org/10.1177/14748851221120120

August, V., & Westphal, M. (2024). Theorizing democratic conflicts beyond agonism. *Theory and Society, 53*, 1119–1149. https://doi.org/10.1007/s11186-024-09565-4

Boix, C. (2003). *Democracy and redistribution*. Cambridge University Press.

Campati, A. (2021). Elite and liberal democracy: A new equilibrium? *Topoi, 41*, 15–22. https://doi.org/10.1007/s11245-021-09762-1

Chen, K. J. (2023). *Re-imagining social-spatial fragmentation: Protests in Latin America during the pandemic*. London School of Economics. Accessed December 24, 2024, from https://blogs.lse.ac.uk/internationaldevelopment/2023/01/18/re-imagining-social-spatial-fragmentation-protests-in-latin-america-during-the-pandemic/#:~:text=Fifty%20years%20on%20from%20late%2D1960s%20counterculture&text=Since%20the%20pandemic%20outbreak%2C%20many,plan%20and%20Chile's%20constitutional%20change

Deneen, P. J. (2018). *Why liberalism failed*. Yale University Press.

Eckstein, S. (2001). *Power and popular protest. Latin American social movements*. University of California Press.

Eslava, F., & Valencia Caicedo, F. (2023). *Origins of Latin American inequality*. Latin American and Caribbean Inequality Review. Accessed December 26, 2024, from https://lacir.lse.ac.uk/en-gb/publications/origins-of-latin-american-inequality

Fergusson, L., & Robinson, J. (2024). *The interaction of economic and political inequality in Latin America*. Latin American and Caribbean Inequality Review. Accessed December 26, 2024, from https://lacir.lse.ac.uk/en-gb/publications/the-interaction-of-economic-and-political-inequality-in-latin-america

Gourevitch, A. (2013). Labor republicanism and the transformation of work. *Political Theory, 4*, 591–617. https://doi.org/10.1177/0090591713485370

Gourevitch, A. (2018). The right to strike: A radical view. *American Political Science Review, 4*, 905–917.

Gutiérrez Aguilar, R. (2015). *Horizonte comunitário-popular: Antagonismo y producción de lo común en América Latina*. BUAP Puebla.

Gutmann, A., & Thompson, D. F. (1997). *Democracy and disagreement. Why moral conflict cannot be avoided in politics and what should be done about it*. The Belknap Press of Harvard University Press.

Habermas, J. (1996). *Between facts and norms. Contributions to a discourse theory of law and democracy*. MIT Press.

Human Rights Watch. (2022). *Latin America: Alarming reversal of basic freedoms*. Human Rights Watch. Accessed December 26, 2024, from https://www.hrw.org/news/2022/01/13/latin-america-alarming-reversal-basic-freedoms

Huntjens, P. (2021). *Towards a natural social contract: Transformative social-ecological innovation for a sustainable, healthy and just society*. Springer Nature. https://www.springer.com/gp/book/9783030671297

Kapoor, I. (2002). Deliberative democracy or agonistic pluralism? The relevance of the Habermas-Mouffe debate for third world politics. *Alternatives, 4*, 459–487. https://doi.org/10.1177/030437540202700403

Lamas, M. (2020). *Dolor y Política. Sentir, pensar y hablar desde el feminismo*. Editorial Océano.

Levitsky, S., & Way, L. (2015). The myth of democratic recession. *Journal of Democracy, 1*, 45–58. https://doi.org/10.1353/jod.2015.0007

Levitsky, S., & Ziblatt, D. (2018). *How democracies die*. Crown Publishing Group.

López Barajas, M. P., & Guerra Favela, T. (2020). *La violencia feminicida en México: Aproximaciones y tendencias*. ONU Mujeres. Accessed December 26, 2024, from https://mexico.unwomen.org/es/digiteca/publicaciones/2020-nuevo/diciembre-2020/violencia-feminicida

Lowndes, V., & Paxton, M. (2018). Can agonism be institutionalised? Can institutions be agonised? Prospects for democratic design. *The British Journal of Politics and International Relations, 3*, 693–710. https://doi.org/10.1177/1369148118784756

Manin, B. (1997). *The principles of representative government*. Cambridge University Press.

Maslin, M. A., Lang, J., & Harvey, F. (2023). A short history of the successes and failures of the international climate change negotiations. *UCL Open: Environment, 5*, 1–16. https://doi.org/10.14324/111.444/ucloe.000059

Merkel, W. (2014). Is there a crisis of democracy? *Democratic Theory, 2*, 11–25. https://doi.org/10.3167/dt.2014.010202

Merlinsky, M. G. (2021). *Toda ecología es política: Las luchas por el derecho al ambiente en busca de alternativas de mundos*. Siglo Veintiuno Editores.

Modonesi, M. (2008). Crisis hegemónica y movimientos antagonistas en América Latina. Una lectura gramsciana del cambio de época. *A Contra Corriente, 2*, 115–140.

Moncau, G. (2024, August 20). *A historic practice of Indigenous peoples in Brazil targeted by agribusiness: Understanding retomadas*. Brasil de Fato.

Mouffe, C. (2000). *The democratic paradox*. Verso.

Mouffe, C. (2013). *Agonistics. Thinking the world politically*. Verso.

Mounk, Y. (2018). *The people vs. democracy: Why our freedom is in danger and how to save it*. Harvard University Press.

NIATERO. (2024). *Twenty years of free land camp*. NIATERO. Accessed December 26, 2024, from https://www.niatero.org/stories/podcasts/twenty-years-of-free-land-camp

Norton, A., & Greenfield, O. (2023). *Eco-social contracts for the polycrisis: Participatory mechanisms, Green Deals and a new architecture for just economic transformation*. Green Economy Coalition. Accessed December 26, 2024, from https://www.greeneconomycoalition.org/assets/reports/GEC-Reports/GEC_Eco-Social-Contracts-Polycrisis-FINAL-Nov23.pdf

OHCHR. (2024). *Brazil must protect Indigenous Peoples' lands, territories and resources, says Special Rapporteur*. OHCHR. Accessed September 30, 2024, from https://www.ohchr.org/en/press-releases/2024/07/brazil-must-protect-indigenous-peoples-lands-territories-and-resources-says

Pabst, A. (2016). Is liberal democracy sliding into 'Democratic Despotism'? *The Political Quarterly, 1*, 91–95. https://doi.org/10.1111/1467-923X.12209

Pabst, A. (2019). *The demons of liberal democracy*. Polity Press.

Pansardi, P. (2016). Democracy, domination and the distribution of power: Substantive Political Equality as a Procedural Requirement. *Revue Internationale de Philosophie, 275*, 91–108.

Paulo, F. S.. (2023). *Marco temporal de terras indígenas é um litígio climático, diz Maurício Terena*. Folha de S. Paulo. Accessed September 30, 2024, from https://www.youtube.com/watch?v=CYMV0WrSVqc

Petras, J., & Veltmeyer, H. (2011). *Social movements in Latin America. Neoliberalism and popular resistance*. Palgrave Macmillan.

Przeworski, A. (2018). *Why bother with elections?* Polity Press.

Retamozo, M. (2005). Movimientos sociales y orden social en América Latina: Sujetos, antagonismos y articulación en tiempos neoliberales. *Desde el fondo, 38*, 27–35.

Rosenblatt, H. (2018). *The lost history of liberalism. From ancient Rome to the twenty-first century*. Princeton University Press.

Selmini, R., & Di Ronco, A. (2023). The criminalization of dissent and protest. *Crime and Justice, 52*, 197–231. https://doi.org/10.1086/727553

Tong, R. (2009). *Feminist thought. A more comprehensive introduction*. Westview Press.

UNRISD. (2021). *A new eco-social contract. Vital to deliver the 2030 Agenda for Sustainable Development*. Issue Brief No. 11. UNRISD. Accessed December 26,

2024, from https://sdgs.un.org/sites/default/files/2021-07/UNRISD%20-%20 A%20New%20Eco-Social%20Contract.pdf

UNRISD. (2022). *Crises of inequality. Shifting power for a new eco-social contract.* UNRISD. Accessed December 24, 2024, from https://cdn.unrisd.org/assets/ library/reports/2022/full-report-crises-of-inequality-2022.pdf

Vegh Weis, V. (2022). *Criminalization of activism. Historical, present and future perspectives.* Routledge.

Young, I. M. (2001). Activist challenges to deliberative democracy. *Political Theory, 5,* 670–690. https://doi.org/10.1177/0090591701029005004

Znoj, M. (2024). How to think about liberal and democratic principles: Three models of illiberal democracy. *Filosofický časopis, 72,* 13–29. https://doi.org/10.46854/ fc.2024.1s13

20

Contestation Movements and the Emergence of Eco-Social Contracts in India and Nepal

Alina Saba and Gabriele Koehler

20.1 Introduction[1]

India and Nepal saw considerable political changes in 2004 and 2007, respectively. In India, a newly elected progressive coalition formed a government in 2004; in Nepal, after a decade-long civil war, a multi-level peace process introduced new forms of governance in 2007. Both new governments addressed stark social and economic cleavages and turned attention to social policies. We interpret this as a move towards rights-based policymaking—a social turn.[2] This relatively progressive phase continued into the 2010s. However, it has now slowed in Nepal due to pushback from majoritarian elites. In India, efforts to create a new eco-social contract are being reversed altogether.

---

[1] Maggie Carter, Deepta Chopra, Manisha Desai, Nicole Harris, Katja Hujo, Isabell Kempf, Cornelia Rühlig and Sachin Siwakoti provided comments. The India sections are based on interviews with experts on India, whom we thank for their expertise.

[2] UNRISD (2016:34) describes the social turn as a gradual shift in ideas and policies which reasserted social issues in development agendas. Koehler and Namala (2020) have a slightly different take.

A. Saba
National Indigenous Women Forum (NIWF), Kathmandu, Nepal

G. Koehler (✉)
UNRISD, Bonn, Germany

© The Author(s) 2025
P. Huntjens et al. (eds.), *Eco-Social Contracts for Sustainable and Just Futures*,
https://doi.org/10.1007/978-3-031-99109-7_20

Citizens and civil society movements have been advocating for policy change and claiming their rights in both countries for decades. These struggles have taken many forms, but the coinciding and emerging coalescence of such movements suggest that UNRISD's (United Nations Research Institute for Social Development) concept of an eco-social contract, understood as an implicit or explicit societal agreement for justice, equal rights and sustainability, can serve as an analytical framework, even if the political and civil society actors do not explicitly use the term.

The purpose of this chapter is to highlight socio-economic and policy similarities between India and Nepal, trace the roles and demands of civil society movements and analyse their impact. The findings build on our activist and advocacy experiences, complemented by desk research. By examining eco-social transformations in specific situations, we conclude that reforms—political, legislative and legal—and behavioural changes are prerequisites for inclusive eco-social contracts, but the former will not fructify without pressure and contestation from civil society movements. Drawing on this, we offer ideas for supporting the emergence of inclusive eco-social contracts.

20.2   Civil Society Contestation and Developments Towards an Eco-social Contract

In India and Nepal, oppression based on class, caste, ethnicity, indigeneity, faith, language, location and—overarching all of these—gender identity (Desai, 2016; Koehler & Namala, 2020) continues to undermine the rights promised to "secure for all the citizens justice, economic, social and political" (Government of India, 2015; Government of Nepal, 2007). Contrary to this commitment, cleavages have worsened over the past decade and are being reinforced by economic inequities and power hierarchies.

In India, the current government under the Bharatiya Janata Party (BJP) has turned to authoritarianism, curtailing the rights and freedoms of marginalised groups and codifying ethnicity- and faith-based exclusions with a view of building a "Hindutva" nation riven with structural systemic violence and social exclusions (Sen, 2021). The original ethos of the Indian Constitution of 1950 is undermined.

Conversely, in Nepal, the political system is moving notionally towards ending discrimination and inequality based on caste, ethnicity, faith and gender identity. There are also efforts to address climate and environmental commitments (Siwakoti, 2024). However, the decentralisation (federalisation) process required for greater representation at different levels of government has been rife with conflicts, often violent. The economic and political power of the dominant ethnic groups appears to be re-entrenching. Nevertheless, policymaking can be understood as a continued, even if challenged, move towards an eco-social contract between various communities and the federal government.

20.2.1 India's Socioeconomic Trajectory

Indian society is deeply fractured (Ambedkar, 2011; Drèze & Sen, 2013:213ff; Roy, 2014; Sen, 2021; Wada Na Todo Abhiyan, 2023). Jean Drèze and Amartya Sen (2013) remark: "India ... has a unique cocktail of lethal divisions and disparities". Even though the Constitution and post-independence governments adopted affirmative action and social justice legislation in support of women, Scheduled Castes, Scheduled Tribes and other oppressed communities, the caste hierarchy in governance, politics and institutions remain. As a result, 75 years after independence, India faces "one of the most extreme increases in income and wealth inequality observed in the world" (Chancel et al., 2022:197). Thus, in 2021, the wealthiest 10 percent of the population owned 65 percent of total household wealth. As of 2020, the female labour income share equalled only 18 percent. In addition, civic space, such as the autonomy of civil society organisations (CSOs) and media freedom, while guaranteed in the Constitution, is under threat (Civicus, 2022, 2023).

The United Progressive Alliance (UPA), a 15-party coalition government under the Indian National Congress (INC), led the government from 2004 to 2014. Building on a tradition of overarching ministerial policy decisions, the UPA government introduced policies to address economic, social and political rights which followed a rights-based logic[3] and can be understood as a "social turn". At the same time, the UPA continued the previous neoliberal government's focus on GDP growth as the easiest solution to overcoming

---

[3] These include inter alia, in chronological sequence: the Right to Information (RTI) Act (2005); the Protection of Women from Domestic Violence Act (2005); the Commissions for Protection of Child Rights Act (2005); the Mahatma Gandhi National Rural Employment Guarantee (MGNREGA) Act (2006); The Scheduled Tribe and Other Forest Dwellers (recognition of forest rights) Act (2006); The Unorganised Workers Social Security Act (2008); the Right of Children to Free and Compulsory Education Act (2009); the Protection of Children from Sexual Offences Act (2012); the Sexual Harassment of Women at Workplace (prevention, prohibition, redressal) Act (2013); the Right to Fair Compensation and Transparency in Land Acquisition, Rehabilitation and Resettlement Act (2013); the National Food Security Act 2013; and the Companies Act (2013).

cleavages. The policies also encompassed social and environmental dimensions, such as the Mahatma Gandhi National Rural Employment Guarantee Act (Chopra, 2014; UNRISD, 2016), and climate-related policies, for example, on energy efficiency and sustainable agriculture. A National Green Tribunal was established in 2010 to oversee cases concerning environmental protection and the conservation of forests and other natural ecosystems and resources (Government of India, n.d.; Desai, 2016:18).

Many of these policies were moderately successful (Ghosh, 2014; Desai, 2016) and one could cast this period as an emerging eco-social contract. However, this (eco-)social policy orientation did an about-face when the National Democratic Alliance government, under the leadership of BJP, formed a new government in 2014. This government introduced discriminatory laws, taking a distinct and deliberate turn away from the more inclusive policy direction pursued since independence. BJP governance advertises the "glory of ancient India" and Hindu nationalism ("Hindutva"). It sees the "world as one single family"—an idea that never applied in India given the caste hierarchy. In fact, the Citizenship Amendment Act (CAA) and the National Register of Citizens 2019 are both antithetical to the "one single family" ideology. The CAA, designed to prevent Muslim refugees from entering India from neighbouring countries and disenfranchising people living in India without recognised documentation, will create large groups of "stateless" people among vulnerable communities, including Scheduled Tribes (particularly vulnerable tribes), nomadic communities, LGBTQI communities, the urban poor, migrants and others (Wada na Todo Abhiyan, 2023).

Violence against Muslim and Christian communities, including physical violence, lynching and murder, has been on the rise since 2014 (Human Rights Council, 2022). The so-called upper-caste Hindus are increasingly controlling cultural symbols, festivals and food habits of minorities and marginalised groups in an attempt to eliminate their traditions. The Foreign Contribution Regulation Act (FCRA) obstructs the renewal of CSO licences needed in order to receive international funding. This is meant to constrict their activities, programmes and networking and undermine their capacity to support vulnerable and marginalised communities.

Independent governance and judiciary institutions are also undermined, consolidating the ruling party's hold on power. Institutions like the Supreme Court, Election Commission and investigation bodies are instrumentalised to promote the interests of the government. Mainstream media often align their reporting with government views and assessments. Posts in important institutions like the Information Commission are left vacant and appointments controlled to encourage the Commissioners to align with government lines

(Drishti, 2021). Many human rights defenders have been arrested and alleged illegal immigrants forced into detention centres (Singaravelu, 2020).

In parallel, the government is dismantling environmental and climate commitments introduced by the UPA government. It is weakening existing provisions when approving investments under the excuse of facilitating business. For example, the Biological Diversity Amendment Bill of 2021 now allows the commercialisation of biological resources and decriminalises what had been offences under the original act (Saldanha, 2023).

This unravelling of the eco-social contract needs an explanation. The well-recognised challenges of poor political will and administrative neglect, or even deliberate discrimination in the implementation of constitutional rights and the reversal of previous progressive policies, stem from the continued hold of the dominant, powerful elites over economy, society and government (Sen, 2021). The Indian Constitution's norms and values of equality—the notion of economic, social and political justice—are not carried through, leading to government capture and the creation of new institutions. The economic liberalisation process since the 1990s favoured the interests of the dominant castes and communities and exacerbated the economic, social and political exclusion of other groups. Arguably, the elite's hold on power facilitated the rise of right-wing majoritarianism and rule (Iqbal, 2021). For example, the Supreme Court de facto promoted right-wing interests in its orders to evict forest dwellers and the Scheduled Castes and the Scheduled Tribes Prevention of Atrocities Act (Sethi, 2019) has been significantly diluted. Such decisions reflect majoritarian caste interests and a departure from constitutional principles.

20.2.2 Nepal's Political and Social Trajectory

Nepal is also a society deeply riven by class, caste, gender, economic and faith-based inequalities, which are exacerbated in more remote or inaccessible areas of the country, notably for Indigenous communities (Lawoti, 2010; Siwakoti, 2024; Human Rights Council, 2021). As a result of structural inequalities in access to employment, land, capital or social assistance, Nepal has extremely high income and wealth inequalities. The richest quintile owns 56 percent of all wealth (Khanal et al., 2019:20). Land—a strategic asset for livelihoods—is highly concentrated: the richest 7 percent of households own 31 percent of agricultural land, while the poorest 20 percent own just 3 percent (Khanal et al., 2019:22).

Given the dire economic and societal cleavages (Lawoti, 2010:7–11), a movement which called itself Maoist presented a set of 40 political, economic and social demands in 1996. These included a secular state, abolition of the caste system, Indigenous rights, gender equality, land reform, and minimum wage and social transfer provisions (Adams, 1998; Lawoti, 2010:19). When the monarchy and government rejected these demands, a violent armed conflict erupted, with brutalities on both sides (Lawoti, 2010:15,19f; Dixit, 2006). The conflict between the royalist army and the Maoist insurgents waged from 1996 until 2006. In total, 15,000 people—combatants and civilians alike—lost their lives (Lawoti, 2010).

In 2006, in a grassroots People's Movement—the Jan Andolan—thousands of citizens took to the streets of Kathmandu demanding to end the hostilities (Sijapati, 2009). In response, the then government on behalf of the monarchy and the Communist Party of Nepal (Maoist) negotiated a Peace Accord. It agreed to a "progressive restructuring of the state to resolve existing class-based, ethnic, regional and gender problems", to convene an interim parliament and to call elections. The parties also agreed on an Interim Constitution which abolished the monarchy (Interim Constitution of Nepal, 2007).

After the 2007 elections, the new government acknowledged that poverty, inequality and social exclusion (Lawoti, 2010; Khatiwada & Koehler, 2014) were the root causes of the decade-long civil conflict, and introduced reforms towards a decentralised—federated—system of local governments, considered more amenable to recognising, addressing and representing Indigenous and oppressed caste interests and concerns.

Building on a considerable history of social policy action (Koehler, 2021), the successive post-conflict governments adopted policies to promote social inclusion and address income poverty, notably in access to health services and income transfers addressing caste- and income-disadvantaged children, single women, pensioners and persons living with a disability or in remote areas (Khatiwada & Koehler, 2014:136ff). Tax concessions with positive discrimination elements, such as tax rebates for the employment of women, Dalits and people with a disability, followed (Chakravarty & Shakya, 2021).

Political governance reform was more demanding. The constitution drafting process (2007–2015) was rife with violent conflict and contentious debates, mainly over restructuring the state into a federal model (Siwakoti, 2024). The concept of a right to social justice was introduced for the first time, ensuring inclusion on the principle of "proportional representation" of marginalised groups in state structures, specifically Dalit, women, Indigenous groups, Madhesi communities,[4] and poor farmers and labourers (Article 21, Interim Constitution of Nepal 2007). These provisions paved the way for amendments to the Nepal Civil Service Act and the Nepal Army and Police Acts to include quotas for marginalised groups.

Dalits, Indigenous peoples and Madhesi groups stressed decentralisation and federalism as a way to attain equal rights as citizens and overcome exclusion by acknowledging identities, ensuring their voice and transparent political representation, giving language rights and access to government office and institutions, and providing equitable access to government resources (Tamang, 2014). The identity issue was central to redressing historical subjugation, for instance, by renaming provinces to reflect the identities of marginalised communities (Lawoti, 2019). In response, the Interim Constitution of Nepal (2007) adopted federalism with a view of ending discrimination based on class, caste, gender, religion, language, culture and region (Article 138.1) (Hachhethu, 2014; Saba & Kohler, 2023).

In contrast to the marginalised groups' demand for ethnicity-based federalism, the traditional political parties—Nepali Congress and the Communist Party of Nepal (Unified Marxist-Leninist, CPN-UML), with the majority of the leaders of both parties belonging to the Hill Hindu or so-called "high" caste—argued for a "realistic model" for administrative ease and accessibility, and to acknowledge challenges in the distribution of natural resources (Saba, 2018). These traditionalist political parties, and some groups in media, administration, judiciary and civil society considered the identity or inclusion agenda as creating competition and animosities among communities. This polarising controversy led to the dissolution of Parliament in 2012. Elections in 2013 enabled the reconvening of the constituent assembly.

After debates in the reconstituted Parliament and contestation from many civil society movements, Parliament succeeded in promulgating the new Constitution of Nepal in 2015. Key provisions which can be interpreted as an eco-social contract include gender equality and the rights of all marginalised communities—two demands of the 1996 Maoist programme—and the right to a clean environment (Government of Nepal, 2015; Siwakoti, 2024).[5]

---

[4] Madhesi People are a community concentrated in southern Nepal, accounting for approximately 19 percent of Nepal's total population. There are class and caste hierarchies within the community, but overall, it faces social and political exclusion.

[5] Clustered by thematic areas, these are: the right against exploitation (Article 29); the rights of women to participate in all bodies of the state on the principle of proportional representation (Article 38.4); the rights of Dalits, which ensures participation of Dalits in all bodies of the state on the basis of the principle of proportional inclusion (Article 40.1); the right to social justice, and specifically the rights of excluded marginalised groups, including women, Indigenous nationalities, Madhesi communities, Muslim communities, persons with disabilities, gender and sexual minorities, poor farmers and labourers, to partici-

However, grievances remained. The Madhesi community objected that electoral constituencies were delineated along ethnic lines (Khanal et al., 2019). Another contentious provision was the question of Khas Aryas, a community of so-called "high" Hindu caste who constitute 31 percent of the population of Nepal and occupy 10 of the highest public offices; of which, all are men except the president (Lawoti, 2019; Hachhethu, 2017:59). The Constitution designated Khas Arya who are economically poor ("indigent") as marginalised, and hence entitled to affirmative action measures such as quotas in elections (Government of Nepal, 2015). This definition diluted the prioritisation of social inclusion of those groups that are truly marginalised in terms of class or caste (Jha, 2017:66; Saba, 2018; Saba & Koehler, 2022). The clause reveals the challenges of addressing historical discrimination and domination based on caste, ethnicity or faith without taking an intersectional approach that includes income/assets, economic class and de facto political power.

Regarding gender and intersectional equality, the Election Commission of Nepal mandated 40.4 percent representation of women in the local elections of 2017. Mayor and deputy mayor/chair and vice chairperson elections required at least one female candidate. At the ward level, one out of two seats reserved for women among a five-member committee was mandated for Dalit women. The election results did see the rise of women and Dalit leaders. However, while the deputy posts were predominantly filled by women, most of the mayor/chairperson officials were male: out of 753 chairpersons and mayors elected, only 18 were women. Dalits constituted only 1 percent of mayors, with the so-called "high" Hindu caste filling 50 percent of the positions; Madhesis 19 percent and Indigenous nationalities 12 percent (Pokharel & Pradhan, 2020). Khas Arya women fill most of the highest decision-making roles instead of a proportional representation of Indigenous, Madhesi and Dalit women. Dalit women's representation was due only to the explicit mandate; the representation of other marginalised women remains nominal.

In other words, barriers to intersectional representation remain significant and provisions in the new Constitution do not necessarily translate into greater equality. The election results show that inclusive policies are implemented only when mandated for a particular group. Nevertheless, it can be argued that in Nepal, despite setbacks, dilution and obstacles, a more inclusive (Sunam & Shrestha, 2019; Saba & Kohler, 2023) (eco)social contract is

---

pate in state bodies on the basis of proportional inclusion (Article 42.1); the establishment of a National Inclusion Commission (Article 258); the right to a clean environment (Article 30); and the right to information (Article 27).

evolving, and that this is primarily due to the persistent contestation—both peaceful and violent—of various communities and identity groups and civil society.

20.3 The Role of Rights-Based Movements for Eco-social Contract Approaches

Civil society has played a significant and driving role in India and Nepal, pushing for what can be understood as a new eco-social contract between citizens and governments.

20.3.1 India

India has a history of protest and a "vibrancy of democratic practice and social movements" (Drèze & Sen, 2013:274). Struggles against the caste system have seen the Dalit community and women in the informal sector at the forefront (Chopra, 2014). The initial struggles focused on the concept of caste hierarchy and untouchability, humiliation, indignity, violence, and the lack of job security and social protection. However, they have since expanded to also demand societal inclusion more broadly—economic inclusion, including in the private sector, decent employment and the effective implementation of government provisions (Thorat & Newman, 2010).

Contestations around the protection of nature have roots that reach back to the 1960s, such as protests against the Narmada Sardar Sarovar Dam, one of the largest in the country (Roy, 2019) and, later, hydropower dam projects (Desai, 2016:48). Other movements protested the Mithi Virdi nuclear power plant (Desai, 2016:17), expansive industrial zones in West Bengal and Gujarat (Desai, 2016:17), and biodiversity theft (Shiva, 1988).

The protests and struggles against caste-based discrimination and for land and ecological rights draw their strength from the active agency and leadership of members of the affected communities (Desai, 2016; Kothari & Joy, 2017; Wada Na Todo Abhiyan, 2023) who stood up against violence and atrocities, referring to their rights under the Scheduled Castes and Scheduled Tribes Prevention of Atrocities Act.

An important campaign in this regard is the National Campaign on Dalit Human Rights (NCDHR). This coalition of human rights activists and academics formed in 1998 promotes community access to justice against discrimination and violence, tracks and monitors budgets mandated by the state for Dalit and Tribal development and promotes Dalit women's rights. The

2018 nationwide protests of Dalit communities (NDTV, 2018) against the Supreme Court's order to dilute the Scheduled Castes and the Scheduled Tribes Prevention of Atrocities Act, testifies to the continued salience of this issue. NCDHR further promotes global coalitions and has brought the issue of caste-based discrimination to the United Nations.

The Shaheen Bagh Protest in Delhi (2019–2020) has become a symbol of citizen action by women against unjust laws and authoritarianism. The movement spread across the country, and hundreds of thousands of women, particularly Muslim women, supported by students, organised peaceful sit-ins against the Citizen Amendment Act and National Register of Citizens, which threatened the citizenship and inclusion of Muslim and other marginalised communities. The movement is inspirational for its dispersed leadership, the pronounced agency of women, and support from other movements and communities.

Civil society organisations (CSOs) have also come together as a platform—Wada Na Todo Abhiyan (Keep Your Promises campaign)—to promote governance accountability at the national and global levels. It monitors the performance of the union government and progress towards meeting global commitments (Wada Na Todo Abhiyan, 2020, 2023). The platform proactively uses the "leave no one behind" principle of the Sustainable Development Goals (SDGs) under the campaign "100 Hotspots: Vulnerable Communities and SDGs in India" (Civil Society Initiative, 2021; Wada Na Todo Abhiyan, 2020). It mobilised massive condemnation for the exemptions the government proposed to forest conservation rules in 2022; in the end, the government agreed not to implement the amendments (Saldanha, 2023).

Students and young people exhibit great courage and leadership, raising their voices for issues of concern and standing up for democratic values. However, many have been jailed for voicing their opinions. In 2020, two students of the Break the Cages movement were arrested on charges of supporting the CAA protests (Sharma, 2021) and only released after 13 months. A student leader from Jawaharlal Nehru University was arrested on charges of plotting communal violence in Northeast Delhi. Sixteen activists were arrested in another case; the late Father Stan Swamy (Front Line Defenders, 2021) died in jail.

20.3.2 Nepal

Nepal experienced an extreme form of contestation in the violent civil conflict of 1996–2006. As mentioned above, the conflict was triggered by massive economic and social inequalities, and notably called for land reform and for caste and gender justice. Post-conflict social protests have been multi-pronged. Today, civil society is more effective and united than in many other countries, and despite some regression compared to the ambitious Interim Constitution of 2007, many elements of an emerging eco-social contract have remained, as witnessed by the provisions in the 2015 Constitution's articles and many policy decisions. For example, in 2007, Nepal became one of the first countries globally to officially recognise the third gender (Sharma, 2015).

The role of rights-based organisations and civil society was instrumental in establishing inclusive policies. The social justice movements of Dalit, Indigenous Nationalities and Madhesis continue to challenge the state to implement inclusive policies.

In 2020, a nationwide independent #DalitRightsMovement protested the killing of a Dalit teenager boy and five other children by a mob of villagers over an inter-caste marriage. The social movement that grew out of this crime continues to advocate against the murder of Dalits and calls for an end to all forms of discrimination and violence based on caste. Similarly, the Samata Foundation, a national network of Dalit human rights defenders, monitors and documents cases of Dalit human rights violations nationally, and runs television programmes and social media campaigns.

The Nepal Federation of Indigenous Nationalities (NEFIN), an umbrella organisation of 59 Indigenous nationalities of Nepal, has been advocating for the rights of Indigenous peoples since the early 1990s. It played a key role during the People's Movement of 2006 and in the constitution drafting and revision processes from 2007–2018 by encouraging the government to ratify international treaties such as the United Nations Declaration on the Rights of Indigenous Peoples (UNDRIP) and the ILO Convention on Indigenous and Tribal Peoples (Convention No. 169). NEFIN mobilises Indigenous peoples nationally and locally for demonstrations and strikes and advocates for environmental and climate rights in line with the Paris Agreement through its Climate Change Partnership Programme. NEFIN also engages in international policy-making at the United Nations and in the Asia Indigenous Peoples Pact (AIPP).

Indigenous peoples have mobilised against various neoliberal development projects implemented aggressively by central and local governments in Nepal, especially in collaboration with multilateral development banks, in hydropower, electricity transmission lines and road expansion (Bhattachan, 2019:369). Lawyers' Association for Human Rights of Nepalese Indigenous Peoples (LAHURNIP) works on free, prior and informed consent (FPIC) and the rights of Indigenous peoples against development projects which have harmed or displaced Indigenous peoples from their homes, locations and livelihoods. In 2021, the movement won a victory against the European Investment Bank's high voltage transmission line project for violating their FPIC rights (IWGIA, 2021).

In 2023, the Tsum Nubri Rural Municipality of Gorkha District adopted the Shagya Act.[6] It will enable the local Indigenous communities to formally govern local natural resources, forests and biodiversity based on Indigenous knowledge (Rights and Resources Initiative, 2023). This was a result of continuous advocacy by the Center for Indigenous Peoples' Research and Development (CIPRED).

Regarding gender justice, there have been many movements (Hewitt, 2018). NIWF and NIWF-Forum, two national-level Indigenous women's rights organisations, advocate for the social, cultural, political and economic rights of Indigenous women within their community and beyond. They advocate for intersectional feminism to recognise women as a heterogeneous group and address the multiple forms of violence Indigenous women face. Problems persist regarding the implementation of constitutional gender provisions and gender-sensitive and gender-responsive legislation, policies and acts, including the intersectional recognition of concerns that affect women based on ethnicity, caste, religion, language, indigeneity, marital status, geographical location, ability and access to health and education (Saba & Hewitt, 2019:1). Discrimination faced by Nepali women goes beyond the gender binary and patriarchal structure and is often coupled with other intersectional barriers, including a deeply entrenched caste system.

In 2018, a consortium of Indigenous women's organisations, including NIWF and NIWF-Forum, submitted a shadow CEDAW (Committee on the Elimination of Discrimination Against Women) report on the "Situation of the Rights of Indigenous Women of Nepal" (NIWF and NIWF-Forum, 2018), bringing attention to the social exclusion and invisibility of Indigenous women in the Constitution, laws, policies, plans, programmes and budget and calling on the state to recognise Indigenous women and Indigenous women with disabilities as distinct legal entities (CEDAW, 2018). Following the submission, the Committee observed the lack of recognition of Indigenous women, and recommended amending the Constitution in line with UNDRIP. This is a historic success of the Indigenous women's movement in the international human rights space.

The street protests of the Indigenous movement in Nepal have slowed in the last few years; we observe growing pessimism among Indigenous groups

---

[6] Shagya is a practice of Indigenous groups to protect their surrounding natural resources, biodiversity and their cultures.

(Chhantyal & Rai, 2020:6). This may be due to increased weariness, the heightened polarisation between Indigenous and non-Indigenous communities as well as some degree of co-optation by political parties (Saba, 2018; Saba & Kohler, 2023). Instead, the movement has taken to other forms of protest. Nevertheless, given the shared intersecting concerns for gender justice, exclusion, discrimination and environmental destruction, contestation in Nepal can be seen as a move towards an eco-social contract.

20.4 Summary and Outlook

In both countries, we observe progress and success. Much of such progress is owed to decades of pressure, action, proposals and pushback against repression, contestation and, in Nepal, armed conflict. These contestations introduced a series of political and societal reforms, such as the many rights-based social policy acts adopted in India under the UPA and the fundamental political reforms in post-conflict Nepal. Agreements were reached between governing elites and the majority of society.

At the same time, our discussion demonstrates the challenges of social change and for lasting, progressive agreements. Regarding the *challenges*, communities marginalised on the basis of class, caste, ethnicity, faith, location and gender, along with their civil society representatives, remain structurally oppressed and often persecuted. The political and socio-economic trajectories of both India and Nepal illustrate the vulnerabilities of just governance. Authoritarian, racist, patriarchal and classist government and police actions continue. Therefore, it is crucial that, at the *level of aspirations*, societal, economic and political inclusion and justice remain at the centre of demands.

Transformations are needed at all levels.

At the *government level*, regulatory reform must be strengthened. Legislation must ensure genuine intersectionality in affirmative action and implementation to realise and consolidate social inclusion (Siwakoti, 2024). Politically, tokenism must be tackled—the practice of nominally electing representatives of disadvantaged groups while highjacking their authority and relegating them to secondary places in decision-making processes. At the *political party level*, cemented groupings and traditions must be overcome (Krämer, 2021). Politicians newer to the scene and/or younger persons need to have space to take political power. In turn, their political positions must reflect and be controlled by independent progressive civil society and democratic institutions.

In the *economic domain*, fiscal policy reform is urgent. Societal inclusion requires progressive fiscal policy for three intertwined reasons: (i) to support adequate social policy expenditures; (ii) to enforce income and wealth redistribution (Chakravarty & Shakya, 2021; Bonnerjee, 2014); and (iii) to enable environmental action (UNRISD, 2022:307ff). Access to decent work must rectify the enormous cleavages between the formal and the informal economy and close the gaping gender-, caste-, ethnicity- and faith-driven employment and wage gaps. This requires fundamental labour law reforms, as a basis for unions, workers and citizens in general, to claim their rights.

At the *interpersonal level*, the identity of people from marginalised communities needs to be valorised. Progressive intersectionality must bring the causes of all marginalised groups together. This requires public messaging, education and radical shifts in employment policies. Additionally, land rights need to be ensured (Kurian & Singh, 2020).

Norms and ideals, too, need transformation. Civil society in India has adopted the concept of the "5 Rs" to promote societal inclusion: marginalised and excluded communities and individuals demand recognition, respect and representation, and there must be access to reparation and reclamation. This is an inherently human rights-based approach (Koehler & Namala, 2020:340ff), and could trigger a move away from Hindutva ideology in India, and in Nepal, facilitate a coming-together of the different identity groups.

However, affirmative action codices, legislation and judicial reforms, behaviour or norm change, or even societal contracts do not on their own lead to genuine economic, political, social and ecological inclusion (Saba & Koehler, 2022; Siwakoti, 2024). Civil society pressure must continue to vigorously claim and reinforce rights (Drèze & Sen, 2013; Piketty, 2020; UNRISD, 2022). Activism is the driver. As one legal expert put it: "Legislative changes come off the back of movements" (Her Forum, 2021). In short, progressive, rights-based civil society activism and contestation remain key for building new, inclusive eco-social contracts for social justice, as the examples of India and Nepal demonstrate.

References

Adams, B. (1998, May 7). Neglect of Bhattarai's demands sparked peoples war. *The People's Review*.

Ambedkar, B. R. (2011). *Annihilation of caste*. Dr. Ambedkar Institute of Social and Economic Change.

Bhattachan, B. K. (2019). Nepal. In D. N. Berger (Ed.), *The indigenous world 2019* (pp. 366–373). The International Work Group for Indigenous Affairs (IWGIA).

Bonnerjee, A. (2014). Social sector spending in South Asia: A mixed bag. In G. Koehler & D. Chopra (Eds.), *Development and welfare policy in South Asia* (pp. 185–197). Routledge.

CEDAW (Committee on the Elimination of Discrimination Against Women). (2018). *Situation of the rights of indigenous women in Nepal.* Shadow report for the sixth periodic report of Nepal. CEDAW/c/NPL/6. 22 October–9 November. Accessed April 2024.

Chakravarty, M., & Shakya, R. (2021). *Taxation policy and gender inequality in South Asia.* Presentation made during a webinar organized by the South Asia Alliance for Poverty Eradication (SAAPE) and the South Asia Tax and Fiscal Justice Alliance (SATaFJA) on Tax Justice and Gender Equality. Online, December 8.

Chancel, L., Piketty, T., Saez, E., & Zucman, G. (2022). *World inequality report 2022.* World Inequality Lab. https://wir2022.wid.world

Chhantyal, G., & Rai, T. B. (2020). *Politics of resistance: Indigenous peoples and the Nepali State.* Nepal Federation of Indigenous Nationalities (NEFIN).

Chopra, D. (2014). The Indian case: Towards a rights-based welfare state. In G. Koehler & D. Chopra (Eds.), *Development and welfare policy in South Asia* (pp. 85–105). Routledge.

Civicus. (2022). *Police disrupt protest in Dhangadhi while attacks on press freedom persist in Nepal.* Accessed November 29, 2023, from https://monitor.civicus.org/explore/police-disrupt-protest-dhangadhi-while-attacks-press-freedom-persist-nepal/

Civicus. (2023). *Nepal: Environmental defenders at risk, journalists targeted and concerns around the new cyber security policy.* Accessed November 29, 2023, from https://monitor.civicus.org/explore/nepal-environmental-defenders-at-risk-journalists-targeted-and-concerns-around-the-new-cyber-security-policy/

Civil Society Initiative. (2021). *Promises & Reality: Citizen's Report on Year Two of the NDA II Government 2020–2021.* Wada Na Todo Abhiyan.

Desai, M. (2016). *Subaltern movements in India: Gendered geographies of struggle against neoliberal development.* Routledge.

Dixit, K. (2006). *A people war: Images of the Nepal conflict 1996–2006.* Nepa-laya Publishers.

Drèze, J., & Sen, A. (2013). *An uncertain glory: India and its contradictions.* Allen Lane.

Drishti IAS (Indian Administration Service). (2021). *Central Information Commission (CIC).* Accessed January 14, 2022, from www.drishtiias.com/daily-updates/daily-news-analysis/central-information-commission-cic

Front Line Defenders. (2021). *Fr. Stan Swamy.* https://www.frontlinedefenders.org/en/profile/fr-stan-swamy

Ghosh, J. (2014). Can employment schemes work? The case of the rural employment guarantee in India. In D. B. Papadimitriou (Ed.), *Contributions to economic theory, policy, development and finance: Essays in Honor of Jan A. Kregel* (pp. 145–171). Palgrave Macmillan.

Government of India. (2015). *The constitution of India.* Ministry of Law and Justice. legislative.gov.in/sites/default/files/coi-4March2016.pdf

Government of India. (n.d.). *National green tribunal: About us.* https://www.greentribunal.gov.in/about-us

Government of Nepal. (2007). *The interim constitution of Nepal 2063*. Nepal Law Commission. constitutionnet.org/sites/default/files/interim_constitution_of_nepal_2007_as_amended_by_first_second_and_third_amendments.pdf

Government of Nepal. (2015). *The constitution of Nepal*. Nepal Law Commission. www.mohp.gov.np/downloads/Constitutionpercent20ofpercent20Nepalpercent202072_full_english.pdf

Hachhethu, K. (2014). Balancing identity and viability: Restructuring Nepal into a Viable Federal State. In B. Karki & R. Edrisinha (Eds.), *The Federalism Debate in Nepal* (pp. 35–54). United Nations Development Programme (UNDP).

Hachhethu, K. (2017). Legislating inclusion: Post-war constitution making in Nepal. In D. Thapa & A. Ramsbotham (Eds.), *Two steps forward, one step back: The Nepal peace process* (pp. 59–63. Accord, Issue 26). Conciliation Resources.

Her Forum. (2021). *Bringing together women in law: Interview with Karuna Nundy*. Her Forum video, 24:06. December 8. www.youtube.com/watch?v=ZIE10DOb6gs

Hewitt, S. (2018). *Nepal: A situational analysis of women's participation in peace processes*. Monash University.

Human Rights Council. (2021). *Universal periodic review. Summary of stakeholders' submissions on Nepal*. Summary submitted to the Human Rights Council. UN Doc. No. A/HRC/WG.6/37/NPL/3. Accessed November 2023.

Human Rights Council. (2022). *Universal periodic review. Summary of stakeholders' submissions on India*. Summary submitted to the Human Rights Council. UN Doc. No. A/HRC/WG.6/41/IND/3. Accessed November 2023.

Interim Constitution of Nepal 2063. (2007). https://constitutionnet.org/sites/default/files/interim_constitution_of_nepal_2007_as_amended_by_first_second_and_third_amendments.pdf

Iqbal, Y. (2021, December 27). How Indian Liberalism Aided the Rise of the Right. *Eurasia Review*. www.eurasiareview.com/27122021-how-indian-liberalism-aided-the-rise-of-the-right-oped/

IWGIA (International Work Group for Indigenous Affairs). (2021, May 4). Nepal Indigenous Communities Vindicated in Rare European Human Rights Victory. *IWGIA News*. https://iwgia.org/en/news/4365-nepal-eib-lahurnip-victory.html

Jha, D. (2017). Comparing the 2007 and 2015 constitutions. In D. Thapa & A. Ramsbotham (Eds.), *Two steps forward, one step back: The Nepal peace process* (pp. 64–66. Accord, Issue 26). Conciliation Resources.

Khanal, D. R., Adhikari, D., Mabuhang, B. K., Deuja, J., Khatiwada, P. P., & Pandey, G. (2019). *Fighting inequality in Nepal: The road to prosperity*. Oxfam International. https://doi.org/10.21201/2019.3903

Khatiwada, Y. R., & Koehler, G. (2014). Nepal, a nascent welfare state? In G. Koehler & D. Chopra (Eds.), *Development and welfare policy in South Asia* (pp. 129–147). Routledge.

Koehler, G. (2021). Effects of social protection on social inclusion, social cohesion and nation building. In M. Loewe & E. Schüring (Eds.), *Handbook on social protection systems* (pp. 636–646). Elgar Publishers.

Koehler, G., & Namala, A. (2020). Transformations necessary to 'leave no one behind': Social exclusion in South Asia. In A. D. Cimadamore, F. Kiwan, & P. M. Monreal Gonzalez (Eds.), *The politics of social inclusion: Bridging knowledge and policies towards social change* (pp. 313–352). UNESCO.

Kothari, A., & Joy, K. J. (Eds.). (2017). *Alternative futures: India unshackled.* AuthorsUpFront.

Krämer, K. H. (2021). Democracy and constitutional crisis in Nepal. *Nepal Observer*, Issue 71. https://nepalobserver.de/archive/0071.pdf

Kurian, R., & Singh, D. (2020). Politics of caste-based exclusion in rural India. In A. D. Cimadamore, F. Kiwan, & P. M. Monreal Gonzalez (Eds.), *The politics of social inclusion: Bridging knowledge and policies towards social change* (pp. 283–312). UNESCO.

Lawoti, M. (2010). Evolution and growth of the maoist insurgency in Nepal. In M. Lawoti & A. K. Pahari (Eds.), *The maoist insurgency in Nepal: Revolution in the twenty-first century* (pp. 3–30). Routledge.

Lawoti, M. (2019). Nepal in 2018: The communist majority and the emergence and decline of hope within a year. *Asian Survey, 59*(1), 133–139. https://doi.org/10.1525/as.2019.59.1.133

NDTV. (2018, April 2). Bharat Bandh protest by Dalit Groups: What you need to know. *NDTV*. https://www.ndtv.com/india-news/bharat-bandh-on-dilution-of-sc-st-act-heres-all-you-need-to-know-1831666

NIWF – National Indigenous Women's Federation; NIWF – National Indigenous Women Forum; NIDWAN – National Indigenous Disabled Women Association Nepal; INWOLAG – Indigenous Women's Legal Awareness Group, Situation of the Rights of Indigenous Women in Nepal; Shadow Report for the Sixth Periodic Report of Nepal CEDAW/C/NPL/6. (2018). https://www.ecoi.net/en/file/local/1451750/1930_1542809703_int-cedaw-css-npl-32567-e.doc

Piketty, T. (2020). *Capital and ideology.* Harvard University Press.

Pokharel, B., & Pradhan, M. S. (2020). *State of inclusive governance: A study on participation and representation after federalization in Nepal.* Tribhuvan University.

Rights and Resources Initiative. (2023, March 1). Indigenous community in Nepal wins recognition for customary practice to protect nature and biodiversity. *The Land Writes Blog* (blog). https://rightsandresources.org/blog/indigenous-community-in-nepal-wins-recognition-for-customary-practice-to-protect-nature-and-biodiversity/

Roy, A. (2014). *The Doctor and the Saint: The Ambedkar-Gandhi Debate: Caste, race, and annihilation of caste.* Verso.

Roy, A. (2019, July 9). Tribespeople in India's Gujarat fiercely resisted a mega dam—but they got a mega statue, too. *Quartz*. https://qz.com/india/1661257/arundhati-roy-reflects-on-narmada-bachao-andolan-statue-of-unity

Saba, A. (2018). *Identifying the challenges of addressing diversity in Nepal through federalism?*. MA Thesis, University of Sheffield.

Saba, A., & Hewitt, S. (2019, March 26). The diversity of Nepal's Women's movement must be recognized. *The Record*. https://www.recordnepal.com/the-diversity-of-nepals-womens-movement-must-be-recognized

Saba, A., & Koehler, G. (2022). *Towards an eco-social contract in Nepal: The role of rights-based civil society activism*. Issue Brief, No. 13. UNRISD.

Saba, A., & Kohler, G. (2023). *Das Neue Nepal: Naya Nepal*. Margit Horváth Stiftung. Accessed November 29, from https://www.margit-horvath.de/das-neue-nepal-naya-nepal/

Saldanha, L. F. (2023, August 5). Modi Government's new environmental laws a threat to India's biodiversity and forests. *Frontline*. https://frontline.thehindu.com/environment/primed-for-plunder-modi-government-new-environmental-laws-biological-diversity-act-forest-conservation-act-a-threat-to-india-biodiversity-and-forests/article67158366.ece

Sen, A. (2021). *Home in the world: A memoir*. Allen Lane.

Sethi, N. (2019, February 21). SC orders forced eviction of more than 1 million tribals, forest-dwellers. *Business Standard*. https://www.business-standard.com/article/current-affairs/sc-orders-forced-eviction-of-more-than-1-million-tribals-forest-dwellers-119022000855_1.html

Sharma, G. (2015, January 7). Nepal to issue passports with third gender for sexual minorities. *Reuters*. https://www.reuters.com/article/idUSKBN0KG0QS/#:~:text=%22We%20have%20changed%20the%20passport,told%20the%20Thomson%20Reuters%20Foundation

Sharma, N. (2021, June 15). Pinjara Tod activists Devangana Kalita, Natasha Narwal granted bail, Delhi HC hails right to protest. *India Today*. www.indiatoday.in/law/story/pinjra-tod-activists-devangana-kalita-natasha-narwal-granted-bail-1814937-2021-06-15

Shiva, V. (1988). *Staying alive women, ecology and survival in India*. Indraprastha Press.

Sijapati, B. (2009). *People's participation in conflict transformation: A case study of Jana Andolan II in Nepal*. Occasional Paper, Peace Building Series No. 1. Future Generations University.

Singaravelu, N. (2020, January 1). Data: Where are detention centres in India? *The Hindu*. www.thehindu.com/data/data-where-are-detention-centres-in-india/article30451564.ece

Siwakoti, S. (2024). *Pathways to a new eco-social contract in Nepal: Judicial interpretations of constitutional guarantees and their implementation*. Working paper 2024–04. United Nations Research Institute for Social Development.

Tamang, S. (2014). Nepal's transition and the weak and limiting public debates on rights. In B. Karki & R. Edrisinha (Eds.), *The Federalism Debate in Nepal: Post peace agreement constitution making in Nepal* (Vol. II, pp. 23–34). UNDP.

Thorat, S., & Newman, K. S. (Eds.). (2010). *Blocked by caste: Economic discrimination and social exclusion in modern India*. Oxford University Press.

UNRISD (United Nations Research Institute for Social Development). (2016). *Policy innovations for transformative change: Implementing the 2030 agenda for sustainable development*. UNRISD.

UNRISD (United Nations Research Institute for Social Development). (2022). *Crises of inequality: Shifting power for a new eco-social contract*. UNRISD.

Wada Na Todo Abhiyan. (2020). Snapshot of socially excluded and vulnerable communities vis-a-vis sustainable development goals in India. In *The 100 hotspots* (Vol. 2). Wada Na Todo Abhiyan.

Wada Na Todo Abhiyan. (2023). *Promise & reality: Citizen review of 4+ years of NDA II Government 2019–2023*. Wada Na Todo Abhiyan.

21

Varieties of Eco-Social Contracts in Japanese Ecovillages and Coliving-Coworking Arrangements

Yogi Hale Hendlin, Yuho Hisayama, and Kazuhiko Ota

21.1 Introduction

Ecovillages are a special case of eco-social work as they combine both coliving and coworking towards ecological ends. In some ecovillages, inhabitants farm together or run other businesses jointly; in others, they live together in certain eco-social arrangements but their income comes from work unrelated to their ecological coliving (e.g., computer coders working remotely subsidising the expenses of those engaged in the operation and maintenance of the ecovillage) (Litfin, 2013). The question of whether and to what degree this activity qualifies as economic activity is subject to the framing of what is necessary to meet needs, and how much of this can be sourced regeneratively through work involving commons, rather than built on a logic of accumulation and

Y. H. Hendlin (✉)
Erasmus University Rotterdam, Rotterdam, Netherlands
e-mail: hendlin@esphil.eur.nl

Y. Hisayama
Kobe University, Kobe, Japan
e-mail: yuhohisayama@pegasus.kobe-u.ac.jp

K. Ota
Nanzan University, Nagoya, Japan
e-mail: otakazu@nanzan-u.ac.jp

© The Author(s) 2025
P. Huntjens et al. (eds.), *Eco-Social Contracts for Sustainable and Just Futures*,
https://doi.org/10.1007/978-3-031-99109-7_21

subjugation to market dynamics. Working together and living together is understood here as on-the-job (and in-the-field and living room) training for best practices in eco-social work.

Ecovillage dynamics rely on attempting to displace as much as possible activities, services, and the production of goods from a money economy, replacing them with both formal and informal obligations of reciprocity. This concurs with how the editors of this volume define eco-social contracts as "implicit or explicit collective agreements across multiple levels of governance, among members of society, aimed at addressing the interconnected polycrisis of the twenty-first century, including inequalities, injustices, climate and ecological breakdown, and faltering trust in institutions" (Huntjens & Kemp, 2025). Ecovillages aim to reimagine and re-materialise the agreements made between residents with each other and the living earth, congruous with how eco-social "agreements [...] rooted in cooperation and the recognition of shared norms and values oriented toward sustainability, equity, and justice" rearticulate "duties of care" (Huntjens & Kemp, 2025; see Chap. 16).

Balancing and attending to needs, rather than amplifying and exacerbating the insecurities and overconsumption of some at the expense of others, fundamentally posits a concept of sufficiency beyond which distortions accumulate (Princen, 2005). Ecologically-oriented coliving and coworking set-ups vary widely in their ability to deliver on this horizon of possibility, but they share the necessity for renegotiating the eco-social contract away from domination and towards partnership—with both fellow humans and the more-than-human world; an asymptote towards which they can near, an ideal to which one can approach but never finally arrive. The success of such experiments depends on the capacity for sustained collective effort, mature consensual conflict resolution protocols, resource mobilisation, and institutional support (whether from existing governmental and capital-accumulated philanthropy, or through *sui generis* commons-created networks), along with the efficacy of achieving temporary autonomous zones of experimentation which preclude corporate or governmental capture (Ostrom, 1990; Gibson-Graham, 2006; Bey, 1991). Finding more joy in relating and encountering directly our food and immediate ecology, and finding camaraderie in neighbours which one can rely on, grieve with, and celebrate together, the ecovillage concept is premised on the notion that most of our unfulfilled needs under the current dominant, atomised and alienated paradigm can be alleviated through rearranging how we live and work together in reference to our local ecology and community (Mychajluk, 2024).

Coliving and coworking in Japan have a long history. The *satoyama* (mountain farming) and *satoumi* (sea village) models provide a historical basis, severed during modernisation, but revived in the postwar period as Japanese sought alternatives to the pressures of individualism, secularism, and consumerism. Our research examines how ecovillages in Japan draw on *satoyama* and *satoumi* principles while navigating the tensions between market forces, ecological ethics, and communal resilience. Some ecovillages align closely with deep ecological commitments, striving for off-grid self-sufficiency and commune-like sharing, while others operate within more mainstream economic and infrastructural systems, incorporating sustainability in a more incremental or symbolic fashion. Similarly, some communities are explicitly spiritual, drawing on Buddhist, Shinto, or animistic worldviews, while others approach sustainability through secular and pragmatic lenses.

Surveying this wide spectrum of ecovillages as eco-social contracts in action, our pilot study highlights the pluralistic ways in which ecological coliving and coworking manifest in Japan. This pilot study employed ethnographic research, interviews, site visits, and news and literature reviews, as well as historical comparisons to generate a sketch of the complexities of ecovillages in the Japanese context. Our analysis directed us particularly to the role of precipitating events, such as the March 11, 2011, Great East Japan Earthquake, in reshaping attitudes towards alternative living arrangements. Through this inquiry, we investigate how ecovillages serve as sites of negotiation, adaptation, and resistance, revealing the broader tensions between tradition and innovation, individual autonomy and collective belonging, and compliance with or departure from dominant sociopolitical structures.

21.2 Ecovillages as Eco-Social Contracts

Ecovillages can be thought of as a post-industrial response to the metacrisis of social and ecological issues facing us. According to the Global Ecovillage Network, an "ecovillage is an intentional, traditional or urban community using local participatory processes to integrate ecological, economic, social, and cultural dimensions of sustainability in order to regenerate social and natural environments" (Global Ecovillage Network, 2025). Ecovillage movements exist around the world, as a loose collection of many thousands of "experiments in living," as John Stuart Mill called them (Anderson, 1991). In the Global North, ecovillages tend to be smaller experimental intentional communities, while in the Global South they are often larger networks of communities preserving local traditions in the face of industrial globalisation (Dawson, 2013). In Japan, both models are present, to a degree. Environmental coliving and coworking projects in Japan take on aspects of either of these templates, as well as develop their own regional models, depending on the motivations, histories, and personalities of each group. By widening the types of organisations engaged in forms of ecological coliving and coworking that only loosely fall into the category "ecovillage," a better understanding of the various forms of their eco-social potential can be obtained.

The Global Ecovillage Network (2025) claims that 97% of the ecovillages they surveyed work actively to restore degraded ecosystems, 90% deliberately sequester carbon through soil and biomass management, and 97% engaged in efforts to replenish natural water systems (Global Ecovillage Network, 2025). Such triumphant claims, however, should be taken with a grain of salt: the levels of restoration, sequestration, or replenishment, are not given; and the data is based on self-reports. While actual measurement is much more difficult, more quantitative studies using lifecycle assessments (LCA) have shown that "Ecovillages achieve large (47–80%) reductions compared to the average" citizen's per capita greenhouse gas emissions (Sherry, 2019).

Quantitative reductions in pollution, waste, and impact are important to track; otherwise, what is the "eco" in ecovillage unless there is some rebalancing between the needs of the natural world and those of humans? Merely having goodwill without it translating into ecological outcomes takes us too easily down the path of "beautiful soul syndrome" (*die schöne Seele*) where thoughts and prayers severed from integrated action in the world serve as avoidance mechanisms (Duvernoy, 2022). At the same time, ecological harmonisation requires a certain amount of reskilling in a world deskilled by assembly line work, Taylorism, Fordism, and the fragmented labour globalisation has brought us. This can often be a long process, rather than an overnight transformation. Thus, multiple ports of entry are necessary to meet people where they are at, and help relax layers of trauma and resistance inhibiting them from eco-social integration. Likewise, multiple pathways of reskilling are also necessary to help us make the transition from total dependency on far-flung commodity chains to self-reliance in community and ecology.

Ecovillages, as efforts to reestablish these eco-social links, exist along several distinct dimensions. These spectra include (a) secular-religious (or spiritual), (b) token versus thoroughgoing ecological commitments, (c) individualist versus collectivist, and (d) default system compliant versus off-the-grid and experimental (Fig. 21.1). While other axes, such as (e) socially diverse versus homogeneous, and (f) economically ambitious versus minimalist/ascetic, or rich capitalist versus communist subsistence could also be introduced, in Japan, those represented in Fig. 21.1 depict those distinctions most relevant for the contemporary Japanese case. Partly, this is due to the overall homogeneity of the Japanese population (98% of the population is ethnically Japanese)

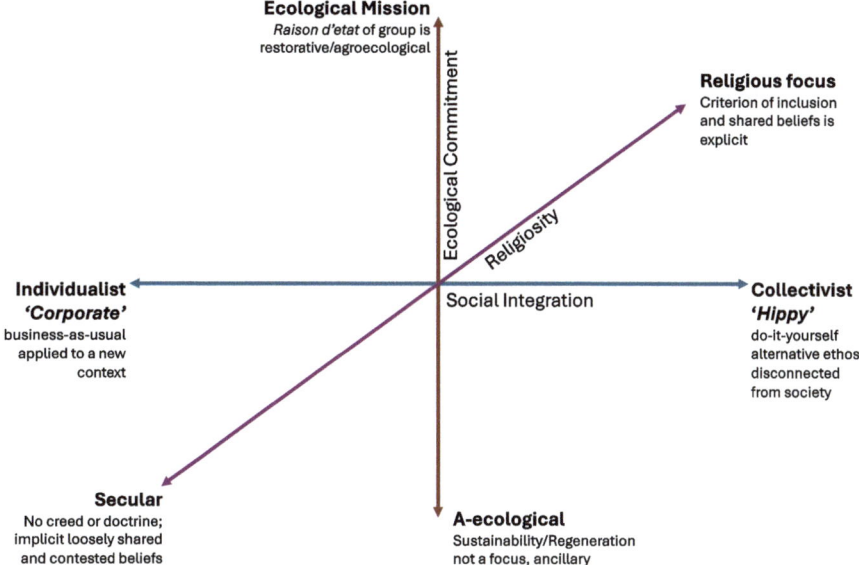

Fig. 21.1   Ecological, religious, and collectivist dimensions of coliving and coworking. Source: Authors

and similarities economically among most ecovillages (with their still outsider alternative status not yet creating gentrifying versions). Diversity along these axes is wide-ranging, and Japanese ecovillages cannot be subsumed into a single category of representation.

Our research examining ecovillages in Japan ranges from corporate head-quarters relocating to beautiful rural settings during the pandemic with 1200 of their employees to provoke a corporate re-ruralisation movement under putative sustainability auspices, to permaculture-inspired traditional *satoyama* farming communities. How "eco" and how "village" each coliving and cowork-ing arrangement is (and how enmeshed these two aspects are) vary widely. The question of ecological commitment and outcomes both in the short term and long term, as well as how this intersects with social welfare improvements and their degree of decision-making decentralisation and communing, is compel-ling for eco-social research, as it is agnostic to the impetus or motivation: whether people form and join ecovillages due to ecological ethics, personal desires for happiness, or the need for belonging. Our preliminary results show that social and ecological commitments track together, suggesting that natural social contracts form mutually reinforcing relationships.

21.3 Contemporary Ecovillages in Japan

There are few commonalities among ecovillages besides their desire to solve some of the most pressing issues surrounding ecological and social sustainability. Some ecovillages, such as As-One, in the city of Suzuka, emphasise social non-domination as their primary virtue. Instead of telling people to do things, or *requiring* people to work, they instead aim to cultivate a self-driven sensibility where members simply recognise things that need to be done, and do them if they have the capacity and desire to do so (Ogawa, 2025). This cultivation of self-motivation is similar to that in any modern household: how to get all household members to take out the trash when they see it needs emptying, do the dishes when they start to pile up, or clean the floor when it gets dirty—without a chore wheel or nagging? While filial relations can sometime solve this through clear communication, in larger communities, when an anonymising aspect enters, the question of "it's not my responsibility" often arises, as we have been conditioned in modernity to not do more than our share without commensurate compensation, and normally are nearly paranoid about the "overcontribution penalty" or "altruistic punishment" of those who have the wherewithal and capacity to do more than their share. Rather than punishing or enforcing members to maintain order, the As-One community aims to spark the intrinsic rewards of service through not mandating work but letting people slowly learn by example, a type of virtue ethics which incidentally, provides the opportunity for self-discovery of the joys of service, bringing people out of their funk of individualism and victimhood which results in anxiety, depression, and overall mood disorders. Since Frank Riessman's (1965) landmark article on helper theory in the journal *Social Work*, research has mounted showing that providing assistance to others and to the community can form the bedrock of re-instilling meaning and purpose in the life of the helper. Combined with recent insights which show a positive correlation between individuals' self-perception and their engagement in pro-environmental behaviours (Capstick et al., 2022), helping others may be the key to unlocking positive self-worth and ecological stewardship.

Many ecovillages in Japan are explicitly anti-capitalist in their rejection of using money internally to compensate for work and their embrace of material minimalism. The modern economic/material minimalism movement has roots in multiple places, but Japanese sources have played a significant role in popularising it, particularly through cultural and philosophical influences like Zen Buddhism, wabi-sabi, and Ma (the concept of space and emptiness). Even outside of ecovillages, minimalism is very popular in Japan, as glorified in Marie Kondo's KonMari method and philosophy of decluttering clothes closets and home items that became internationally famous. As economic pressures have grown in Japan's working class, as well as the cultural shift where the salarymen no longer have a guaranteed life path of working hard and being able to secure long-term stable employment, as women now also compete in the job market, and as Japan's economic growth has stagnated for three decades, the life model previous generations could depend on no longer applies. Economic precarity and cost-of-living crises have acted as drivers for Japanese to pursue alternative models multisolving for various needs. Thus, material/economic minimalism and alternative social recognition structures—both, which are routinely found in ecovillages offer normal people the opportunity to sidestep the social judgement, economic insecurity, and anxious alienation they might otherwise face if they hadn't sloughed off the hegemonic model of social success.

For this reason, early ecovillage movements such as the Yamagishi commune have strong ascetic qualities, with social norms including hard work, communal meals, and decision-making, but no financial compensation for work rendered. The Yamagishi commune also, despite their low consumerism, do not actively adopt an explicitly environmental or sustainability ethos. While their shift in orientation from endless dopamine addiction from consumerism to intrinsic dignity and mutual recognition as belonging to a larger group dedicated towards a common good does of course lead to substantial sustainability gains, the instantiations are uneven, as this is not the principle value selected for. Instead, their anti-consumerist customs are based on their notion of what actually makes humans happy—which includes consensus decision-making and mutual recognition for contributions (Spiri, 2008).

A major motivation for ecovillages is to revive meaning in life. And generally, there is the notion, in Japan and worldwide, that in order to have a purpose, and rediscover the wonder of the world, reestablishing our connecting to the natural world is crucial (Sherry, 2019; Litfin, 2013). Just as we have seen the degradation of our natural environments through ecocide, we have also been dislocated from the sustaining communities of meaning with both other humans and the more-than-human world through semiocide—the destruction of our meaningful relations (Hendlin, 2024). Ecovillages serve as attractors, spaces for experimentation not just in terms of social arrangements or ecological engagement, but also for experimenting with how to find purpose in an alienated world.

Minori Ogawa, founder of Dana Village in Fukushima Prefecture, became galvanised to start an ecovillage after the Great East Japan Earthquake, also known as the Tohoku Earthquake, which caused the Fukushima Daiichi nuclear disaster on March 11, 2011. Indeed, given the historical agony of the nuclear bombing of Hiroshima and Nagasaki, after the Fukushima Daiichi nuclear disaster, many Japanese became heavily involved in the anti-nuclear movement, including looking for ways to live outside of an economy and society that could create the circumstances for such a predictable accident, as a way of personally processing the trauma incurred ecologically and socially (Perrow, 1999; Park et al., 2015). In response, Ogawa decided to found an ecovillage in Fukushima prefecture, to demonstrate that contrary to perceptions in the news media, that this (very large) prefecture is not terminally contaminated. This act sought to counter the repeat of the treatment of *hibakusha* (those radiation-affected survivors of the nuclear bombing during WWII), who were seen as 'tainted' and discriminated against post-bombing. This act of peace and ecological restoration by founding Dana (Eco)Village was understood as a way to take personal action to create alternative lifestyles, narratives, and forge relations with each other and the planet, rather than contributing to the factors which lead to the disaster.

Ogawa is one of few female ecovillage founders. As a result, she estimates that 70–80% of visitors and incoming (temporary) residents are women (Ogawa, 2025), presumably attracted to not just the fact of a female founder (which is not overtly apparent in online materials), but through the presentation and description of the ecovillage in the public sphere as detailed by Ogawa. Dana Village, although renting the principal building and land (hence, making major upgrades to the poorly insulated architecture of the main building unfeasible), cultivates an immense amount of their food, between 90 and 95% of all food consumed, thanks to greenhouses, rice paddies, vegetable rows, and orchards—all on only a hectare of land. This small ecovillage (5–15 members at a time), reduces food dependency, gleans and shares food that elders in the village otherwise could not pick for themselves, and aims to make their own soy sauce from the soy beans they grow. Housed in an old schoolhouse, one of the many that have become vacant as Japan's population has decreased and the average resident age—especially in rural areas—has increased considerably, Dana Village has also built several tiny houses, and recently purchased a neighbouring piece of land for expansion of the ecovillage. In their mountainous location in Nozawa, they have, through family members and neighbourhood committees, begun to integrate with the larger community. Tensions between the vegan, breakfast abstaining, almost monastic life of the ecovillage and the typical postwar diets and expectations of the surrounding community are indicative of the perennial issue for ecovillages in Japan and elsewhere of how to bridge the building of an alternative world in the confines of a given location, while adapting and integrating with the surrounding globalised habits of the local population.

As ecovillages adopt experimental modes of living which challenge traditional economic, political, and cultural organisational models, tensions with dominant social structures and those still adhering to them abound. But how these tensions are navigated indicate their level of success or failure. Ecovillages that are too rigid fail, as do those lacking a focus on values and virtues with which inhabitants feel identification and agency. Reorienting towards commons while in the midst of a capitalist system also presents challenges. Finding forms of coworking and coliving that benefit the commons and are socially as well as ecologically sustainable takes resources, and although many resources can be found outside the money economy, the need for capital and funding for the most part remains. As the end goal of many ecovillages includes material self-sufficiency through reskilling, the path towards approaching this requires compromises and negotiation. As long as viable economic options within the default system continue to reduce, however, out of necessity as well as idealism, alternative modes of coliving and coworking—together in the form of ecovillages—will likely become increasingly attractive, in Japan and elsewhere.

21.4 *Satoyama* and *Satoumi* Traditional Models

The *satoyama* (mountain farming) and *satoumi* (sea village) models of ecological sustainability are well-known in Japan, as they are historical traditions that provided almost complete self-sufficiency for the families who would have their farms, rice fields, bamboo groves, and forests (or fishing and seaweed harvesting for *satoumi*) mostly sustainably managed for generations. These models of low-impact living, often consisting of clusters of families living together simply but providing for all of their material needs autonomously, are understandably praised. Traditional mountain and sea village life, however, is often idealised in Japan, a phenomenon that has received critical scrutiny in the past decades (Matsumura, 2012). Especially with regard to the social limits of these proto-ecovillages, patriarchy and strict gender and role hierarchies have begun to be questioned as viable models (without modification) for current eco-social coworking and coliving arrangements (Matsumura, 2012). Nonetheless, the traditional (Buddhist) virtue of caretaking of mountaintops and seasides, which has led to preserving the mountaintops in Japan, carries elements of ecological coliving and coworking worth propagating reflexively.

The Japanese word *satoyama*, found in pre-modern literature, was not widely used as a term until the second half of the twentieth century. It subsequently became a keyword in nature conservation activities and was sometimes associated with anti-modern utopian ideas. The more recent word *satoumi* denotes the sea version of *satoyama*. These terms can be interpreted to mean either "mountain" (*yama*) or "sea" (*umi*) in the "neighbourhood" (*sato*) of humans, and are generally understood to emphasise mutualistic relationships between nature and humans. In this sense, *satoyama* are made use of by humans; however, research has demonstrated that *satoyama* and *satoumi* often provide habitat for more abundant biota than even primary forests and wild meadows due their careful curation of different species which provide ecological niches for various plants and animals (Kankyosho, 2001, 2021).

Satoyama arrangements have repeatedly been subject to threats during modernisation. Since the 1960s, on one hand, the shift from firewood and charcoal for heating and cooking to fossil fuels, the introduction of chemical fertilisers, and the decline of livestock for domestic use resulted in the economic decline of *satoyama* forests. Devalued under a single-metric economic system that no longer valued the ecosystem services of these cared-for areas, these traditionally protected forests were instead in many instances developed for urban and housing expansion, including golf course construction. A notable example of this transformation is the Tama New Town development in Tokyo (Matsumura, 2012), where they decimated a *satoyama* area that provided the backdrop for the Studio Ghibli film *Pom Poko* (1994).

Since the 1970s, on the other hand, there has been a notable influx of the worldwide eco-movement into Japan, precipitating a marked increase in the legitimacy and renaissance of nature conservation activities. The concept of *satoyama*, first introduced in its modern form by the Japanese forest ecologist Tsunahide Shidei, in the context of nature conservation has become a staple of the conservation movement mixed with nostalgia and the tensions in modern Japanese society for preserving ancestral land-use practices and social relations, including artisanal activities and crafts (such as making baskets, cups, and other objects from bamboo, or traditional soy sauce cultivation and brewing methods). The novelty of Shidei's concept lies in its focus on the sociocultural relationship between nature and humans, between *satoyama* and the people living there, in contrast to the prevailing nature conservation theory of the time, which prioritised "pure" wild nature (Okada, 2017). The 1990s saw the establishment of the Satoyama Research Network, and the term "*satoyama*" gained widespread popularity through the efforts of Mitsuhiko Imamori who collected photographic artefacts on the subject, the Japanese Broadcasting Corporation (NHK) specials entitled Satoyama Stories, and the 2005 Achi

Expo themed "Nature's Wisdom." Around that time, the Japanese Ministry of the Environment started promoting the "Satoyama Initiative" to preserve forest areas to benefit both biodiversity and human livelihoods. This initiative asserts that *satoyama*-like landscapes are not exclusive to Japan, but are also present all over the world. In this regard, the initiative underscores the significance of the Asian perspective that nature and human beings are inextricably linked.

It is important to acknowledge that the concept of *satoyama* is historically dynamic, used sometimes arbitrarily, and has shifted over time. Actual *satoyama* landscapes have been shaped by various social, historical, and cultural contexts, and cannot be perceived as a static, non-temporal utopia. Research has demonstrated that *satoyama* have actually undergone continuous transformation in response to historical and social shifts; however, there is a pervasive tendency to depict *satoyama* in overly idealised forms within wider society, symbolising the utopian coexistence of humans and nature. This tendency must be critically evaluated, as it is frequently informed by a nostalgic eco-nationalist perspective and is susceptible to the influence of various political actors (Matsumura, 2022). Maintaining a nuanced and flexible understanding of *satoyama* allows resisting the tendencies towards historical revisionism while avoiding any association with nationalism or auto-orientalism. Refining our perceptions of the concept to align with the evolving nature of how it is operationalised can allow this guiding principle to not smother experimentation (such as the permaculture of Masanobu Fukuoka), but offer an ideal recognised as such.

21.5 Discussion and Conclusion

The diversity of ecovillages in Japan generate a number of conceptual as well as practical challenges. On one hand, there are large-scale developments led by corporations or local governments—such as Panasonic's Fujisawa Sustainable Smart Town (2014–)—in which state-of-the-art technologies and capital are deployed to manage energy and water resources efficiently while actively reducing CO_2 emissions and enhancing disaster preparedness (Sakamoto, 2021). On the other hand, there exist small-scale communities based on traditional *satoyama* wisdom and permaculture principles—such as the As-One Community Suzuka and the Sankaku Ecovillage "Saihate"—that emphasise harmonious coexistence with nature and the sharing of an internal vision, combining organic agriculture and off-grid technologies to achieve higher levels of self-sufficiency. These smaller models are often operated from the bottom-up through deliberative decision-making by the members, fostering strong communal bonds and flexible lifestyles (Katayama, 2016; Nakabayashi, 2016; Hirayama, 2020).

One interesting example of a for-profit company taking-up the mantle of ecovillages and their eco-social claims is the major Japanese human resources company, Pasona, which transferred the vast majority of its employees (1200 of 1800 total) including management away from Tokyo to Awaji island, starting in 2020 during the pandemic and completed in May 2024 (Pasona, 2020). Their reasoning for the move includes environmental sustainability, but also higher quality of life for their employees to enjoy. As a for-profit company, their insistence on branding their company's move as more than just adding another company campus, rather with the specific ambition of establishing an eco-social contract is a hybrid case. At the same time, Awaji Island (Awajishima) has a population of 129,000 people, so Pasona's arrival represents an increase of the island's population by 1%, but their alteration of the island culture is many times magnified due to their outsized economic power. Despite their professed eco-social intentions, since moving their headquarters there, the price for rental housing increased 20–30% on the island, making it difficult for residents to afford housing (Kozo, 2022). This outsized power of the company has generated elements of gentrification—an issue which may be more widespread with improving the social and ecological quality of life for business employees, when it is not integrated with the already pre-existing local ecology and communities. The overall environmental impact from Pasona is not yet known (and will be part of our future study). Taking into account the full range of ecological and social impacts of the move is important to get a better idea of how top-down corporate models of sustainability coliving and coworking work, and how they differ from other models, such as ecovillages which usually do not have profit motives and often are structured as eco-social entrepreneurships, cooperatives, or B-corporations. Pasona is a limit case for what could possibly be included under the rubric of an ecovillage, not necessarily falling neatly inside the category. Nonetheless, it serves as an interesting example of the tensions between ecological coliving and coworking aspirations and capitalist system embeddedness.

The technological approaches and philosophy about infrastructure differ widely between corporate-led and *satoyama*-based models. Regarding energy self-sufficiency, many ecovillages employ local resources such as solar power, wood-burning stoves, and wood pellets. In contrast, large-scale smart towns establish advanced infrastructures like battery storage systems and Internet of Things (IoT)-controlled smart grids. Some new, more "solarpunk" ecovillages aim to leverage blockchain technologies, such as decentralised autonomous organisations (DAOs). The non-profit initiative AkiyaDAO, for example, raises funds via blockchain to renovate abandoned houses known as *akiya*, repurposing them into creative residencies and interactive habitats for community gathering. In other words, corporate-led projects offer a sophisticated environmental urban model on a large scale, while traditional *satoyama*-based ecovillages provide grassroots alternatives which reimagine societal systems. Both aim to achieve a sustainable way of life, and highlight the challenges inherent in existing systems.

In recent years, not only successful cases but also instances of dissolution have been reported. The closure of Saihate Village (2011–2023) in Kumamoto after approximately 12 years can be attributed to difficulties in governance that emerged between freedom and effective management. One of the founding members of Saihate summed up the situation: the political context of anti-nuclear power and other issues that arose in the wake of the 2011 Tohoku earthquake eventually faded away, and instead, multi-level marketing and conspiracy theories began to take hold, creating a climate in which the community was prone to becoming overly spiritualistic. Through conflicts claiming that the "ecovillage" was nothing more than a name, and that the emphasis was on inner spirituality rather than on environmental and social issues, eventually this artificial separation between inner peace and actively working to achieve outer peace and sustainability, lost its appeal (Djpolypical, 2024). In contrast, communities like As-One Community Suzuka (2011-present) in Mie, and the Yoichi Ecovillage (2012-present) in Hokkaido, have continued their activities by actively engaging with external partners, incorporating non-profit organisational structures, and cooperating with local communities to secure a stable financial foundation (Kurushima, 2019; Hirayama, 2020). Psychological, social, and ecological isometry seem crucial to creating and maintaining successful ecovillages.

In Japan, although support policies and legal frameworks specifically labelled for "ecovillages" are not yet fully established, related initiatives—such as the Sustainable Development Goals (SDGs) Future Cities programme and the Organic Village Creation project—have been increasingly utilised to make this general direction of ecological and social focus more legible to the state (Cabinet Office, 2019; Ministry of Agriculture, Forestry and Fisheries, 2023). The more ecovillages are formally recognised, however, the more it is expected that there will be difficulties in trying to make them consistent with the Agricultural Land Act, the Building Standards Act, and other regulations, leading to the probability of creating a special zoning system to account for their unique structure and land use characteristics. Thus, there is a trade-off between formalised recognition of ecovillages as an institutionalised category: the more they are institutionally recognised, the less experimentation is possible, and the more conformity to pre-existing categories they will be subjected to.

Ecovillages present a unique case study for understanding experiments in creating various new eco-social contracts. In Japan, the full array of ecovillage configurations currently exist. Without creating hierarchies of more versus less prescriptive models, we can nonetheless distinguish between different aspects of ecovillages, serving different constituencies and their needs. While an upscale ecovillage based on luxury ecotourism may be able to garner more resources for further ecological work, the additional carbon spent may balance this out. The ascetic rice farm ecovillage may use fewer exogenous resources, but require high amounts of gas and wood for heating in the winter due to poor insulation. Ecovillages with work required from all members and visitors according to a strict schedule may repel certain people, while those that emphasise more voluntary labour for non-core members may be able to attract a larger swath of the public romanticising back-to-land movements. These pluralistic models exist because there are different needs and capacities for different demographics and psychographics. The commonality is that they create liminal spaces where pre-existing economic and environmental relations are put aside to certain degrees, providing a common space for exploring other ways of relating with self, other, and nature.

This chapter provided a brief overview of the various forms of eco-social contracts involved with the aspiration of ecovillages and coworking and coliving models in Japan, from large companies to small DIY (do-it-yourself) ecovillages, focusing on the *satoyama* history of eco-social work and contracts which undergird the ecovillage imaginary in Japan. There's also the recognition the precipitating events, such as the March 11, 2011, Great East Japan Earthquake, can also galvanise latent misgiving about dominant social contracts, and spur action to rethink and reorganise eco-social defaults through the creation of alternative structures (ecovillages, in this case). Thus, the dual affordances of drawing on traditional but depreciated modes of relating, such as *satoyama* eco-social relating, combined with calamities jolting people out of the status quo, can provide the impetus for action and the blueprint for how to experiment. By themselves, traditional models and shocks are not enough to produce eco-socially favourable outcomes (Huntjens, 2021; Klein, 2008; Ito & Sugiura, 2021). Together, however, inherited traditions offer scaffolding for emergent practices, and on this basis, crises catalyse the reconfiguration of social and ecological relations in ways empowering care relations. In Japan, norm creation from the bottom-up, based on spiritual connection with nature and community (*inochi*) does not easily map on to Sustainable Development

Goals or other institutionally legible outside-in ecological metrics. Nonetheless, by prioritising communal well-being over individual competition, and developing practices to feed evolutionary needs of recognition and belonging without exacerbating the need for strict hierarchy, Japanese ecovillage principles demonstrate approaches towards creating partnership rather than dominating societies (Eisler & Fry, 2019). Thus, the various ecovillage models in Japan reflect both a reactivation of latent eco-social imaginaries and a response to socio-environmental contingencies, suggesting how agency is distributed across historical, material, and relational dimensions.

References

Anderson, E. S. (1991). John Stuart Mill and experiments in living. *Ethics, 102*(1), 4–26. https://doi.org/10.1086/293367

Bey, H. (1991). *T.A.Z. the temporary autonomous zone, ontological anarchy, poetic terrorism*. Writing in Book edition. Autonomedia.

Cabinet Office, Regional Revitalisation Promotion Division. (2019). *Case studies of SDGs future cities and municipal SDGs model projects*. https://www.chisou.go.jp/tiiki/kankyo/

Capstick, S., Nash, N., Whitmarsh, L., Poortinga, W., Haggar, P., & Brügger, A. (2022). The connection between subjective wellbeing and pro-environmental behaviour: Individual and cross-national characteristics in a seven-country study. *Environmental Science & Policy, 133*(July), 63–73. https://doi.org/10.1016/j.envsci.2022.02.025

Dawson, J. (2013). From Islands to networks: The history and future of the ecovillage movement. In J. Lockyer & J. R. Veteto (Eds.), *Environmental anthropology engaging ecotopia: Bioregionalism, permaculture, and ecovillages* (pp. 217–234). Berghahn Books. https://doi.org/10.1515/9780857458803-018

Djpolypical. (2024). *On the conclusion of Saikate: Our failures, let's talk about survival strategies, shall we? note*. https://note.com/djpolypical/n/n9a06ded3e81f

Duvernoy, R. J. (2022). The 'beautiful soul' and 'religious consciousness': Deleuze and Nishida. *Comparative and Continental Philosophy, 14*(1), 30–43. https://doi.org/10.1080/17570638.2022.2098560

Eisler, R., & Fry, D. P. (2019). *Nurturing our humanity: How domination and partnership shape our brains, lives, and future*. Oxford University Press.

Gibson-Graham, J. K. (2006). *A postcapitalist politics*. University of Minnesota Press.

Global Ecovillage Network. (2025). *Global ecovillage network*. February 5, 2025. https://ecovillage.org/

Hendlin, Y. H. (2024). Semiocide as negation: Review of Michael Marder's Dump philosophy. *Biosemiotics, 17*(1), 233–255. https://doi.org/10.1007/s12304-024-09558-x

Hirayama, M. (2020). Case study of eco-villages (final report). *Meiji Gakuin University Institute for International Studies Annual Report, 23*, 59–78.

Huntjens, P. (2021). *Towards a natural social contract: Transformative social-ecological innovation for a sustainable, healthy and just society.* Springer Nature. https://www.springer.com/gp/book/9783030671297

Huntjens, P., & Kemp, R. (2025). The transformation flower approach for eco-social contracting: Comparative insights from eight case studies in the Global South and North. In P. Huntjens, N. Mohamed, K. Hujo, & M. Desai (Eds.), *Eco-social contracts for sustainable and just futures* (Chapter 16, pp. 283–312). Springer Nature.

Ito, T., & Sugiura, M. (2021). Satoyama landscapes as ecological mosaics of biodiversity: Local knowledge, environmental education, and the future of Japan's rural areas. *Environment: Science and Policy for Sustainable Development, 63*(5), 14–25. https://doi.org/10.1080/00139157.2021.1953911

Kankyosho. (2001). 環境省「日本の里地里山の調査・分析について(中間報告」 https://www.env.go.jp/nature/satoyama/chukan.html

Kankyosho. (2021). 環境省自然環境局生物多様性センター「生物多様性レポート2021」 https://www.biodic.go.jp/moni1000/findings/reports/pdf/2021_satoyama.pdf

Katayama, H. (2016). Urban, open-type eco-village: The As-One Suzuka community's attempts. *Bio-City, 68*, 72–81.

Klein, N. (2008). *The shock doctrine: The rise of disaster capitalism* (1st ed.). Picador.

Kozo, K. (2022). "パソナの淡路島移転から2年 見えてきた成果と宿題." 産経新聞:産経ニュース. 地方・西. October 29, 2022. https://www.sankei.com/article/20221029-6BYYR4MBXNN7RJYDKVKCZ3SE6A/.

Kurushima, R. (2019). A place to live peacefully and "Yuruyuru": A visit to a "waste material ecovillage". *Colocal* (pp. 171–179, 197–204). https://colocal.jp/topics/lifestyle/ecovillage/20190508_124752.html

Litfin, K. T. (2013). *Ecovillages: Lessons for sustainable community* (1st ed.). Polity.

Matsumura, M. (2012). 松村正治「多様な人びとと多様な里山、その多様な関係性」『多摩ニュータウン研究』 = Studies on Tama newtown』 14 (2012), 8–16.

Matsumura, M. (2022). The Satoyama movement and its adaptability: Beyond ideology and institutionalisation. In M. Taisuke & F. Mayumi (eds.), *Adaptive participatory environmental governance in Japan: Local experiences, global lessons* (pp. 67–92). Springer.

Ministry of Agriculture, Forestry and Fisheries. (2023). *Towards the creation of organic villages.* https://www.maff.go.jp/j/seisan/kankyo/yuuki/organic_village.html

Mychajluk, L. (2024). Learning sustainability through enterprise work in ecovillages. *Studies in Continuing Education, 46*(2), 246–265. https://doi.org/10.1080/0158037X.2023.2222072

Nakabayashi, Y. (2016). The community called an eco-village. *Design Studies Special Issue, 23*(4), 48–51.

Ogawa, M. (2025). Interview. By Yogi Hendlin. Dana Village ecovillage, Fukushima Prefecture. (14 March).

Okada, W. (2017). 岡田航「「里山」概念の誕生と変容過程の林業政策史」『林業経済研究』63–1 (2017), 58–68.

Ostrom, E. (1990). *Governing the commons: The evolution of institutions for collective action*. Cambridge University Press.

Park, D. J., Wang, W., & Pinto, J. (2015). Beyond disaster and risk: Post-Fukushima nuclear news in U.S. and German Press. *Communication, Culture & Critique*, July, n/a–n/a. https://doi.org/10.1111/cccr.12119

Pasona. (2020). 1200 Employees to Awaji by May 2024 towards 'A Truly Rich Lifestyle & Workstyle' Pasona Group Headquarters' decentralisation and relocation to Awaji Island begins | NEWS & TOPICS Detail. パソナグループ. 2020. http://www.pasonagroup.co.jp/english/news/tabid770.html?itemid=4031&dispmid=1331.

Perrow, C. (1999). *Normal accidents: Living with high-risk technologies* (2nd ed.). Princeton University Press.

Princen, T. (2005). *The logic of sufficiency*. MIT Press.

Riessman, F. (1965). The 'Helper' therapy principle. *Social Work, 10*(2), 27–32. https://doi.org/10.1093/sw/10.2.27

Sakamoto, M. (2021). Planning, design and practice of the intermediate zone in Fujisawa SST. *Ie to Machinami, 40*(2), 22–27.

Sherry, J. (2019). The impact of community sustainability: A life cycle assessment of three ecovillages. *Journal of Cleaner Production, 237*(November), 117830. https://doi.org/10.1016/j.jclepro.2019.117830

Spiri, J. (2008, February 1). Whatever happened to Yamagishi? Idealism, nature and the environment in Japan's cooperative Agrarian Community. *Asia-Pacific Journal: Japan Focus* (blog). https://apjjf.org/john-spiri/2666/article

22

The Economy We Want: Exploring the Potential of Participatory Mechanisms to Catalyse Economic Transformations

Najma Mohamed

22.1 Introduction

Transforming our economic systems is the most ambitious and urgent transformation the world has ever known. But a transition on this scale requires broad and deep social support. Support for rapid and widespread climate actions *hinges* on public understanding of policy (Bo et al., 2023), and the support of and demand for transition policies and processes that place the well-being of people and the planet above profit. If trust is the currency of the transition (Cameron-Smith, 2023), people should be at the heart of the processes where the terms of economic transitions are being negotiated. Economic transitions will not find the support they need to accelerate and endure unless it can give people a stake in delivering an economy in which they feel included, listened to and rewarded. Ultimately, we need eco-social contracts that both give people a say in the economic decisions being made in their name and drive just economic reforms.

Calls for a fairer, sustainable, safer world—and for the systemic changes that can help humanity achieve this, are no longer radical (Hopkins &

This chapter draws on a policy report (Mohamed, 2023). The author acknowledges the permission of the Green Economy Coalition to reproduce parts of the report in this chapter.

N. Mohamed (✉)
UN Environment Programme World Conservation Monitoring Centre,
UNEP-WCMC, Cambridge, UK
e-mail: najma.mohamed@unep-wcmc.org

© The Author(s) 2025
P. Huntjens et al. (eds.), *Eco-Social Contracts for Sustainable and Just Futures*,
https://doi.org/10.1007/978-3-031-99109-7_22

Greenfield, 2022). As the United Nations Environment Programme's (UNEP) Inger Andersen states "Only a root-and-branch transformation of our economies and societies can save us from accelerating climate disaster" (UNEP, 2022). And people agree. Seventy-four per cent of people in the richest economies in the world want economic transformation, with three in four people wanting to transform economic systems to prioritise health, well-being and the protection of the planet over a singular focus on profit and economic growth (EarthforAll, 2024). The Green Economy Coalition's (2024) survey of global attitudes to green economy found that 71% of people would choose "stronger environmental protection even at the cost of slowed economic growth" and that "contrary to many established narratives, demand for green action is considerably stronger in poorer countries than wealthier ones". And young people agree. Over half of the surveyed groups in the youth climate movement identified an economic "system that puts profit over people and planet" as the root cause of the climate and ecological crisis (Herfort et al., 2023).

Wealth and income inequality is growing in almost every country and in every economy. Over the last decade the richest 1% have amassed around half of all the new wealth created, while extreme poverty and unemployment continue to rise (Oxfam, 2023). Rising costs of living is affecting households across the world, with people unable to afford essential needs, such as food, water and energy. At the same time, democracies are in decay, with the democratic institutions that form the "scaffolding of civic space" being eroded by both the decline in democratic values and the ascent of authoritarian regimes (Civicus, 2021). Trust is being lost as people are left increasingly vulnerable and unprotected, with mounting fears about economic security and their economic future. As deliberations on the climate and nature agenda continue, almost all the economies of the world are off track to meet climate and biodiversity targets (Oceanographic, 2020), and action and ambition remain far below what is needed to meaningfully address the required economic reforms driving this crisis. These issues cannot be tackled alone, because they all arise from the same underlying problem: how our economies and societies are organised, ruled and managed.

Economic changes need to be anchored in restored trust and in a renewed social mandate for economic transformation. In this chapter, 'economic reform' refers to the process of rethinking the economic assumptions, models and institutions that govern the economy. This reform process is difficult—and often avoided—because structural economic change involves not just complex technical challenges, but also intensely contested questions of politics and power. Only structural economic reform that reorganises how our economies and societies are designed and managed, makes nature visible in economic decision-making and engages *with* people is a commensurate solution for the task ahead. Mobilising people as economic reform activists means supporting (all) their routes of agency. This includes their role in negotiating eco-social contracts for economic transformation.

22.2 Why a New Social Contract for Economic Reform?

An eco-social contract not only requires a reconfiguration of rights, responsibilities and duties of care to secure the well-being of people and the planet, it needs a different economy (Huntjens & Kemp, 2022). An economy that enables all people to create and enjoy prosperity and promotes equity within and between generations, an economy that safeguards, restores and invests in nature and supports sustainable consumption and production, an economy that is rights-based and an economy guided by integrated, accountable and resilient institutions.

Eco-social contracts are not only about the why and the what, but also the how. It should be catalysed by participatory mechanisms and processes that recognise and respect diverse values, perspectives and the rights of all stakeholders. It should be rooted in societal mandate and enable people's participation and engagement in the design of policies, projects and investments, ensuring that the outcomes, trade-offs, costs and benefits of interventions are distributed equitably and with transparency. What is meant by a social mandate for eco-social change? In simple terms it is a justification and permission structure for action—by the government, filtered through legitimate political or democratic institutions, or by society at large, through additional deliberative mechanisms. And negotiating eco-social contracts requires an enabling environment—civic and democratic space and an engaged and informed citizenry with access to diverse participatory mechanisms and processes to engage with and hold power to account.

As Fig. 22.1 shows, social mandate acts as part of a fly-wheel of self-perpetuating momentum for economic reform—as necessity of reform leads to creation of a mandate, which builds trust and spurs action, demonstrating change is possible, unlocking yet wider avenues for economic reform. For example, youth climate activists who have self-organised and catalysed greater ambition around the world are already beginning to create a mandate for economic change, now based on their position as the imminent inheritors of

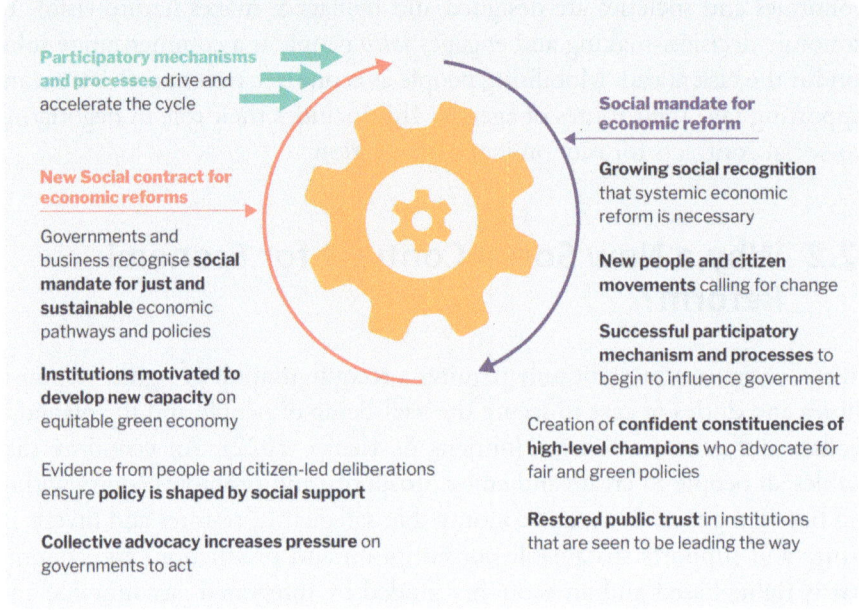

Participatory mechanisms and processes drive and accelerate the cycle

Social mandate for economic reform

New Social contract for economic reforms

Governments and business recognise **social mandate for just and sustainable** economic pathways and policies

Institutions motivated to develop new capacity on equitable green economy

Evidence from people and citizen-led deliberations ensure **policy is shaped by social support**

Collective advocacy increases pressure on governments to act

Growing social recognition that systemic economic reform is necessary

New people and citizen movements calling for change

Successful participatory mechanism and processes to begin to influence government

Creation of **confident constituencies of high-level champions** who advocate for fair and green policies

Restored public trust in institutions that are seen to be leading the way

Fig. 22.1 Building the social mandate for economic reform. Source: Author (based on Mohamed, 2023)

economic and environmental decline. They have pushed governments into greater ambition on climate and nature as growing national litigation rooted in children's rights show (Child Rights International Network, 2024) and have also engaged deeply with questions of structural economic transformation (Herfort et al., 2023). Another example, though now a missed opportunity, are COVID-19 green economic recovery efforts, which presented an opportunity to align climate and nature policies with economic recovery efforts. Most of these were underscored by a strong social mandate that sought to prioritise jobs and livelihoods, and health and well-being above profit in economic recovery (UNEP, 2021).

Negotiating eco-social contracts is a two-step process—a goal in and of itself—and a process to centre people in driving economic transformation at multiple levels. It offers a means by which people are actively engaged in decision-making processes ensuring that their values, visions and priorities are considered. Eco-social contracts could help define the goals and shape of the economy and provide legitimacy and accountability to the twenty-first-century economic transition we're now a part of. It is a tonic for declining public trust and a way for people to articulate what they want from their government (Hopkins et al., 2020).

There is a growing consensus from diverse players working on social, economic and climate justice that a new social contract fit for the challenges of the twenty-first century requires governance transformations that enable citizen deliberation, action and voice to be integrated in decision-making. Examples are emerging where the frame of new social contracts provides the lens to reconfigure the relationship between people and power, and where diverse mechanisms for citizen deliberation, voice and influence are catalysing economic reform. A wide range of participatory mechanisms, *the institutions and processes to engage people in policies and processes*, are being used to shape new social contracts. South Africa's Presidential Climate Commission is forging a social compact to integrate the interests and the voices of ordinary citizens in the country's just transition strategy (Presidential Climate Commission, 2021). Citizen assemblies for climate change have been held across Europe to deliberate upon and help shape and build legitimacy for climate policy (Hierlemann et al., 2022). And globally, climate litigation has doubled since 2015 with an increasing number of citizens and civil society challenging government and corporate inaction on climate goals and commitments using legal mechanisms (Setzer & Higham, 2021). These mechanisms are not only testing the temerity of policy ambitions but helping to restore decaying democracies.

It is not only our economies, but also our democracies that are in recession. Democracy is under threat globally—from reactionary and nationalist politics, from declining public trust and loss of political legitimacy and from a shrinking democratic and civic space. In 2021, nearly 75% of the world's population lived in a country that faced democratic deterioration, a decline "broad enough to be felt by those living under the cruellest dictatorships, as well as by citizens of long-standing democracies" (Freedom House, 2021). The expansion of authoritarian rule, declining civic space, elite capture and the brutal repression of civil liberties are hitting hard as the impacts of the cost-of-living crisis and COVID-19 economic recovery push millions into hunger, poverty and unemployment. In 2022, 101 out of 198 countries saw an increase in civil unrest (Verisk Maplecroft, 2022). This crisis of political legitimacy, the mounting disillusionment with governments and declining trust will be an obstacle to economic transitions. Reviving democracies and establishing principles of inclusive governance is a central part of one of the biggest societal projects of the twenty-first century: reimagining the purpose and structure of our economies.

22.3 An Overview of Institutional Mechanisms for People-Centred Governance

There have been multiple and interconnected reasons for putting people at the core of decision-making processes. Foremost among these have been the failings of representative democracy and the inability of elected representatives to adequately represent and speak for people. Those in power are mainly working "to preserve and perpetuate a system that benefits the few at the expense of the many" (UNRISD, 2023), and more often than not, elected politicians, whose policy choices favour the affluent, are instruments of elite capture that result in entrenching economic and social privilege and deepening social exclusion. With the gap between citizens and governments, evidenced in declining trust, civil unrest and loss of political legitimacy, a deliberative wave (OECD, 2020) has been growing at all government levels for more than a decade. Among the multitude of reasons (Bherer et al., 2016) for including citizen participation, the most highly cited are: restoring public trust, opening up the decision-making process, providing information, fostering social change, including citizens, holding decision-makers to account, increasing transparency, achieving public support and restoring legitimacy and acceptance of policy decisions. An equally diverse repertoire of participatory mechanisms and arrangements exist to advance citizen participation.

Innovative citizen participation and democratic innovations address diverse policy questions, from social and environmental issues such as health, housing and climate change to envisioning and planning economic pathways and models (OECD, 2020). But a recent review shows that people's participation in economic governance has received comparatively little attention (Thorpe & Gaventa, 2020). If people agree that the world needs systemic economic change, how do they become part of the process to find alternative means to govern and measure the economy? How can they challenge the short-termism of economic decision-making that ignores social and environmental demands? How do they challenge elite capture and privilege? How do they centre well-being in economic policymaking? How do they ensure that nature is visible in economic decisions? And how can they use existing and emerging institutional mechanisms to hold their governments to account?

Participatory mechanisms can play a catalytic role in driving and accelerating the social mandate for systemic economic reform. Citizens, acting as individuals or represented through civic associations, now voluntarily and regularly contribute to decision-making throughout policy processes, reinforcing democratic legitimacy and enabling greater transparency and accountability, as this study on participatory budgeting illustrates (De Soysa, 2022). These mechanisms show how people not only engage in but also have opportunities to shape, influence, review and monitor the purpose and substance of policies and processes.

Most countries now aspire to a level of social inclusion where they seek to design 'win-win' policies that deliver economic, environmental and social outcomes. Direct and deliberative democracy, as illustrated through the mechanisms described above, can complement representative democracy and restore trust and political legitimacy in decision-making processes. They can build public understanding and social support for economic policy ensuring that these are socially relevant and are delivered with social mandate.

An analysis of participatory mechanisms, drawn from the Organisation for Economic Co-operation and Development's (OECD) report on innovative citizen deliberation (OECD, 2020) and other sources, forms the basis of this chapter (Mohamed, 2023). This analysis found that citizen juries and deliberative polls offer the opportunity for the public to engage in focused policy discussions, but are often single-issue focused and were not regarded as appropriate mechanisms to deliberate on economic reform questions. Strategic litigation, while itself not a deliberative process, is often rooted in citizen and people movements and could be a follow-on strategy from a deliberative process and a mechanism to ensure accountability. And collective bargaining processes are very stakeholder-focused, i.e. trade unions, but can be a key conduit particularly for trade union confederations to engage in economic policy and governance and become part of alliances for economic reform. This section will focus on the mechanisms that were found to hold the most promise for reimagining eco-social contracts for economic transformation.

Citizen participation mechanisms for negotiating eco-social contracts for economic reform, can deliver critical outcomes such as (a) the construction of citizen knowledge, awareness and action on economic reform (b) the strengthening of mechanisms and practices that enable people to participate in economic reform, (c) the strengthening of responsive and accountable states and other powerful actors such as corporations in economic decision-making, (d) the development of inclusive and cohesive societies that support economic transition policies and processes, and (e) the social mandate for economic reform that builds a just and equitable world (Gaventa & Barrett, 2010).

But these mechanisms are not separate to and neither immune to the challenges of polarisation, elite capture and the influence of power, nor of democratic decline and backsliding. Ensuring that the democratic wave does not crash and lead to the dissipation of the inclusive and participatory governance frameworks essential for eco-social contracting will require maintaining the trust and integrity of the policies and institutions that safeguard and protect political rights. Moving beyond nominal, instrumental and representative participation to transformative participation where citizens have more power to negotiate and change the status quo, where "their voices are heard *and* responded to" (White, 1996) is also needed. This could also require shifting participatory mechanisms from being top-down processes that come into being through efforts at inclusive decision-making to processes initiated at the behest and demands of citizens and social movements. The recent Climate Emergency Social Contract (CESC) for South Africa's Government of National Unity (Climate Justice Charter Movement, 2024) is one such example where a civil society coalition put forward a position and a mechanism for a "deep just transition" challenging the government's just transition policy framework, prompting it to consider "life and ecocentric" concepts to reorganise the economy and deliver an economy in service of all life. In some places, participatory mechanisms are becoming an enduring part of political and social systems and institutions, shifting from once-off experiments and reactive responses to democratic innovations that are a permanent and evolving mechanism in the institutional landscape (Hierlemann et al., 2022).

Which of these participatory mechanisms are best-placed to enable greater engagement of people in economic reform? Mechanisms for direct public participation are being operationalised in very different ways across the world. For instance, citizen assemblies are widely adopted in Europe, the United Kingdom and United States (Sortition Foundation, 2025) while citizen dialogues, just transition deliberation and participatory budgeting have a more global application. The most effective mechanisms are those that are not one-off exercises to achieve greater participation and inclusion in the short term, but ones that become part of the fabric of decision-making where citizen and people-led positions and recommendations are meaningfully engaged in policy- and decision-making processes.

The following criteria have been applied to assess which participatory mechanisms could boost people's participation in decision-making around economic reform: they engage in complex issues and produce broad solutions; they present extended opportunities for focused discussions at various stages of decision-making; they engage with experts and evidence; they create opportunities for alternative framings and interpretations; and they enable the devolution of power for transformational change. They should also engage a wide spectrum of the population, young and old for instance, in the deliberations. A final criterion is that they offer the opportunity to nurture intersectional alliances, coalitions and movements for economic reform—engaging with frontline policy questions and ensuring that the agenda is amplified by civil society, people's movements, trade unions and others who champion economic and social justice.

Figure 22.2 represents the five key mechanisms that offer a conduit for amplifying the social mandate and building a movement for a just and sustainable economy: citizen assemblies, participatory budgeting, just transition processes, citizen dialogues and climate commissions and advisory councils.

Fig. 22.2 Key participatory mechanisms for people's engagement in economic reform. Source: Author (based on Mohamed, 2023)

While the latter is a representative, rather than deliberative mechanism, climate commissions engage in public deliberation, and are key catalysts for bringing peoples' voices into climate and economy deliberations. Illustrative examples of these are provided below.

Citizen deliberation in net zero governance, through *citizen assemblies* for instance, has been extensively employed in the United Kingdom (UK) to feed into policy formulation, with thirty-nine climate assemblies convened between 2019 and 2023 at various scales (local to national). While they are seen as an avenue for holding the UK government to account to meet its 2030 climate targets, they have not met all the objectives set (Ainscough et al., 2025), notably in providing an oversight and accountability role. While the overwhelming majority of citizen assemblies have been held in the UK, Europe and the United States (Sortition Foundation, 2025), there are a growing selection of assemblies being convened globally (Bürgerrat, 2025). A key recommendation emerging from Ainscough et al.'s (2025) analysis is the need to embed citizen assemblies in governance institutions. *Participatory budgeting* has seen widespread adoption globally, and involves citizens participating in decision-making around the allocation of public resources (OECD, 2020). While the city of Lisbon in Portugal has adopted participatory budgeting since 2008, it transformed this process into a green participatory budgeting exercise in 2020, adopting a diverse range of participatory mechanisms from in-person to web-based platforms to engage citizens in discussion and debate around a green budget (Centre for Public Impact, 2021).

Just transitions are not only about placing social justice at the centre of policies and plans that address economic transitions, but also about the *just transition processes* that engage citizens and workers (Mohamed & Montmasson-Clair, 2022; Montmasson-Clair et al., 2022). The Just Transitions Aotearoa Group (2023) in Aotearoa New Zealand developed a guide for communities, rooted in *talanoa*, a Pacific Island form of deliberative dialogue, to "develop positive visions for change, transform unfair systems, draw on diverse strengths and worldviews, and come together to solve problems in ways that work better for everyone". This guide draws on diverse case studies of *talanoa* and citizen deliberations that address economic, climate and ecological challenges in Aotearoa New Zealand. The Green Economy Coalition's seven *citizen-led policy dialogues* (Mohamed, 2020) in Asia, Africa, Latin America and the Caribbean illustrate how people—small and informal enterprises, women, local communities, marginalised populations—have been able to get a stake and have their say in green economic policy processes. It shows how inclusion can be mainstreamed through all stages of the economic policy cycle, creating a new dynamic between local citizens, small businesses and communities, and policymakers further 'upstream'. The process produced national and regional Green Economy Barometers which presented, from citizens' perspectives, what a green and fair economy should look like.

Finally, even as *climate commissions* lead from a climate and not an economic transformation mandate, they are playing a critical role in climate governance oversight and accountability, ensuring that governments uphold their commitments to climate action. These institutions often facilitate public deliberation, acting as catalysts for bringing peoples' voices into climate and increasingly, economic transformation deliberations. This is illustrated in South Africa's Presidential Climate Commission that is tasked with building a social compact to guide the country's transition to a low-carbon, climate-resilient economy, and developing a shared understanding and agreement among different stakeholders to ensure the transition is equitable and inclusive, particularly for climate vulnerable groups (Grice, 2021).

These participatory mechanisms offer an opportunity for people to engage in negotiating and setting the terms of economic transformation policies, plans and investments to shape the purpose, form and structure of economic reform pathways. They need further exploration and testing at multiple levels. Key steps in testing of these mechanisms include:

- Identifying territories and regions where there is the opportunity to support participatory mechanisms that can advance people's participation in economic reform.
- Finding the entry point or signature issues to negotiate new social contracts for economic reform.
- Selecting the suitable level of deliberation (global, regional, national, local).
- Ensuring that citizens are informed, supported and able to engage in deliberations.
- Assessing the extent of civic space as a critical enabling factor and where constrained, identifying alternative strategies for citizen deliberation.

Beyond participation in the deliberation mechanisms, civil society and social movements play a key role in ensuring transparency and accountability post-deliberation with recourse to wide-ranging strategies including campaigning, lobbying and advocacy, strategic litigation, protest action and building solidarity around the agenda of economic reform.

22.4 Seven Reflections on People-Centred Economic Transformation

Participatory mechanisms give people more power to challenge, negotiate and change the status quo, but also creates opportunities to engage with planning, designing and monitoring interventions for structural economic reform, i.e. reforms or measures that are intended or aim to be systemic, long-lasting and applicable in most geographic and economic contexts (Hopkins & Greenfield, 2022). While this chapter has sought to highlight the potential of participatory mechanisms for ensuring greater social inclusion and public engagement in economic transformations, there are several challenges that impinge democratic, inclusive and participatory decision-making processes for economic governance more broadly.

The current rise of far-right populism in all countries and the growing corporate capture of governance in fact make inclusive governance approaches now seem pretty radical. For example, the United States' withdrawal from the Paris Agreement (that could be regarded as a global eco-social contract for the climate) could decimate climate action as governments and corporations reduce or even renege on the transformative ambition of the economic reforms that are central to address climate change, such as the just transition from fossil fuels to clean energy. This is already visible in the European Union (EU) and many parts of the world where populist parties that oppose strong and inclusive climate action are gaining support (Dirkx & Wettengel, 2024). Another challenge constraining citizen participation is ensuring that this participation is actually meaningful. As shared in the example of citizen assemblies in the UK (Ainscough et al., 2025), these mechanisms must move from instruments of consultation and engagement to processes that people trust, where they can hold leaders and decision-makers to account and influence decision-making. Binding, collective and citizen-wide agreements do exist, but many participatory mechanisms still operate at the nominal levels of citizen participation. Rising political inequality, diminishing civic space and elite capture at multiple levels can impact people's ability, capacity and willingness to participate in these institutional mechanisms.

How can we build collective and shared visions for a just transition to fairer and greener economies, grounded in a broad consensus between different stakeholders? How can economic decisions be made "democratically, collectively, for the benefit of all—and not dictated by the vested interests of powerful corporations, authoritarian states or the super-rich" (War on Want, 2025)? As suggested by other chapters in this book, forging alliances across struggles by "connecting workers, Indigenous land defenders, climate activists in a shared vision, which centres the perspectives and experiences of those living on the frontlines of these crises" will be key (War on Want, 2025). This section explores cross-cutting insights that can help address the barriers to inclusive economic governance and ensure that participatory mechanisms are designed and delivered in a way that it achieves green and fair economic transformations.

- *Broaden the geographical spread of mechanisms.* Participatory mechanisms should be applied beyond countries and places that have civic space to places where people's participation in economic reform is equally, if not more, critical. This includes frontline communities where climate, environmental and economic inequalities are hitting hardest, e.g. citizen assemblies in the Global South, in climate vulnerable countries and regions, and with informal workers (GRID-Arendal, 2022) and refugee communities. It should be held in fossil fuel-producing countries and regions where workers and communities need to transition to new industries and jobs. It should include more city- and place-level deliberations such as the deliberations around economic change. Examples include the Place-Based Climate Action Network (2025), the Transition Towns Network (Rapid Transition Alliance, 2019) and Global Green New Deals being designed at city level (C40 Cities, 2021).
- *Involve people early and often, recognising that there are structural and political, and at times, cultural, barriers to citizen participation in economic decision-making.* Inclusion of people in decision-making needs to be embedded in governance systems, with engagement from the outset of a policy or a process to ensure that they are part of the debate, the diagnosis, the decision and the monitoring and review of interventions (Mohamed, 2020). Highlighting society's visions, values and priorities creates a 'demand-pull' on policy that makes otherwise 'top-down' decision-making more robust, sustainable and relevant. Participatory mechanisms offer working approaches to foster meaningful inclusion through all stages in the policy and planning cycles, but also face multiple barriers. These could include structural barriers such as lack of access to information and technology or the time to participate in governance processes, political barriers such as mistrust in policymaking processes or gatekeeping by powerful elites or cultural barriers such as social bias that could lead to discrimination against certain groups or values and beliefs that limit equal participation of women or young people. Participatory mechanisms need to be context-specific and designed in a way that all visions, values and views can

be represented to arrive at a collective vision for an economy that works for all.

- *Identify, engage and include all stakeholder communities.* Deliberative processes must assess who is missing from the conversation. Whose voices are not being heard? Who is not getting a seat at the table? Are young people, minority groups, informal workers, refugee communities and others engaged and brought into processes? Is the language and style of engagement inclusive and empowering? Are we rebuilding a social contract with nature (Mohamed & Huntjens, 2023) as the children and youth in Ireland are (Children and Young People's Assembly on Biodiversity Loss, 2022)? Are there opportunities for cross-stakeholder deliberations and alliances, between workers, youth, women, entrepreneurs and farmers for instance? Are complementary structures needed to engage invisible voices and perspectives? The assumption that participatory mechanism engages all voices must be challenged if these are to reflect central principles of participatory justice (Montmasson-Clair et al., 2022). Diverse and blended public engagement methods need to be enlisted, such as holding deliberations in-person and on-line or in plenary and in smaller groups to ensure that efforts are exerted to achieve inclusivity.

- *Facilitate knowledge and exchange* to reflect on and learn about what works and what doesn't. For example, the European Knowledge Network on Climate Assemblies (KNOCA, 2025) offers a platform for sharing best practice on the design and implementation of climate assemblies. The OECD's (2025) Paris Collective on Green Budgeting works in close partnership with governments globally to co-design practical and pragmatic approaches to green budgeting. An important feature of these networks is that they give a central role to practitioners alongside researchers, policy-makers and other stakeholders ensuring that concepts and ideas are rooted in practice and real-world experiences. Learning and exchange can ensure that deliberative processes are refined and build upon good practice. There is also a need to facilitate South-South engagement on economic governance in the current economic climate to "promote the exchange of best practices and support...in the common pursuit of their broad development objectives" and "as an expression of South-South solidarity and a strategy for economic independence and self-reliance of the South based on their common objectives and solidarity" (South Centre, 2009).

- *Foreground and engage with the challenge of shrinking civic space and actively support transparency and accountability.* Globally, there is an increase in civil unrest and growing concerns that the watchdog role of civil society is being shut down. Civic space that "allows civil society and individuals to organise, participate and communicate freely and without discrimination" enables people's participation in new social contract formulations (Civicus, 2021). But in most societies, this space is constrained, necessitating actively seeking spaces for civil society engagement, most especially in authoritarian regimes (O'Driscoll, 2018). For instance, engagement in less controversial, more technical oversight processes, such as participatory budgeting or governance devolution, can offer opportunities for holding decision-makers to account. And in many places, civil protest remains a means where people can demonstrate their concerns, and thereby demand the institution of deliberative processes. For example, the citizens' climate assembly in France was a direct response to the *gilet jaune* protests against the fuel tax increases, prompting the government to bring citizens into climate policy deliberations (Giraudet et al., 2022).

- *Step up intersection with global economic reform agendas.* The People's Assembly—Global Call to Action Against Poverty (GCAP) (GCAP, 2024), the Global Assembly to Address the Climate and Ecological Crisis (Global Assembly, 2021) and the People's Charter for an Eco-Social World (International Federation of Social Workers and others, 2022) provide examples of ways in which deliberative mechanisms can not only be run at global level, but also feed people's visions into global economic governance institutions helping to make them become more inclusive (Chatham House, 2021). A new global social contract requires an alignment between the array of negotiated agreements that envisions the world we want, such as the Sustainable Development Goals (SDGs), Kunming-Montreal Global Biodiversity Framework and Paris Agreement, amongst others. For citizens, voters and communities, the SDGs offer a way to hold decision-makers to account—a framework to think strategically across a range of variables over a fifteen-year period, and a guide on reaching negotiated and agreed goals. The SDGs and other global 'eco-social contracts' need to be assessed against the realities of economic governance as experienced by ordinary people, many of whom are impacted by global agreements, on trade and finance for instance, that need to be negotiated.
- *Build a movement for economic reform.* Finding meaningful common ground between civil society, social movements and others who are working in very different fields can be tricky. What unites, for example, a youth assembly on new economics, a city-level deliberation on doughnut economics and a global organisation campaigning for global green deals? At first glance—a vision for a just and sustainable economy. The underlying concerns common to each of these initiatives, and many more besides, are that the very purpose, structure and form of our economies must be reformed. This

shared diagnosis forms the basis for building a critical mass and movement for economic reform as illustrated in this extract from an essay on a transformational new social contract (Phillips, 2022):

> People are coming together in growing numbers in connected movements to ensure that their wages go up, their healthcare is provided and they are not burdened by debt; to at last break the hold of white supremacism and structural violence; and to win a Green New Deal to protect their environment, provide quality public transport and create millions of jobs. It's not just that this social contract is worth fighting for—it's that it can only be won through millions of people fighting for it.

Participatory mechanisms are a central plank in the overarching ambition of building a movement for economic reform. Inclusion in decision-making thus becomes part of a broader societal project for inclusive, accountable and transparent governance. Participatory mechanisms enable people's needs to be heard and taken into consideration. It empowers people to track outcomes, hold decision-makers to account and if needed, to take recourse to justice in the face of inaction. And it is emerging as catalyst for negotiating eco-social contracts, as the case of the civil society-led Climate Emergency Social Contract in South Africa (Climate Justice Charter Movement, 2024) or the People's Plan for Nature in the UK (National Trust, Royal Society for the Protection of Birds and WWF-UK, 2023) which both present people's visions for a just and sustainable world.

22.5 Conclusion

Several aspects of the broken social contract point towards the need for negotiating eco-social contracts that spur the transformation of economies and societies. The current economic system is driving the interlinked climate, biodiversity, inequality and democracy crises. Transparent, participatory and accountable decision-making where people have a say in the economic policies and pathways that affect their lives should be core to eco-social contracts.

Involving people in reimagining the economies they want and contributing to economic decision-making is already happening. Citizen engagement and deliberation on economic reform is among the frontiers for expansion of the deliberative wave, boosting the agenda of engaging people in economic governance and restoring trust, credibility and legitimacy for policy and political action. But there's a cautionary side to this tale. In the same way that the trust in a legitimate, accountable and transparent state needs to be restored, trust in the knowledge, creativity and innovation of people to envision economic futures and engage in economic governance must be a given.

Economic reform at the scale needed is unlikely to attract broad public support unless it is shaped and informed by the needs and concerns of ordinary people. Without societal mandate and demand, governments and corporations are unlikely to act with the legitimacy, ambition, transparency and accountability needed to avoid the social, economic and climate catastrophes we're heading towards (Amnesty International, 2021).

Participatory mechanisms are key conduits for people to define and negotiate eco-social contracts for economic reform. They offer a means for amplifying people's engagement in formulating new and diverse pathways for economic reform, shaped by varied contexts, agendas and approaches. They build and strengthen the relational dimensions of eco-social contracts, restoring people's trust in public institutions and amplifying the social mandate for just and sustainable economic pathways and policies. And they can connect civil society organisations and social movements working on interrelated challenges such as climate, inequality, nature and democracy around a shared mission of achieving economic transformation in an era of rising economic inequality, political instability and democratic decline.

The multiple, escalating and interconnected crisis cannot be tackled alone, because they all arise from the same underlying problem: the rules that govern our economies. Eco-social contracts for economic reform must deliver fairer economies and just societies. A democratic and intersectional alliance, rooted in active citizen engagement and participation in economic policies and processes at multiple levels, can help drive societal demand for just and sustainable economies. It already is.

References

Ainscough, J., Killeen, L., Lewis, P., Shepherd, A., & Willis, R. (2025). Citizen deliberation in Net Zero governance: Learning lessons and looking forward. *Journal of the British Academy, 131*. https://doi.org/10.5871/jba/013.a13

Amnesty International. (2021). *Stop burning our rights! What governments and corporations must do to protect humanity from the climate crisis.* Amnesty International.

Bherer, L., Dufour, P., & Montambeault, F. (2016). The participatory democracy turn: An introduction. *Journal of Civil Society, 12*(3), 225–230. https://doi.org/10.1080/17448689.2016.1216383

Bo, L., Dabla-Norris, E., & Srinivasan, K. (2023, February 9). Support for climate action hinges on public understanding of policy. *IMF Blog*. Accessed

March 14, 2025, from https://www.imf.org/en/Blogs/Articles/2023/02/09/support-for-climate-action-hinges-on-public-understanding-of-policy?utm_medium=email&utm_source=govdelivery

Bürgerrat. (2025). *Citizens' Assemblies Worldwide*. Bürgerrat. Accessed March 14, 2025, from https://www.buergerrat.de/en/citizens-assemblies/citizens-assemblies-worldwide/

C40 Cities. (2021, July 21). *Green new deals in action. Nine cities commit to simultaneously tackle inequality, social justice and climate change*. C40 Cities. https://www.c40.org/news/global-green-new-deal-pilots/

Cameron-Smith, A. (2023). The currency of the transition: Why grand visions need trust. *Unlock Net Zero*. Accessed March 14, 2025, from https://www.unlocknetzero.co.uk/home/introducing/the-currency-of-transition-why-grand-visions-need-trust

Centre for Public Impact. (2021). *Green participatory budgeting: Lisbon, Portugal*. Accessed April 17, 2025, from https://centreforpublicimpact.org/public-impact-fundamentals/green-participatory-budgeting-lisbon-portugal/#Background-and-Context

Chatham House. (2021). *Reflections on building more inclusive global governance. Ten insights into emerging practice*. The Royal Institute of International Affairs.

Children and Young People's Assembly on Biodiversity Loss. (2022). *Call to action*. Accessed March 15, 2025, from https://cyp-biodiversity.ie/wp-content/uploads/2022/11/CYPABL-Calls-to-Action.pdf

Civicus. (2021). *What is civic space and why is it shrinking?* Accessed March 14, 2025, from https://www.youtube.com/watch?v=Fm-KeAbhVCk

Climate Justice Charter Movement. (2024). *Towards a Climate Emergency Social Contract (CESC) for South Africa's Government of National Unity*. Accessed April 17, 2025, from https://cjcm.org.za/download/5f669fe6-5af9-4abb-9812-09e9e3f749ce

CRIN (Child Rights International Network). (2024). *Recent developments on children's rights and climate litigation*. Accessed April 17, 2025, from https://home.crin.org/readlistenwatch/stories/recent-developments-on-childrens-rights-and-climate-litigation

De Soysa, A. (2022). *Participatory budgeting. Public participation in budgeting processes*. Transparency International.

Dirkx, M., & Wettengel, J. (2024, June 7). Right-wing populists challenge Europe's climate efforts. *Clean Energy Wire*. Accessed April 17, 2025, from https://www.cleanenergywire.org/news/right-wing-populists-challenge-europes-climate-efforts

EarthforAll. (2024). *Uncovering global attitudes to system change*. Earth for All Global Survey. Accessed March 14, 2025, from https://earth4all.life/global-survey-2024/

Freedom House. (2021). *Freedom in the World 2021. Democracy under Siege*. Freedom House.

Gaventa, J., & Barrett, G. (2010). *So what difference does it make? Mapping the outcomes of citizen engagement.* IDS Working Paper 347. Institute of Development Studies.

Giraudet, L.-G., Apouey, B., Arab, H., Baeckelandt, S., & Bégout, P. et al. (2022). "Co-construction" in Deliberative Democracy: Lessons from the French Citizens' Convention for Climate. *Humanities and Social Sciences Communications, 9*(207).

Global Assembly. (2021). *Global assembly to address the climate and ecological crisis.* Accessed March 14, 2025, from https://globalassembly.org/

Global Call to Action Against Poverty (GCAP). (2024). *The peoples assemblies.* Accessed March 14, 2025, from https://gcap.global/peoples-assembly/

Green Economy Coalition. (2024). *Global attitudes to green economy: 2024.* Accessed April 17, 2025, from https://www.greeneconomycoalition.org/news-and-resources/polling-survey-global-attitudes-green-economy-2024-results

Grice, J. (2021, November 30). Building a social compact around the climate transition. *South African Labour Bulletin.* Accessed April 17, 2025, from https://www.southafricanlabourbulletin.org.za/building-a-social-compact-around-the-climate-transition/#:~:text=The%20PCC%20plans%20to%20meet,activity%20to%20support%20their%20livelihoods

GRID-Arendal. (2022). *A seat at the table: The role of the informal recycling sector in plastic pollution reduction, and recommended policy changes.* GRID-Arendal.

Herfort, N., Johnson, J., Macintyre, C., & Born, S. (2023). *Mapping the global youth climate movement: Towards a green economic mandate.* Climate Vanguard and Green Economy Coalition.

Hierlemann, D., Renkamp, A., & Demidov, A. (2022, June 24). *From innovation to permanence: Citizens' assemblies in the EU.* BertelsmannStiftung. https://www.bertelsmann-stiftung.de/en/our-projects/democracy-and-participation-in-europe/project-news/from-innovation-to-permanence-citizens-assemblies-in-the-eu

Hopkins, C., & Greenfield, O. (2022). *Setting a structural agenda for a green economic recovery from COVID-19.* Discussion Paper. Green Economy Coalition.

Hopkins, C., & Greenfield, O., & Mohamed, N. (2020). *Is the moment for a new social contract here?* Green Economy Coalition. Accessed March 14, 2025, from https://www.greeneconomycoalition.org/news-and-resources/is-the-moment-for-a-new-social-contract-here

Huntjens, P., & Kemp, R. (2022). The importance of a natural social contract and co-evolutionary governance for sustainability transitions. *Sustainability MDPI, 14*(5), 1–26. https://www.mdpi.com/2071-1050/14/5/2976

International Federation of Social Workers and others. (2022). *People's charter for an eco-social world.* Accessed March 15, 2025, from https://newecosocialworld.com/the-peoples-charter-for-an-eco-social-world/

Just Transitions Aotearoa Group. (2023). *A guide to just transitions for communities in Aotearoa New Zealand.* Motu Economic and Public Policy Research.

Knowledge Network on Climate Assemblies. (2025). *About KNOCA.* Accessed March 15, 2025, from https://www.knoca.eu/about-knoca

Mohamed, N. (2020). *Inclusion matters: Policy insights and lessons from the green economy coalition's national dialogues.* Green Economy Coalition.

Mohamed, N. (2023). *Building new social contracts: An overview of participatory mechanisms for economic governance.* Green Economy.

Mohamed, N., & Huntjens, P. (2023). *Dismantling the ecological divide: Toward a new eco-social contract.* UNRISD Issue Brief 15. https://cdn.unrisd.org/assets/library/briefs/pdf-files/2023/ib15-a-new-contract-with-nature.pdf

Mohamed, N., & Montmasson-Clair, G. (2022). Normative framework to assess to the just transition to a net zero carbon society. In N. Xaba & S. Fakir (Eds.), *A just transition to a low carbon future in South Africa low carbon futures.* Mapungubwe Institute for Strategic Reflection (MISTRA).

Montmasson-Clair, G., Patel, M., & Wolpe, P. (2022). *People's voices: Participatory justice for a just transition in South Africa.* TIPS and UK Pact.

National Trust, Royal Society for the Protection of Birds and WWF-UK. (2023). *The People's Plan for Nature.* Accessed April 17, 2025, from https://peoplesplanfornature.org/sites/default/files/2023-03/PPFN-Executive-Summary-FINAL.pdf

O'Driscoll, D. (2018). *Civil society in authoritarian regimes. Knowledge, evidence and learning for development.* Helpdesk Report. Institute for Development Studies.

Oceanographic. (2020). World remains "far off track" to meet climate change and biodiversity targets for 2020. *Oceanographic Magazine.* Accessed March 14, 2025, from https://oceanographicmagazine.com/news/biodiversity-targets-2020/

OECD. (2025). *Paris collective on green budgeting.* Accessed March 15, 2025, from https://www.oecd.org/en/about/programmes/paris-collaborative-on-green-budgeting-.html

OECD (The Organization for Economic Cooperation and Development). (2020). *Innovative Citizen Participation and New Democratic Institutions: Catching the Deliberative Wave.* OECD Publishing. https://doi.org/10.1787/339306da-en

Oxfam. (2023). *Survival of the richest: How we must tax the super-rich now to fight inequality.* Briefing Paper Oxfam.

Phillips, B. (2022). *How can a transformational new social contract be won?* Christian Aid. Accessed March 15, 2025, from https://www.christianaid.org.uk/news/policy/how-can-transformational-new-social-contract-be-won

Place-Based Climate Action Network. (2025). *Delivering energy and climate policy at the city level in the UK.* Accessed March 15, 2025, from https://pcancities.org.uk/about

Presidential Climate Commission. (2021, December 7). *PCC calls for a solid social compact for South Africa's climate change just transition.* Presidential Climate Commission. Accessed March 14, 2025, from https://www.climatecommission.org.za/news-and-insights/presidential-climate-commission-calls-for-a-solid-social-compact-for-south-africas-climate-change-just-transition

Rapid Transition Alliance. (2019). *Transition towns: The quiet networked revolution.* Accessed March 15, 2025, from https://rapidtransition.org/stories/transition-towns-the-quiet-networked-revolution/

Setzer, J., & Higham, C. (2021, July 21). *Global trends in climate litigation: 2021 snapshot.* London School of Economics and Political Science. Accessed March 14, 2025, from https://www.lse.ac.uk/granthaminstitute/publication/global-trends-in-climate-litigation-2021-snapshot/

Sortition Foundation. (2025). *Citizens' assemblies & sortition around the world.* Sortition Foundation. Accessed March 15, 2025, from https://www.sortition-foundation.org/where

South Centre. (2009). *South-south cooperation principles: An essential element in south-south cooperation.* Accessed April 17, 2025, from https://www.southcentre.int/wp-content/uploads/2013/09/South-South-cooperation-Principles_EN.pdf

Thorpe, J., & Gaventa, J. (2020). *Democratising economic power: The potential for meaningful participation in economic governance and decision-making.* IDS Working Paper 535. Institute of Development Studies.

UNEP. (2021). *Are we building back better? Evidence from 2020 and pathways to inclusive green recovery spending.* UNEP. Accessed April 17, 2025, from https://wedocs.unep.org/bitstream/handle/20.500.11822/35282/AWBBB_ES.pdf

UNEP. (2022, October 27). *Inadequate progress on climate action makes rapid transformation of societies only option.* UNEP Press Release. Accessed March 14, 2025, from https://www.unep.org/news-and-stories/press-release/inadequate-progress-climate-action-makes-rapid-transformation

UNRISD. (2023). *Crises of inequality. Shifting power for a new eco-social contract.* Research and Policy Brief 39. UNRISD https://cdn.unrisd.org/assets/library/briefs/pdf-files/2023/rpb-39-crises-inequality.pdf

Verisk Maplecroft. (2022). *Political Risk Outlook 2022.* Verisk Maplecroft.

War on Want. (2025, February 28). *Building power when the far-right is on the rise.* War on Want. Accessed April 17, 2025, from https://waronwant.org/news-analysis/building-power-when-far-right-rise

White, S. (1996). Depoliticising development: The uses and abuses of participation. *Development in Practice, 6*(1), 6–15. https://doi.org/10.1080/0961452961000157564

23

Epilogue: Five Elements for Advancing a Climate for Humane Conversations

Carlos Alvarado Quesada

In recent years, I've participated in various forums—multilateral, national, and local—where I've observed well-meaning groups tackling issues like the climate crisis, the need for an energy transition, migration or the fragmentation of the multilateral system. Yet, despite shared goals and similar data, these conversations often end in misunderstanding, paralysis, mistrust, and even polarisation. We might all see the same scientific reports and economic analyses, but that rarely means we converge in our actions.

I don't claim to have absolute answers to these complex challenges. However, I'd like to share a set of attitudes and mindsets that, in my experience, can help us break through gridlocks and move discussions—and actions—into more productive territory. Here are five elements I believe are essential for effective and humane climate discussions that lead to improved understanding and accelerated implementation of what the editors of this volume call a new eco-social contract for people and planet.

C. A. Quesada (✉)
Fletcher School of Tufts University, Medford, MA, USA
e-mail: Carlos.Alvarado@tufts.edu

© The Author(s) 2025
P. Huntjens et al. (eds.), *Eco-Social Contracts for Sustainable and Just Futures*,
https://doi.org/10.1007/978-3-031-99109-7_23

23.1 Exploring Our Own Positionality

In a recent panel discussion, for purposes of economy of time, I summed up an argument by concluding "we must urgently transition away from fossil fuels."

A young woman in the audience responded, "Mr. President, I like what you say, but I'm from Nigeria. How can Nigeria do that if 87% of our exports are fossil fuel exports, and those account for 45% of our government revenue?" (Saha et al., 2023).

This was a powerful reminder. I realised I was speaking from my own positionality, shaped by my experience as the former leader of a country where over 90% of electricity is generated from renewable sources, and which launched the first national decarbonisation plan after the Paris Agreement was signed by 196 parties in 2015.

My country's context emphasises the urgent need to address sea-level rise, which threatens Pacific Island nations in addition to communities across the Caribbean and Central America. However, for someone from Nigeria, fossil fuels are inseparable from immediate economic survival. Both are legitimate and important concerns.

Positionality, as Queen's University (n.d.) in Canada defines it, is "where one is located in relation to their various social identities (gender, race, class, ethnicity, ability, geographical location, etc.)." To this, I would add the dimension of livelihoods. We each bring our unique identities and experiences to discussions, but we tend to present our singular views as if they are universal truths. By recognising our own positionality, we're reminded that others' journeys, challenges, and solutions might differ from ours. In doing so, we make room for empathy.

23.2 Exercising Empathy

Empathy, as the *Oxford Dictionary* (2025) defines it, is "the ability to imagine and understand the thoughts, perspectives, and emotions of another person." It serves as a bridge to others' experiences, allowing us to move beyond awareness of our own positionality and truly appreciate where others are coming from.

Empathy is deeply tied to memory—both individual and collective. Our experiences of joy, pain, anxiety, love, and hope shape how we imagine these emotions in others. For instance, my love for my son was a driving force behind my commitment and promotion of Costa Rica's *National*

Decarbonization Plan. That same love, I know, resonates in others, even when our paths may differ. Recognising our positionalities helps us find empathic connections, even with those whose circumstances or priorities may contrast with our own.

As we look to the future, we all share a common goal: to thrive and, above all, to survive on this planet. What we often leave out are the details of how our pursuit of well-being might affect that of others. Framing discussions from this perspective fosters collaboration—and it was this shared vision that made the Paris Agreement of stopping global warming possible.

However, if we focus on our short-term interests, on our quick wins, then we will be limited in our ability to build a different future. Especially around climate change we see profound trade-offs between short-term benefit and long-term sustainability.

Through empathy, we can transcend and connect with a common human identity. This shared humanity enables us to engage in dialogue with open minds, fostering understanding even when points of view conflict.

23.3 Complexity and Overcoming Dualistic Thinking

Another essential element is moving beyond dualistic thinking—the tendency to see the world in dichotomies. Often, we use opposing pairs to simplify complex realities: North and South, male and female, human and nature, mind and body, us and them, rich and poor. While these categories help us make sense of the world, they risk oversimplification or faulty framing of situations.

For instance, when we talk about the Global North and Global South, we imply a sharp divide between wealthy, historically colonising countries and those less wealthy, often formerly colonised. Yet, that is a misleading and contested generalisation. Just to mention one reason, there are aspects of the "Global South" in the North, represented by marginalised communities, and small but powerful pockets of the "Global North" in the South, seen in elite classes within developing countries.

Reality is nuanced; the world is filled with complexities that when cast against dualistic terms appear contradictory. This is how we can miss the most crucial insights, and ignore the clear levers of action, or the places where leadership is needed most. Philosopher Val Plumwood (1993) argues that dualistic pairs create a false dichotomy, hiding overlaps and shared experiences. Concepts like *Ubuntu* from South Africa (I am because we are), described in this volume as Indigenous cosmovisions that underpin non-Western social contracts, offer a non-antagonistic perspective, blending individual and community identities. Similarly, in Indigenous American cultures, statements like "I am with the river, I am with the land" integrate nature and humanity, avoiding separation. Moving beyond dyads allows us to see the world in shades of complexity that are more trustworthy, and so fosters richer, more fruitful dialogues.

23.4 Understanding History and Decolonising Our Assumptions

Empathy is deeply rooted in memory. Both personal and collective memories shape our perspectives and understanding of the world. Our interactions with others are an interplay between past, present, and future. However, many societies tend to treat the past as irrelevant or inconvenient, especially when discussing colonisation and its legacy.

In my view, it is crucial to study and understand history and histories. This knowledge helps us grasp our positionality and cultivate empathy, allowing us to move beyond the anger, resentment, guilt, or indifference that history can evoke. For example, the impacts of colonialism linger, affecting resource distribution, power dynamics, and environmental policies.

Acknowledging this history enables us to approach discussions with greater awareness of the systemic inequalities that persist today. Recognising the weight of historical context helps us address present issues more equitably, overcoming inherited assumptions and honouring the full complexity of what brought us to this point.

23.5 Leadership

Finally, what do we do with these insights? The answer lies in leadership and agency of the people, by the people, and for the people. Scientific scenarios suggest that if we continue our current path, by the year 2100, Earth will be vastly different, presenting harsh and even unlivable conditions for future generations. The *2023 State of the Climate Report* (WMO, 2024) identified trends that could result in 3 to 6 billion people being "confined beyond the livable region." But this future isn't inevitable. It will be the product of the decisions we take today and tomorrow. This leadership by all and human willpower can shape a different outcome.

When we talk about the climate crisis, the urgent need for an energy transition, and the transformation of food systems, we often view these changes as external forces. Yet the reality is that it's not only the world around us that's changing—we, too, are undergoing a profound transformation. This is a comprehensive human transition.

True leadership is more than just vision; it demands empathy, humility, and the capacity to unite people. It means embodying these five elements: learning about positionality, exercising empathy, overcoming dichotomous thinking, decolonising our assumptions, and channelling these qualities into action. These attitudes may also help leaders to create much stronger forms of a common human identity—a bond that is direly needed in these times of rising geopolitical tensions.

Cultivating these five areas does not guarantee success, but in my experience, it strengthens the skills needed to foster productive, compassionate discussions. These attitudes open pathways for constructive agreements among diverse groups, helping us move forward together in a world that urgently needs unity and action.

References

Oxford Dictionary. (2025). *Empathy*. Oxford University Press.

Plumwood, V. (1993). *Feminism and the mastery of nature*. Routledge.

Queen's University. (n.d.). *Positionality statement*. Centre for Teaching and Learning. URL: Positionality Statement I Centre for Teaching and Learning

Saha, D., Walls, G., Waskow, D., & Bergen, M. (2023, January 26). *To shift away from oil and gas, developing countries need a 'just transition' to protect workers and communities*. World Resources Institute. https://www.wri.org/insights/just-transition-developing-countries-shift-oil-gas

WMO. (2024). *State of the global climate 2023*. World Meteorological Organization.